Diseases of Field Crops and their Management

Mr. Shivam Kumar presently pursuing his Ph.D. in Department of Plant Pathology from C.S.A. University of Agriculture & Technology, Kanpur, U.P. (India) under the guidance of Prof. S. K. Biswas. He obtained his B.Sc. (Ag.) Honours from ANDUA&T, Ayodhya, U.P.& M.Sc. (Agr.) in Plant Pathology from C.S.A.U.A. &T, Kanpur, U.P. He qualified ASRB-NET. He published 12 research papers, 03 book chapters, and 05 popular articles. He was received 'Best Poster Presentation Award in 2024 by Indian Phytopathological Society, New Delhi, and 'Young Plant Pathologist Award' in 2023 by ATDS, Ghaziabad (U.P.).

Mr. Ravi Kumar working as Ph.D. research scholar in Deptt. of Plant Pathology, C.S.A. University of Agriculture & Technology, Kanpur (U. P.) under the guidance of Prof. S. K. Biswas. He completed his B.Sc. (Hons) Agriculture from B.U.A.&T, Banda and M.Sc. in Plant Pathology from CSAUA&T, Kanpur. He qualified ASRB-NET. He has authored 9 research articles in national and international journal, 3 Book chapters, 15 popular articles.

Dr. Samir Kumar Biswas working as Professor & Head, Deptt. of Plant Pathology, C.S.A. University of Agriculture & Technology, Kanpur (U.P.). He obtained his B. Sc. (Ag.) degree from BCKV, West Bengal in 1994, M. Sc. (Plant Pathology) in 1997 and Ph.D. (Plant Pathology) in 2001 both from I.A.R.I, New Delhi. He has published more than 150 research papers in National & International Journal, 04 books (Sustainable disease management of Agricultural Crops & Organic farming and management of biotic stresses, Diseases of pulse crop and their sustainable management), 19 book chapters and 10 articles. He was received the Young Scientist Award-2003 by UPCAR (Lucknow), M. K. Patel Memorial Young Scientist Award by Indian Phytopathological Society, New Delhi in 2011 and PR Verma Award 2003 by Indian Society of Mycology & Plant Pathology.

Ms. Anjali Arya working as Senior Technical Assistant (Group A) in Department of Agriculture, Uttar Pradesh. She completed B.Sc. (Hons) Agriculture and M.Sc. (Agr.) in Plant Pathology degrees from SardarVallabhbhai Patel University of Agriculture and Technology, Meerut, U.P. (India). She has been awarded Vice Chancellor's Silver Medal in her graduation.

Dr. Vishnu D. Rajput working as Head of 'Soil Health Lab' at Southern Federal University, Rostov-on-Don, Russia. He has published 331 scientific publications, 70 chapters, 18 books, H-Index: 37. He is holding national and international projects including Mega-Grant, BRICS. In year 2023, he is included in the list of 2% world's potential scientists by Elsevier BV publishing house.

Diseases of Field Crops and their Management

– *Editors* –

Mr. Shivam Kumar

Mr. Ravi Kumar

Dr. S.K. Biswas

Miss. Anjali Arya

Dr. Vishnu D. Rajput

2025

Daya Publishing House®

A Division of

Astral International Pvt. Ltd.

New Delhi – 110 002

ISBN 978-93-5919-510-0

Published by : **Daya Publishing House®**
A Division of
Astral International Pvt. Ltd.
– ISO 9001:2015 Certified Company –
4736/23, Ansari Road, Darya Ganj
New Delhi-110 002
Ph. 011-43549197, 23278134
E-mail: info@astralint.com
Website: www.astralint.com

Printed at : **Replika Press Pvt Ltd**

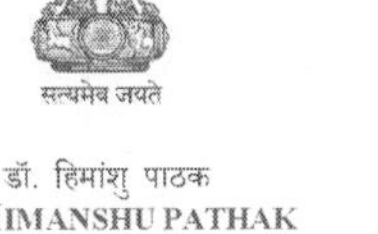

सत्यमेव जयते

डॉ. हिमांशु पाठक
DR. HIMANSHU PATHAK
सचिव (डेयर) एवं महानिदेशक (आईसीएआर)
Secretary (DARE) &
Director General (ICAR)

भारत सरकार
कृषि अनुसंधान और शिक्षा विभाग एवं
भारतीय कृषि अनुसंधान परिषद
कृषि एवं किसान कल्याण मंत्रालय, कृषि भवन, नई दिल्ली -110 001

GOVERNMENT OF INDIA
DEPARTMENT OF AGRICULTURAL RESEARCH AND EDUCATION (DARE)
AND
INDIAN COUNCIL OF AGRICULTURAL RESEARCH (ICAR)
MINISTRY OF AGRICULTURE AND FARMERS WELFARE
Krishi Bhavan, New Delhi 110 001
Tel: 23382629 / 23386711 Fax: 91-11-23384773
E-mail: dg.icar@nic.in

FOREWORD

India proudly boasts the second-largest technical workforce globally, trailing only behind China. In the realm of agriculture, estimates from agricultural scientists suggest that the country faces substantial losses, amounting to several crores, in terms of food grains due to pests and diseases. The disease management is the practice of minimising disease in crops to increase quantity or quality of harvest yield .Therefore, the crop management requires a multidisciplinary approach because pathogens (fungi, bacteria, viruses. nematodes) not only interact with each other, but with other biotic and abiotic factors to create heavy yield loss. The best way to ensure success of a disease-management program is to adapt it to the diseases expected and to use integrated disease-control measures.

The publication entitled **"Diseases of Field Crops and their Management" is** a commendable endeavor to provide comprehensive understanding of field crop diseases. The book systematically gathers a plethora of information on field crop's diseases and their effective management. The book is designed to cater to the needs of undergraduate and postgraduate students, researchers, educators, planners, and cultivators,

I would like to compliment the authors for the meticulous efforts for the publication and hope the book will be useful for various stakeholders.

(Himanshu Pathak)

Dated the 20th November, 2023
New Delhi

Preface

As we stand at the intersection of agriculture and scientific inquiry, it becomes increasingly evident that safeguarding our global food supply demands a comprehensive understanding of the challenges faced by those working the fields. In today's time, just as the entire world of humans is suffering from diseases, in the same way the world of plants is also suffering from diseases. We can't even imagine the microbes that are invisible so serious and dangerous for us. The speed with which the population has exploded is shocking and providing food to this population is a huge challenge. To save humanity, it is very important for us to save our trees and plants because ultimately they are the source of our food and nutrition. A human being is able to share his sorrow and pain but plants cannot speak. Because of this we have to observe them very closely. There have been many tragedies in history such as the Irish Famine, the Begal Famine, which shows that single plant disease can kill millions of people.

A big problem faced by growers, researchers and students to identify crop diseases and their eco-friendly management. There are so many literatures available about crop diseases but not systematically compiled to cover all the field crops. Keeping all these things in mind, the book "**Diseases of Field crops and their Management**" has been designed and compiled in such a way that it covers all the diseases of field crops and people can easily get information about the diseases of their crops, symptoms and eco-friendly management. This volume is not merely a compilation of facts and figures but a concerted effort to bridge the gap between scientific research and practical agricultural application. We have endeavored to distill the complexities of crop diseases into a format accessible to agronomists, researchers, students, and all those with a vested interest in the health and resilience of our crops. Each chapter delves into the nuances of specific diseases, offering insights into their etiology, symptoms, and management strategies. We explore not only the traditional methodologies but also the cutting-edge advancements in the field of plant pathology. The aim is to empower the reader with the knowledge needed to identify, understand, and combat these diseases effectively. In assembling this work, we have drawn upon the expertise of seasoned researchers, agricultural scientists, and practitioners from diverse corners of the globe. We extend our gratitude to the agricultural community whose ceaseless dedication ensures that our fields remain productive. It is our hope that this book serves as a valuable tool in their arsenal, aiding in the quest for sustainable and resilient crop management. As we embark on this exploration of diseases affecting field crops, we invite readers to delve into the intricate world of plant pathology, recognizing that a deeper understanding is the first step toward securing a future where our fields flourish, and our harvests thrive.

Mr. Shivam Kumar

Mr. Ravi Kumar

Dr. S.K. Biswas

Miss. Anjali Arya

Dr. Vishnu D. Rajput

Contents

Chapter 1

Diseases of Rice Crop

Mehjebin Rahman[1], Mateti Gayithri[1] and Merajul Hussain[1]

[1]Department of Plant Pathology, Assam Agricultural University, Jorhat – 785013, Assam

**Corresponding author: mehjebinr@gmail.com*

Introduction

Rice (*Oryza sativa*) is the one of the most important cereal crop, cultivated and consumed daily by more than half of the world's population. Due to large population depends on this tropical crop to meet their daily dietary needs, big scale production occurs in various parts of the world. Rice diseases, which are mostly caused by bacteria, fungus, and viruses, cause severe crop damage and production loss. The extent of plant damage induced by these diseases is determined by the intrinsic capacity of the plant. This chapter provides a comprehensive review of the numerous diseases that restrict rice crop growth caused by bacteria, fungus, and viruses.

Diseases of Rice Crop classified as below

Fungal Diseases

1. Blast (leaf, neck, nodal and collar): *Pyricularia oryzae* (P.S. = *Magnaporthe grisea*)
2. Brown spot: *Bipolaris oryzae* (P.S = *Cochliobolus miyabeanus*)
3. False smut: *Ustilaginoidea virens* (P.S. = *Villosiclava virens*)
4. Kernel smut: *Tilletia barclayana* (P.S. = *Neovossia horrida*)

5. Sheath blight: *Rhizoctonia solani* (P.S. = *Thanatephorus cucumeris*)
6. Sheath rot: *Sarocladium oryzae* (P.S. = *Acrocylindrium oryzae*)
7. Stackburn (Alternaria leaf spot): *Alternaria padwickii*
8. Stem rot: *Sclerotium oryzae* (P.S. = *Magnaporthesalvinii*)

Bacterial Diseases

1. Bacterial leaf blight (BLB): *Xanthomonas oryzae* pv. *oryzae*
2. Bacterial leaf streak: *Xanthomonas oryzae* pv. *oryzicola*
3. Grain rot: *Burkholderia glumae*

Viral Diseases

1. Rice dwarf virus
2. Rice grassy stunt virus
3. Rice hojablancatenuivirus
4. Rice ragged stunt virus
5. Rice tungro bacilliform virus
6. Rice tungro spherical virus

Nematode Diseases

1. Rice root-knot nematode: *Meloidogynegraminicola*
2. White rice tip nematodes: *Aphelenchoidesbesseyi*
3. Ufra disease of rice: *Ditylenchusangustus*

Physiological Disorders

1. Khaira disease of rice: Zn deficiency
2. Akiochi disease: Excess of H_2S

1. Bacterial Leaf Blight (BLB) of Rice

Causal organism: *Xanthomonas oryzae* pv. *oryzae*

Systematic Classification

The causal organism belongs to Domain: *Bacteria,* Phylum: *Pseudomonadata,* Class: *Gammaproteo bacteria,* Order: *Xanthomonadales,* Family: *Xanthomonadaceae,* Genus: *Xanthomonas,* and Species: *oryzae.* The pathogen is single-celled, rod-shaped gram-negative bacteria with dimension of 1.2 x 0.3–0.5 µm. They lack capsules and do not generate spores. On nutrient agar, their colonies are round, smooth, and pale yellow with an entire border.

Symptoms

There are three phases of bacterial leaf blight disease: Leaf blight, Kresek (The seedling blight or wilt phase) and Pale-yellow leaf. **Leaf blight:** Bacterial slime on newly formed lesions, resembling a drop of milky dew, is visible in the early morning. Later, behind the leaf, the slimy ooze hardens up and turns into tiny golden beads. Infected seedling leaves curl up and turn a greyish green colour. The illness spreads, causing the leaves to wilt and turn yellow to straw-colored, which causes the entire seedling to dry out and perish. **Kresek (the seedling blight or wilt phase):** Kresek damage to seedlings can occasionally be mistaken for early rice stem borer damage. Squeeze the bottom end of infected seedlings between your fingers to separate Kresek signs from stem borer damage. Yellowish bacterial ooze pouring out of the cut ends is a sign of kresek. Rice plants with kresek cannot be plucked from the ground like plants with stem borer infestations. **Pale-yellow leaf:** Lesions typically appear as water-soaked to yellow-orange stripes on leaf blades, leaf tips, or on leaf portions that have been damaged mechanically on older plants. The edge of the lesions is wavelike, and they advance to the leaf base. As a result of the development of numerous saprophytic fungi, old lesions change from yellow to greyish white with black spots. Lesions may reach the leaf sheath on severely diseased leaves.

Figure 1.1: BLB Symptoms in Rice Leaves (*Source*: IRRI Rice knowledge bank).

Etiology

Both tropical and temperate habitats are susceptible to its occurrence, particularly in lowland regions that get irrigation and rainfall. Weather conditions that favours the disease is relative humidity above 70 per cent and temperatures between 25 and 34 °C. Because bacteria from afflicted plants can enter other rice plant's roots and leaves through water, *Xanthomonas oryzae*.pv.

oryzae (Xoo) can spread swiftly throughout farms. Due to wind and water, other crops and rice paddies may contract the Xoo bacteria. In addition to common grasses and weeds, Chinese sprangletop, rice cut-grass (*Leersia oryzoides*), and other plants outside rice may also be impacted by Xoo. Xoo can survive in rice seeds, straw, other living hosts, water, and, for a brief period of time, soil when the climate is unfavourable for growing crops.

Disease Cycle

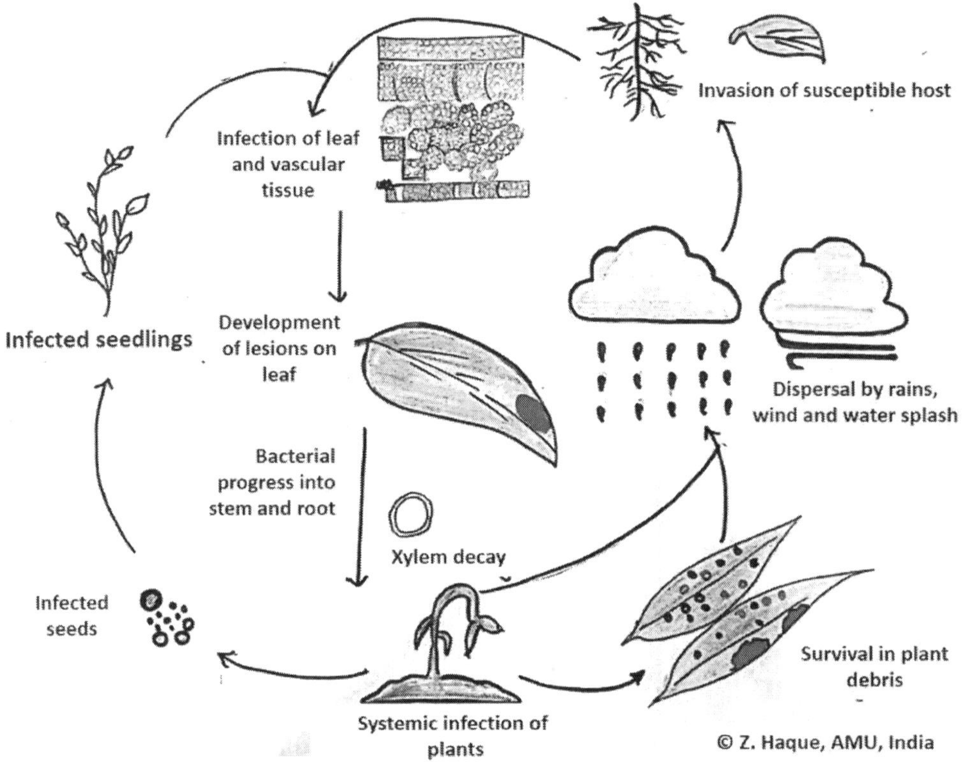

Figure 1.2: Disease Cycle of BLB. (X. oryzae. pv. orzae)

Primary source of inoculum are infected seeds with *Xanthomonas oryzae*. pv. *oryzae* or previous crop plant debris. Bacteria multiply through vascular tissue and develop symptoms of lesions on leaf surface further progresses into stem and root causing system infection. Secondary infection is caused by irrigation water or through dispersal by rain splashes. Bacteria survive in plant debris after harvesting of infected plants which serve as primary inoculum for next season crop.

Management

1. Nurseries to be maintained in upland conditions.
2. Planting resistant varieties such as IR 20, TKM 6.
3. Seeds should be treated with hot water.
4. A stringent commitment to sanitation is required, including the eradication of weed hosts and the plowing of straw, stubble, rice ratoons, and volunteer seedlings that may harbor bacterial infections.
5. With the right fertilizer treatment and plant spacing, the disease can be managed.
6. Ensuring good drainage of fields and nurseries.
7. When yield gets affected by severe illness, a copper-fungicide may be sprayed at the heading stage.

2. Bacterial Leaf Streak of Rice

Causal organism: *Xanthomonas oryzae* pv. *oryzicola*

Scientific Classification

The causal organism belongs to Domain: Bacteria, Phylum: Pseudomonadata, Class: Gammaproteo bacteria, Order: Xanthomonadales, Family: Xanthomonadaceae, Genus: *Xanthomonas*, and Species: *oryzae*. X. oryzae pv. *oryzicola* are aerobic gram-negative rod-shaped bacteria. Both spores and capsules are absent in the bacteria. Their only polar flagellum assists with their movement. They can grow positively at 28 °C. On nutritional agar, the bacterial colonies are circular, smooth, convex, viscid, and pale yellow.

Etiology

Areas with high temperatures and high humidity are prone to bacterial leaf streak in respect to growth stage of the plant, the disease is susceptible from tillering to initialial stages of panicle formation. With little loss in grain yield, mature rice plants may quickly recover from leaf streak.

Symptoms

Initially dark green streaks are observed which later turn light brown to yellowish grey. When held up to the light, the lesions are translucent. In really severe cases, entire leaves may turn dark and die. The lesions lengthen, infect larger veins, and turn brown. Under humid conditions, microorganisms on the lesions surface seep out and create tiny, yellow bands-like exudates. In extreme circumstances, the leaves dry out. The illness initially emerges as small, linear lesions within leaf veins that are soaked in water. Narrow brown spot and bacterial leaf streak can be mistaken for one another.

Figure 1.3: Bacterial Leaf Streak Symptoms in Rice Leaves (*Source*: IRRI rice knowledge bank).

Disease Cycle

Primary source of inoculum are infected seeds with *Xoo*or previous crop plant debris. Bacteria multiply through vascular tissue and develop symptoms of lesions on leaf surface further progresses into stem and root causing system infection. Raindrop splashes or irrigation water might disseminate secondary infections. Bacteria survive in plant debris after harvesting of infected plants which serve as primary inoculum for next season crop.

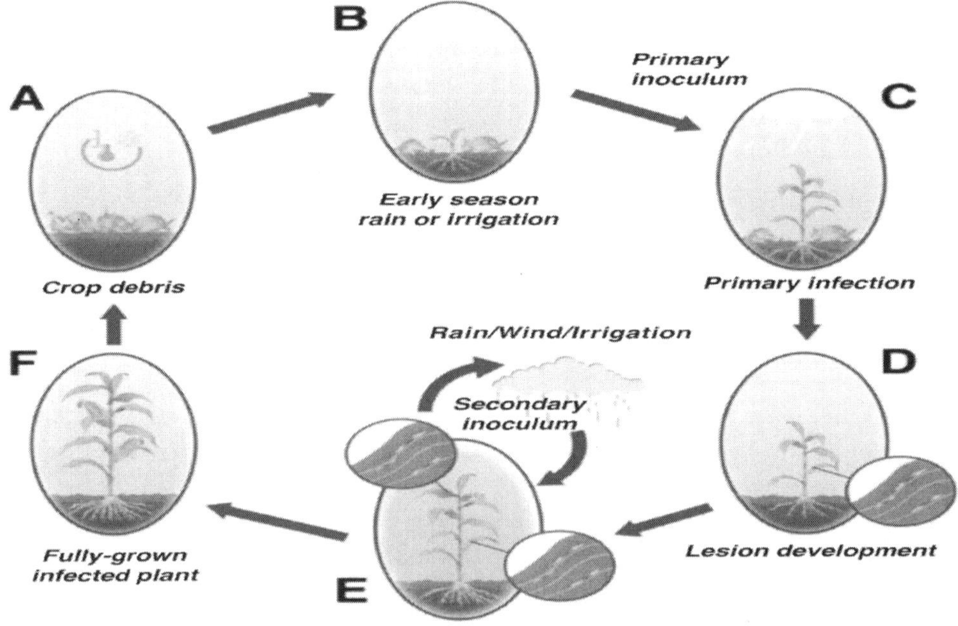

Figure 1.4: Disease Cycle of Bacterial Leaf Streak of Rice.

Management

1. Nurseries to be maintained in upland conditions.
2. Planting resistant varieties such as IR 20, TKM 6.
3. Seeds should be treated with hot water.
4. A stringent commitment to sanitation is required, including the eradication of weed hosts and the plowing of straw, stubble, rice ratoons, and volunteer seedlings that may harbor bacterial infections.
5. With the right fertilizer treatment and plant spacing, the disease can be managed.
6. Ensuring good drainage of fields and nurseries.
7. When yield gets affected by severe illness, a copper-fungicide may be sprayed at the heading stage.
8. Spraying of botanicals *viz.*, lemon grass or mint extract at 20 per cent also found to be effective.

3. Sheath Rot

Causal organism: *Sarocladium oryzae*

Scientific Classification

The fungi belongs to Order: Hypocreales, Phylum: Ascomycota, Family: Sarocladiaceae, Genus: *Sarocladium*, and Species: *oryzae*. The fungus produces white mycelium which issparsely branched andseptate. Conidiophores that form from the mycelium are branched once or twice, with 3–4 branches in each whorl, and are significantly thicker than vegetative hyphae. Conidia produced by the final branches, phialides, are single-celled, hyaline, fusiform, smooth, cylindrical to slightly typically moderately curved, and measure 4-9 x 1-2.5 m. In rare instances, the fungus infects the sheath in addition to bacterial infections (such as *Pseudomonas fuscovaginae*) that damage the sheath and discolour the grain. The fungus enters the rice plant by its stomata and wounds and spreads intercellularly in the mesophyll and vascular bundle tissues. Mycelium from the sheath rot fungus can be found on seeds and infected debris.

Etiology

Sheath rot is more prevalent in rainy seasons than dry ones and is present across majority of the world's paddy-growing countries, particularly in areas that are rainfed. When crops are starting to grow, it can also occur in areas with high relative humidity, substantial nitrogen fertilizer application, and temperatures between 20 and 28 °C. Higher planting densities favours spread of the disease. Plants which serve entrance points for the fungus such as wounds and injuries caused by stem borers during the onset of panicles are more prone to the disease. The fungus usually causes the uppermost leaf sheaths that surround newborn panicles to decay, which delays or prevents the emergence of the panicles.

Figure 1.5: Sheath Rot Symptoms in Rice Panicles (*Source*: IRRI rice knowledge bank).

Symptom

Lesions with grey centres and dark reddish brown edges develop in leaf sheath. Sheath of the flag's leaf gets discoloured and the leaf sheath covering the newborn panicles begins to rot. They get larger, frequently merge, and may completely cover the leaf sheath. Young panicles might become completely or partially encased in the sheath due to severe infection. Panicles that are not submerged decay, and the flowers change from red to dark brown. Inside the afflicted sheaths and juvenile panicles is a whitish powdery growth. Panicles and grains that have been infected are infertile, discoloured, and shrivelled. The disease is significant when the rice crop is getting ready to mature. Sometimes sheath rot lesions are mistaken for sheath blight lesions. Sheath rot, which is more common than sheath blight, interferes with panicle exsertion and lowers grain output by delaying or stopping panicle emergence, producing empty seeds, and producing sterile panicles leading to panicles rotting and discolored grains. Sheath blight affects the bottom parts of the plant and its leaves; it doesn't impede panicle exsertion. Check for lesions emerging on the topmost leaf sheaths enclosing the panicles, partially or fully opened panicles, sheath rot, and the formation of whitish powdery mildew growth on diseased leaf sheaths in order to confirm a sheath rot infection.

Disease Cycle

Sowing of infected seeds serves as primary inoculum. Germinated seedlings are unhealthy infected by sheath rot disease and dies early. The dead plants remains as plant debris in the field carrying the inoculum for next season. Airborne conidia serves as secondary inoculum and infecting plants at booting stage affecting sheath, panicles and grains.

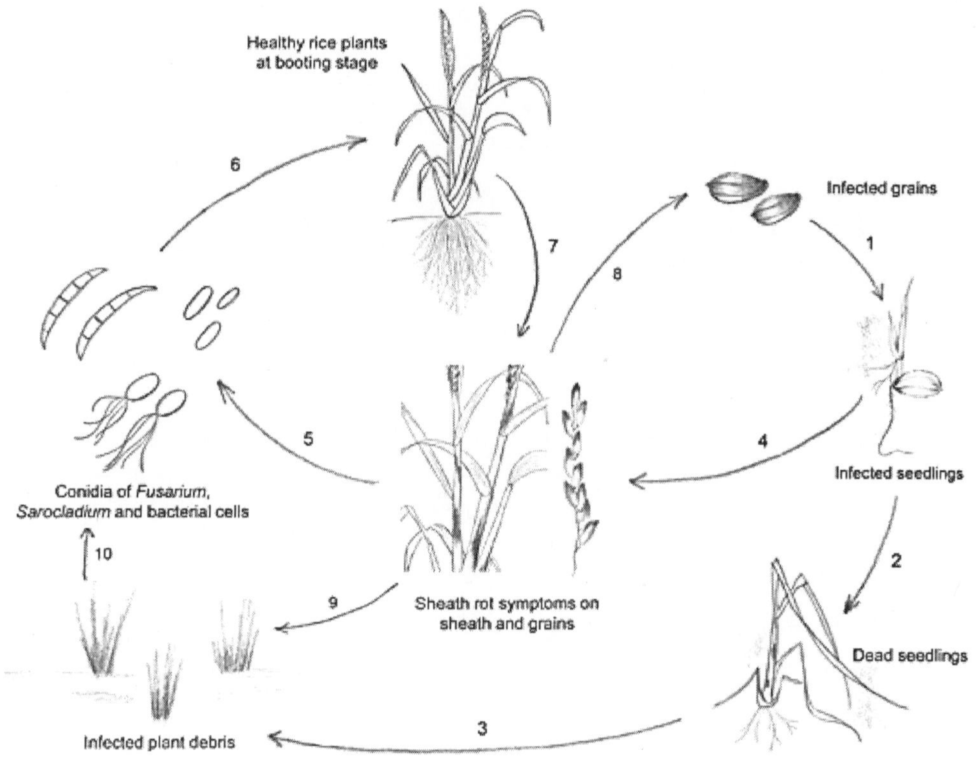

Figure 1.6: Disease Cycle of Sheath Rot of Rice.

Management

1. Biocontrol agents improve crop development and grain yield while reducing the severity of sheath rot by 20–42 per cent. Seed treatment with Bacillus subtilis @10g/kg followed by seedling dip treatment for 30 minutes can be done. 30 days after transplanting, foliar application of *Pseudomonas fluorescens* can be done by mixing 2.5kg with 50kg of FYM. Depending on the severity of the illness, a foliar spray of Bacillus subtilis at a 0.2 per cent concentration might be given 45 days after transplanting, every 10 days for a total of three applications,
2. Since it is a seed borne disease, healthy seeds must be used. Seeds must be treated with Mancozeb or Edifenphos before sowing.
3. Steer clear of dense spacing. Plant three seedlings per hill, spaced 20 cm apart in rows.
4. Using fertilizer containing potash during the tillering stage.

5. Maintenance of field sanitation and keeping away weeds.
6. Reduce the amount of insects in rice fields. Plants are injured by insects, which makes it possible for fungi to infiltrate the plant and spread infection.
7. 7. The fungi can cause infection in crops sown in field affected in earlier seasons as disease inoculum remains in crop residues. After harvesting, remove contaminated stubble.
8. Application of $CaSO_4$ and Zn SO_4 or benomyl by foliar spraying

4. Stem Rot

Causal organism: *Sclerotium oryzae*

Scientific Classification

The fungi belongs to Phylum: *Basidiomycota*, Subphylum: *Hymenomycotina*, Class: *Agaricomycetes*, Subclass: *Agaricomycetidae*, Order: *Agaricales*, Family: *Typhulaceae*, Genus: *Sclerotium*, and Species: *oryzae*. The sheath's exterior tissues contain globular, black perithecia. By the time the spores develop, the asci are narrowly clavate, with walls that are virtually imperceptible, and they are deliquescing. Sclerotia are smooth, nearly globose, black, and have a rounded shape. The presence of diseased rice plant remains in the field after harvest is more frequently linked to sclerotia than to free-living sclerotia in the soil. Sclerotia float to the water's surface and infect immature rice plant's leaf sheaths near the water's edge the next season, after the field has been flooded. Throughout the growing season, new infections keep popping up. Dark, upright, and septate features characterise the conidiophores. Conidia grow on pointy, fusiform, three-septate, curved, and erect.

Symptoms

When rice is water-sown, stem rot first occurs with tiny, black lesions on the leaf sheaths along the water line during the tillering stage. The condition progresses as the lesions grow. Infected sheaths peel off and die as the disease worsens. The culm may eventually get infected and stem rots. Inside the infected culms, there are countless microscopic white and black sclerotia and mycelium that are visible. Culm lodges became infected, resulting in empty panicles and chalky grain. Tiller death is a result of severe infection. The plants become lodged as a result of the disease.

Etiology

Infestation of leaf hoppers and stem borer makes fungal attack favourable. High doses of nitrogenous and potash fertilizers also make susceptible condition for the disease to occur. Stem rot symptoms in rice stems

Figure 1.7: Symptoms of Stem Rot of Rice (*Source*: IRRI rice knowledge bank, Plantix).

Disease Cycle

Small, black resting structures called sclerotia are how the fungus overwinters. Sclerotia can exist naturally in the soil, although they are more frequently connected to sick rice plant leftovers that are left in the field after harvest. After the field is flooded the following season, sclerotia float to the water's surface and infect new rice plants' leaf sheaths there. Throughout the growth season, new infections keep happening.

Management

1. Sclerotia in the field can be reduced by burning straw, stubble or any crop leftover after harvest.
2. Increasing soil pH with split applications of high potash and lime to balance fertiliser use minimises stem rot infection and boosts production.
3. Chemicals like Thiophanate-methyl and Fentin hydroxide, which are applied during the onset of the disease and mid-tillering, respectively, might lower the incidence of stem rot in rice fields.
4. Other fungicides, like Validamycin A and Ferimzone 600-800 ml/acre are also effective against the fungus.

5. Foot Rot or Bakane Disease

Causal organism: *Gibberella fujikuroi*

Scientific Classification

The pathogen belongs to Kingdom: Fungi, Division: Ascomycota, Class: Sordariomycetes, Order: Hypocreales, Family: Nectraceae, Genus: *Gibberella*, and Species: *fujikuroi*. The pathogen creates ascospores sexually, and they develop inside an ascus sac. The asci are 90-102 x 7-9 m in size, cylindrical, piston-shaped, and flattened above. They have four to six spokes, but eight are uncommon. Hyphae have septa and are branching. The fungus has microconidiophores that bear microconidia and macroconidiophores that bear macroconidia. Sclerotia are 80 × 100 m in size. They are round and dark blue. More often than not, the stroma areplectenchymatous and yellowish, brownish, or violet in colour.

Symptoms

Bakanae is both seed and soil borne fungal disease. Root rot occurs through soil infection. Through the roots or crowns, the fungus infects plants. Pink-white fungus grows on these plants from the soil level and works its way up the stems. The plant gets systemically infected by the fungus.Under the leaf sheaths, the nodes on diseased plants may be pink to purple, and the lower nodes may sprout roots. At flowering, airborne spores called 'conidia' and 'ascospores' grow on the tillers (stems and leaf sheaths) contaminate the grain. After being planted, infected seed creates elongated seedlings known as bakanae. Infected seedlings with lesions on their roots die in the seedbed. After transplanting, infected plants have very tall, pale, thin leaves, generate fewer tillers, and only partially or completely fill their grains. Additionally, spores are produced in high quantities on the straw and twigs, contaminating the panicles after harvest.

Figure 1.8: Foot Rot or Bakane Disease of Rice (*Source*: IRRI rice knowledge bank).

Etiology

High temperatures (>30°C) and dryland farming are favourable for bakanae infection. Root rot is favoured by lower temperatures and extremely dry soil.

Spores are carried by the wind or through water. On and in contaminated seed, long-distance spread occurs. The fungus survives in seeds for around 1-2 years and in soil for about a year. Up to 25 per cent of seed borne levels have been observed.

Disease Cycle

Sowing of infected seeds serves as primary inoculum. Germinated seedlings are unhealthy infected by sheath rot disease and dies early. The dead plants remain as plant debris in the field carrying the inoculum for next season. Air borne conidia serve as secondary inoculum and infecting plants at booting stage affecting sheath, panicles and grains.

Management

1. After every two years, rotate rice with resistant crops like vegetables or root crops.
2. Use seed that has been certified as being free of contamination. Infected seeds can be separated using salt water.
3. Fungicide seed treatment before planting with Thiram, Thiophanate-methyl, or Benomyl at 2g/kg is successful.
4. For dry seed coating, Benomyl or Benomyl-t should be used at 1-2 per cent of seed weight.
5. Steeping seeds in 1 per cent CuSO4 solution or 2 per cent Formalin is also advised.
6. Organomercurials such as Agrosan GN and Ceresan @ 2 g a.i./kg seed are particularly effective seed treatments.
7. The Trycyclazole or Azoxystrobilurin at 0.3 per cent is also efficient to manage disease.
8. As soon as you notice any sick plants, remove them.
9. As soon as possible after harvest, burn straw and stubble or till soil.

6. Blast of Rice

Causal organism: *Pyricularia oryzae* (**Sexual stage:** *Magnaporthe grisea*)

Scientific Classification

The pathogen belongs to Domain- *Eukarya,* Kingdom- *Fungi,* Phylum- *Ascomycota,* Class- *Sordariomycetes,* Order- *Magnaporthales,* Family- *Pyriculariaceae/ Magnaporthaceae,* Genus- *Pyricularia/Magnaporthe,* and Species- *grisea.* The conidia of pathogen are formed in clusters, conidiophores are long septate and olivaceous. They are pyriform to ellipsoid, hyaline to plae olive green in color, 3 celled attached at the broader base by a hilum. Ascospores inside the perithecia are hyaline, fusiform, 4 celled and slightly curved.

Etiology

Since the fungus produces spores all year round, the disease is mainly disseminated through airborne conidia. The infected straw and seeds include mycelium and conidia, serve as important sources of inoculum. The conidia may be spread to different fields by irrigation water. In just six days more spores can be produced when spores settle on leaves, germinate, enter the leaf, and cause a lesion. Primary infections are those caused by spores that travel great distances. Typically, leaves with primary infections have a few, widely dispersed spots on them. The original infections can give rise to spores that can start many other infections. This cycling is called as secondary spread.The serious outbreaks of blast in fields and localized areas are caused by secondary spread. *Panicum repens, Digitaria marginata, Brachiaria mutica, Echinochloa crusgalli,* and *Leersia hexandra* act as collateral hosts.

Favourable Conditions

Rainy days, gloomy sky, sporadic drizzles, extended dew period, and high relative humidity (93–99 per cent) with Low temperature at night (between 15-20°C or less than 26C) are favourable for the disease. The presence of collateral hosts and excessive amounts of nitrogen additionally contribute to the disease. Rice blast forecasting requires minimum nighttime temperatures between 20 and 26°C. and high relative humidity levels of 90 percent or higher for at least one week during any of the three susceptible crop growth stages–seedling stage, post-transplanting tillering stage, and neck emergence stage. Japan developed the first leaf blast forecasting model, called BLAST. Later on, a number of additional models, including PYRICULARIA, PYRIVIEW, BLASTAM, EPIBLA, and PBLAST, were also developed.

Symptoms

Leaf Blast

On the leaves, there are tiny lesions that have a greenish-blue fleck. These lesions become larger in wet conditions and resemble spindle-shaped spots with a greyish centre with a dark brown border. The spots converge and a significant portion of the leaves dry out and wither as the condition worsens. Sheath also develops spots. A severely infected field and nursery appear burnt.

Nodal Blast

Black lesions girdle on the nodes. All plant parts beyond the infected nodes could die as a result of the damaged nodes split apart.

Collar Blast

When spores undergo overwintering and infect the flag leaf's collar, the resultant symptoms include collar rot. The infection first appears close to the leaf sheath at the base of the flag leaf. Later on, the disease spreads upward to the leaf surrounding the flag leaf, which eventually turned brown, dried, and dropped off.

Collar Blast　　Leaf Blast　　Neck Blast

Figure 1.9: Blast of Rice (*Source*: Simkhada *et al.*, 2021).

Neck Blast

Fungi attack the peduncle during flower emergence, resulting in a brownish-black lesion called rotten neck, neck rot, or panicle blast. Grain filling is not caused by early neck infections; rather, partial grain filling is caused by late infections. Little brown to black spots may also be visible on the glumes of the panicles that are extremely infected. Yield losses from the pathogen could reach 61 per cent, depending on the infection stage.

Disease Cycle

It is theoretically possible that contaminated seeds initially colonise roots, which could lead to lesion formation and conidia spreading through the air. During a growth season, a single lesion can yield 2000–6000 conidia every day for up to 14 days. Several cycles of infection and reproduction serve as a source for secondary spread. However, a number of variables, such as temperature, rainfall, water depth in the paddy, amount of nitrogen applied to the rice, and degree of genetic resistance in the infected cultivar, can affect the number of cycles and number of spores produced on each individual lesion.

Management

1. Grow resistant to moderately resistant varieties like CO 47, IR 20, ADT 36, ADT 39, ASD 18, and IR 64.
2. It is not recommended to cultivate extremely susceptible types like IR 50 and TKM 6.
3. Eliminate the weed hosts from the channels and bunds in the field.

4. Pseudomonas fluorescens at 10 g/kg of seed or Captan, Thiram, Carbendazim, or Tricyclazole at 2 g/kg are used to treat the seeds.
5. Apply 300 mg/L of tricyclazole or 500 mg/L of carbendazim to the nursery.
6. Apply 500 ml/ha of Edifenphos to the main field.

7. Brown Spot

Causal Organism: *Helminthosporium oryzae* (**Sexual stage:** *Cochliobolus miyabeanus*)

Scientific Classification

The belongs to Domain- *Eukarya,* Kingdom- *Fungi,* Phylum- *Ascomycota,* Class- *Dothideomycetes,* Order- *Pleosporales,* Family- *Pleosporaceae,* Genus- *Cochliobolus,* and Species- *miyabeanus.* Mycelium is brown in colour and septate. Conidiophores are geniculate, brown in colour occur singly or in small groups. Conidia are pale to golden brown in color generally curved with a bulged centre with tapered ends and are 6-14 septate.

Symptoms

From seedling to milky stage, the fungus attacks the crop in the main field. The leaf blade and glumes show the clearest symptoms, which take the form of tiny spots on the coleoptile, leaf blade, leaf sheath, and glume. The dark brown with golden halo specks eventually become cylindrical or oval. The leaf dries out as several areas coalesce. Affected nurseries are frequently identifiable from a distance by their burnt look and the death of the seedlings. On glumes, dark brown or black dots also appear, discolouring the grain. It results in seed germination failure, seedling death, and decreased grain quality and weight.

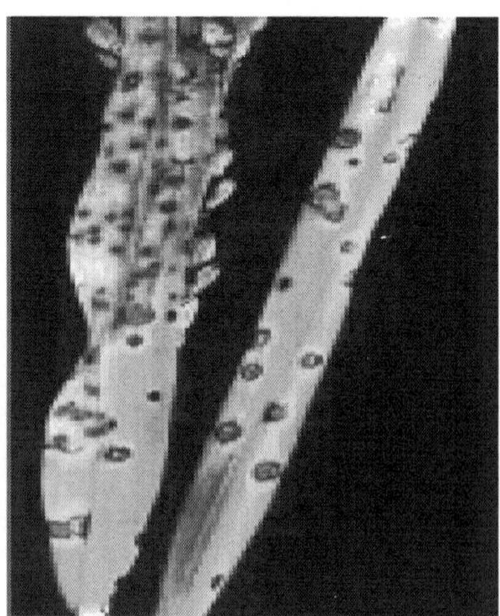

Figure 1.10: Brown Spot of Rice
(*Source*: Orillo *et al.,* 2014).

Favourable Conditions

1. Ideal conditions are between 25 and 30 oC and a relative humidity of more than 80 per cent.
2. An abundance of nitrogen worsens the condition.

Etiology

The primary infection is caused by infected seeds and stubbles. Conidia on contaminated grain and mycelium in diseased tissue are both still viable after two to three years. Conidia which are airborne infect the plants both in nursery and in main field. The brown spot fungus is typically found in regions where rice cultivation has a long history. Older lesions and infected debris release airborne spores that can cause disease. *Leersia hexandra* and *Echinochloa colonum* are examples of collateral hosts on which the fungus survives.

Disease Cycle

The principal source of inoculums for the fungus is considered to be the soil and infected plant parts, such as grains, straw, and stubbles, where it may remain in for two to three years. It has also been noted that some weed hosts act as reservoirs for inoculum. Necrotic lesions on the coleoptile and sheath of the initial leaves typically indicate the onset of the primary infection, which is caused by the infected seed. The secondary infection caused by airborne spores from the main lesions is mostly responsible for the subsequent lesion on leaves. Although B. oryzae spores were found in the air throughout the year, the amount found were higher in the cooler, more humid months.

Management

1. Field sanitation: removal of unwanted hosts and infected debris from the field.
2. Apply slow release nitrogenous fertilisers
3. Cultivate resistant cultivars like Bhavani and Co 44.
4. Use disease free seeds.
5. Apply 2.5 g/kg of Thiram or Captan to the seeds. For a nursery that is 20 days old, spray Edifenphos 40 ml or Mancozeb 80 g.
6. When the grade reaches 3, spray the crop in the main field with 500 ml of Edifenphos or 2 kg/ha of Mancozeb. Repeat after 15 days if necessary.

8. Sheath Blight

Causal organism- *Rhizoctonia solani* (**Sexual stage:** *Thanetophorus cucumeris*)

Scientific Classification

The pathogen belongs to Domain- *Eukarya,* Kingdom- *Fungi,* Phylum- *Basidiomycota,* Class- *Agaricomycetes,* Order- *Cantharellales,* Family- *Ceratobasidiaceae,* Genus- *Rhizoctonia* and Species- *solani.* It produces large number of spherical brown sclerotia. When the septate mycelium of the fungus is young, it is hyaline; as it ages, it turns yellowish brown. It produces a lot of brown, round sclerotia. The disease is soil borne. Fungus produces long usually long cells of sepatate mycelium which are hyaline when young, yellowish brown when old. Mycelia are formed in 3 distinct types. They are moniloid cells, lobate hyphae, and runner hyphae. Sclerotia are thick masses of mycelia. They are irregular, hemispherical, flattened at the bottom, and while young, they are white. As they grow older, they turn brown or dark brown. Sclerotia have a diameter of 1-6 mm. They might combine to create a bigger mass. Large sclerotia are more virulent.

Symptoms

The crop is affected by the fungus from the tillering to heading stages. On leaf sheaths close to the water's surface, the first signs are observed. Oval, elliptical, or irregular greenish-gray dots develop on the leaf sheath. With an uneven blackish brown or purple brown border, with the central part of the spots gradually turns greyish white as they become larger. From the water line to the flag leaf, entire tillers are covered by lesions on the upper regions of plants that spread quickly and coalesce with one another. When a leaf sheath has numerous big lesions, the entire leaf usually dies, and in extreme circumstances, the entire plant's leaves may be blighted. The infection extends to the inner sheaths resulting in death of the entire plant. Older plants are even more vulnerable. Strongly infected plants yield poorly filled grain, especially in the bottom portion of the panicle, throughout the early heading and grain filling growth stages.

Figure 1.11: A. Irregular lesions on the rice sheaths during tillering, B. Many lesions on the leaf gives the appearance of snake skin, C. Field view of sheath blight affected field (Turaidar *et al.,* 2018).

Favourable Conditions

1. High temperature (30–32 °C) and high relative humidity (96–97 per cent).
2. Closer planting.
3. High dosage of nitrogenous fertilizers.
4. Frequent rainfall
5. The presence of infection bodies or sclerotia.

Etiology

In dry soil, the pathogen may persist for about 20 months as sclerotia or mycelium, but only 5-8 months in moist soil. The water used for irrigation spreads sclerotia. The fungus has a broad host range.

Disease Cycle

While the production of basidiospores by *Thanatephorus cucumeris* on the host plant has the potential to initiate infection, this aspect of the epidemiology of rice sheath blight is often regarded as insignificant. In tropical habitats, the pathogen lives through sclerotia and mycelia in infected plant debris/straw, as well as through weed hosts; in temperate regions, sclerotia serves as the main source of inoculum. The fungus spread in double-cropped regions as a result of ratoons serving as a key inoculum source. Ploughing, levelling, transplanting, and weeding are examples of agricultural practises that assist the remaining sclerotia in reaching the water's surface and establishing their first contact with the host. The main sources of infection are rainwater runoff and flood irrigation, which allow floating sclerotia to spread widely (Singh *et al.*, 2016). 250 plant species, including commercially cultivated crops, are host to the pathogen (Chahal *et al.*, 2003).

Management

1. Grow resistant cultivars, such as Pankaj, Swarau Dhan, and Mansarovar.
2. Use organic amendments, such as FYM 12.5 tons/ha or neem cake at 150 kg/ha. Avert the transfer of irrigation water from contaminated to uncontaminated areas.
3. Deep ploughing in summer and stubble burning.
4. After 30 days of transplanting, apply 2.5 kg/ha of P. fluorescens to the soil (mix the product with 50 kg of FYM/sand and apply).
5. At the boot leaf stage, apply a 0.2 per cent folar spray of P. fluorescens, and then repeat ten days later.

9. False Smut

Causal Organism: *Ustilaginoidea virens* (**Sexual stage:** *Claviceps oryzae sativa)*

Scientific Classification

The pathogen belongs to Domain- *Eukarya,* Kingdom- *Fungi,* Phylum- *Ascomycota,* Class- *Sordariomycetes,* Order- *Hypocreales,* Family- *Incertae sedis,* Genus- *Ustilaginoidea* and Species- *virens.* Chlamydospores develop as warty, olivaceous, spherical to elliptical spore balls. Smaller, whiter, and smoother, younger spores. The centres of only a few of the green spore balls develop one to four sclerotia.

Symptoms

Figure 1.12: False Smut Symptom (*Source*: Akos *et al.,* 2019).

Individual ovary or grains are converted by the fungus into velvety-looking green spore balls. A panicle's impacted spikelets are few and far between. Each every grain of rice converted into a cluster of golden fruiting structures. Infected grain features greenish, velvety-looking smut balls. The smut balls appear small and steadily increase in size until they reach a diameter of 1 cm. In a panicle, only a few grains are diseased while the majority are healthy. The smut ball bursts and changes colour from orange to yellowish-green or greenish-black as the growth progresses. Stages of reproduction and ripening are typically impacted.

Favourable Conditions

1. Cloudy and rainy conditions throughout flowering and maturity.
2. Soils that are high in nitrogen.
3. The existence of fungi that overwinter as chlamydospores and sclerotia.

Etiology

Alternate hosts include grasses and varieties of wild rice. The primary source of inoculum is spores in the air. Chalmydospores are a secondary source of infection, whereas ascospores made from sclerotia serve as the primary source. Chlamydospores are air borne, abundant at heading stage.

Disease Cycle

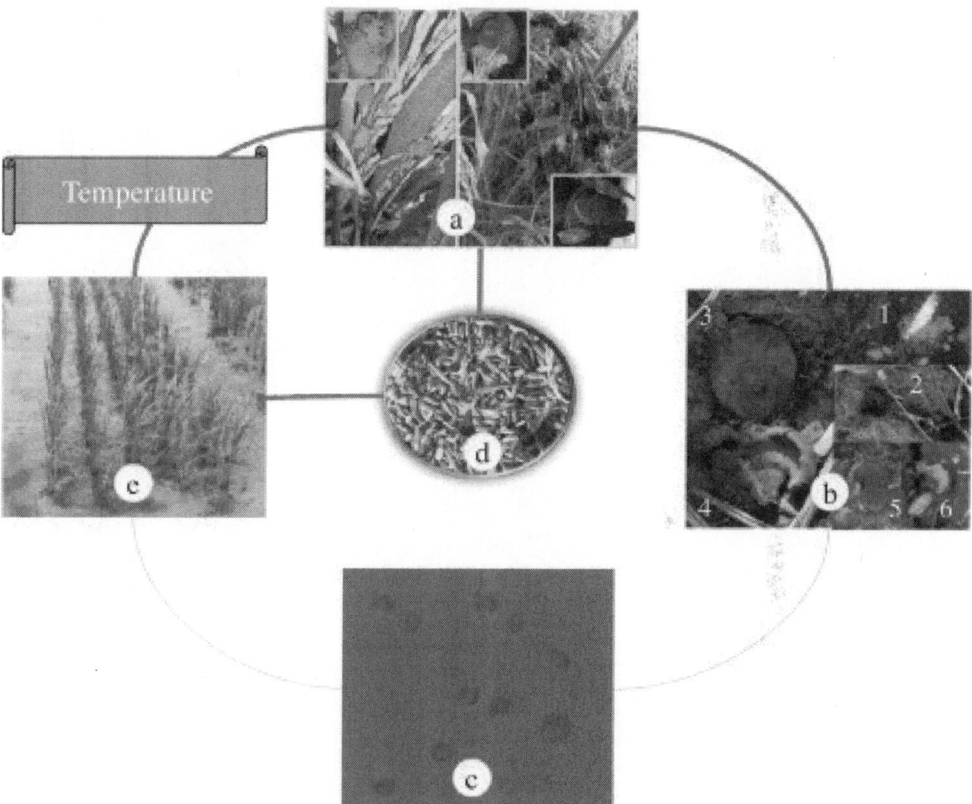

Figure 1.13: Disease Cycle of False Smut of Rice (*Source*: Moin U. Salam).

Due to limited mobility brought on by the lack of spores, the pathogen survives in unfavourable conditions by developing latent mycelia or sclerotia (Anees *et al.*, 2010; Summer, 1996). According to Brooks (2007), sclerotia can endure in soil for a maximum of two years and can be distributed during field preparation and flood irrigation. By moving to nearby fields with regularly flooded irrigation and attaching themselves to accessible plants, the Sclerotia could spread and cause false smut disease. Plus, ideal circumstances facilitate the pathogen's growth. Early signs appear as water-soaked ulcers on the leaf sheath near the waterline.

Management

Preventive Methods

1. Use disease free seeds, control insect pests, nitrogen in split application is suggested, removal of infected plant debris.

Cultural Methods

1. Destruction of straw and stubble from infected plants.
2. During harvesting, diseased plants should be removed and destroyed so that sclerotia do not fall in the field.
3. Early planted crop has less smut balls than the late planted crop.

Chemical Methods

1. Seed treatment with Carbendazim 2.0g/kg of seeds.
2. Spraying of Blitox-50 or Propiconazole at 2.5g/lt.
3. Spray hexaconazole at 1ml/lt or Chlorothalonil 2g/lt.

10. Udbatta Disease

Causal Organism: *Ephelis oryzae* **(Sexual stage:** *Balansia oryzae-sativa)*

Scientific Classification

The pathogen belongs to Domain- *Eukarya,* Kingdom- *Fungi,* Phylum- *Ascomycota,* Class- *Sordariomycetes,* Order- *Hypocreales,* Family- *Clavicipitaceae,* Genus- *Ephelis* and Species- *oryzae. Pycnidiospores of pathogen are hyaline, needle shaped and 4-5 celled.*

Figure 1.14: Udbatta Disease of Rice (*Source*: Rice Knowledge Management Portal).

Symptoms

Appear at the time of emergence of the panicles. Since the fungal mycelium has matted the infected panicle together, the entire ear head is transformed into a straight, compact, cylindrical, black spike-like structure. The size is noticeably smaller and the spikelets are bonded to the core rachis. Greyish stroma encases the entire spike, with convex pycnidia submerged in it.

Favourable Conditions

1. Presence of bacteria on leaves and in the water or those surviving in debris.
2. High humidity and warm temperature.
3. Early stage of planting from maximum tillering to panicle intiation.

Etiology

Fungus is seed borne externally and systemic. Presence of lustrous grayish white films of fungal growth in young leaves of the infected seedlings suggest the entry of the pathogen during germination of seeds.

Disease Cycle

The fungus is seed borne, infection being initiated at the time of emergence of the panicle. Seedling infection is the principle mode of infection.

Management

1. The pathogen is internally seed borne.
2. Solar treatment of seeds is effective in killing the pathogen carried in the seeds, if any.
3. Hot water seed treatment at 45 °C for 10 min. effectively controls the disease.
4. Removal of collateral hosts.

References

Akos, I. S., Yusop, M. R., Ismail, M. R., Ramlee, S. I., Shamsudin, N. A. A., Ramli, A. B. and Chukwu, S. C. (2019). A review on gene pyramiding of agronomic, biotic and abiotic traits in rice variety development. *International Journal of Applied Biology*, **3**(2): 65-96.

Anees, M., Edel-Hermann, V. and Steinberg, C. (2010). Buildup of patches caused by Rhizoctoniasolani. *Soil Biology and Biochemistry*, **42**(10): 1661-1672.

Brooks, S. A. (2007). Sensitivity to a phytotoxin from Rhizoctoniasolani correlates with sheath blight susceptibility in rice. *Phytopathology*, **97**(10): 1207-1212.

Chahal, K. S., Sokhi, S. S., and Rattan, G. S. (2003). Investigations on sheath blight of rice in Punjab. *Indian Phytopathology*, **56**(1): 22-26.

Orillo, J. W., Cruz, J. D., Agapito, L., Satimbre, P. J. and Valenzuela, I. (2014). Identification of diseases in rice plant (oryzasativa) using back propagation Artificial Neural Network. In *2014 International Conference on Humanoid, Nanotechnology, Information Technology, Communication and Control, Environment and Management (HNICEM)*, pp. 1-6.

Simkhada, K. and Thapa, R. (2022). Rice blast, a major threat to the rice production and its various management techniques. *Turkish Journal of Agriculture-Food Science and Technology*, **10**(2): 147-157.

Singh, R., Sunder, S., and Kumar, P. (2016). Sheath blight of rice: current status and perspectives. *Indian Phytopathology*, 69(4): 340-351.

Sunder, S., Singh, R. A. M., and Agarwal, R. (2014). Brown spot of rice: an overview. *Indian Phytopathology*, **67**(3): 201-215

Turaidar, V., Reddy, M., Anantapur, R., Krupa, K. N., Dalawai, N., Deepak, C. A. and Kumar, K. H. (2018). Rice sheath blight: major disease in rice. *International Journal of Current Microbiology and Applied Sciences*, **7**: 976-988.

eagri.org

www.agritech.tnau.ac.in

plantvillage.psu.edu

ipm.ucanr.edu

Chapter 2

Diseases of Wheat Crop

***Vishwa Vijay Raghuvanshi*[1]*, Archana Kushwaha*[2]*, Prabha Siddharth*[3] *and Girijesh Kumar Jaisval*[3]**

[1]Research Scholar, Department of Plant Pathology, Acharya Narendra Deva University of Agriculture and Technology, Kumarganj, Ayodhya, U.P.
[2]Research scholar, Department of Plant Pathology, Govind Ballabh Pant University of Agriculture and Technology, Pantnagar, Udham Singh Nagar, Uttarakhand
[3]Research Scholar, Department of Plant Pathology, Chandra Shekhar Azad University of Agriculture and Technology, Kanpur, U.P.
**e-mail: raghuvanshi550vv@gmail.com*

Introduction The most important staple food in the world is wheat. It ranks second only to rice as one of the major food crops in India. On approximately 26 million hectares, or million tonnes of grains, it is grown. *Triticum aestivum* L., which makes up the majority of the Indian wheat crop despite being planted primarily in Northern and Central India, accounts for around 20 per cent of the land under cereals and about 20 per cent of the yearly production. A little amount of durum wheat is cultivated in South India. They can be grown everywhere from the arid Deccan Plateau to an altitude of roughly 2800 meters in the Himalayas. Over a period of five to six months, the Indo irrigated the environment. It may be grown as a dry crop. The finest area for growing wheat is the Gangetic tract, where it is grown primarily wherever there is sufficient rainfall as well as the primary crop in areas with irrigation capabilities. In India, there are more than one hundred varieties of wheat with various botanical and agronomic characteristics; about 50 of these are now being cultivated in various regions. The two most significant diseases that can affect wheat are rusts and smuts. Additionally, to fungus-related diseases, there are a few bacterial, viral, and nematode-related conditions. There is a quick description of the major diseases affecting wheat in India.

Table 2.1: List of Major Diseases of Wheat and their Causative Agent

Sl.No.	Name of Disease	Causal Agent
A. Fungal diseases		
1.	Black rust/Stem rust	*Puccinia graministritici*
2.	Brown rust/Leaf rust	*Puccinia recondita*
3.	Yellow rust/Stripe rust	*Puccinia striiformis*
4.	Loose smut	*Ustilago nuda tritici*
5.	Flag smut	*Urocystisagropyri*
6.	Karnal bunt/Partial bunt	*Neovossia indica*
7.	Hill bunt	*Tilletia indica*
8.	Powdery mildew	*Erysiphe graminis tritici*
9.	Leaf blight	*Alternaria triticina*
B. Nematode diseases		
1.	Ear cockle/Seed gall/Wheat gall	*Anguina tritici*
2.	Molya/Cereal cyst nematode	*Heterodera avenae*
C. Nematode and bacterial associated disease		
1.	Tundu/Sehu/Yellow ear rot	*Anguina tritici* and the bacterium *Clavibacter tritici.*
D. Viral diseases of wheat		
Sl.No.	Name of Viral Diseases	Vector
1.	Wheat streak mosaic	Eriophyid mite (*Aceria tulipae)*
2.	Wheat spindle streak mosaic	*Polymyxa graminis*
3.	Soil borne wheat mosaic virus	*Polymyxa graminis*

1. Rust

One of the earliest crop plant diseases is rust. Wheat rust is mentioned in the Bible. The rust disease of wheat and barley was well known to ancient Romans and Greeks. Since then, man has had to coexist with it while continuously attempting to stop its spread through a variety of tactics. In India, the rust fungus result in an annual loss of 8-15 per cent of grain productivity. There are three main rust infections that damage wheat. They are: Black or stem rust (*Puccinia graminis* f. sp. *tritici*), Leaf or brown or orange rust (*Puccinia recondite*) and Yellow or stripe rust (*Puccinia striiformis*).

A. Black Rust/Stem Rust

Occurrence and Economic Importance

In the Central provinces, it was severe in 1839 and nearly completely destroyed the wheat crop. Over two million tonnes of grain were lost to the rust epidemics in Madhya Pradesh, Maharashtra, Rajasthan, and Uttar Pradesh in 1946-1947. West Bengal, Bihar and the eastern portions of Uttar Pradesh saw severe rust in 1956–1957, rendering some tracts of grain unusable for harvest, not even for seed. According to estimates, wheat rust in India will cause a loss in grain yield of at least 15 per cent, costing farmers close to Rs. 1 billion. In practically every region of India where wheat is produced, black rust can be found. In the northern parts of India, the disease typically appears in March as the crop is approaching maturity and only slightly reduces grain output, however in the southern regions, it typically appears in November or December and causes significant losses. This rust is also capable of attacking barley.

Symptoms

The flecking of leaves, leaf sheaths, culms, and floral structures is the first sign of rust infection. These flecks are the precursors to uredosori, which quickly grow into rectangular, reddish-brown pustules that frequently converge before exploding to reveal a mass of brown uredospores. The entire leaf blade and other afflicted sections will seem brownish from a distance when several uredosori burst and discharge their spores. Teleutosori are created in the later part of the season. They are noticeable, oblong or linear, dark brown to black, and frequently converge to form linear patches of black lesions, hence the name black rust.

Figure 2.1: Black Rust of Wheat (*Puccinia graminis* f. sp. *tritici*).

Causal Organism

Black rust is an obligate parasite obtain their food from only living organisms. The mycelium that makes up the fungus's body is well-branched. Primary mycelium is septate, endophytic, intercellular, branching and monokaryotic. Dikaryotization causes the secondary mycelium, which emerges later in the life cycle, to be dikaryotic. Haustoria are produced by mycelium having an apex that resembles a knob via which it takes nutrients from the host cell. *Puccinia graminis* f.sp. *tritici* is a macrocyclic rust fungus that produces all five types of spore stages: Stage- 0 (Pycniospores/spermatia), Stage- I (Aeciospores), Stage- II (Urediospores), Stage- III (Teliospores), Stage- IV (Basidiospores). Karyogamy and meiosis occur in stage- III (Teliospores).

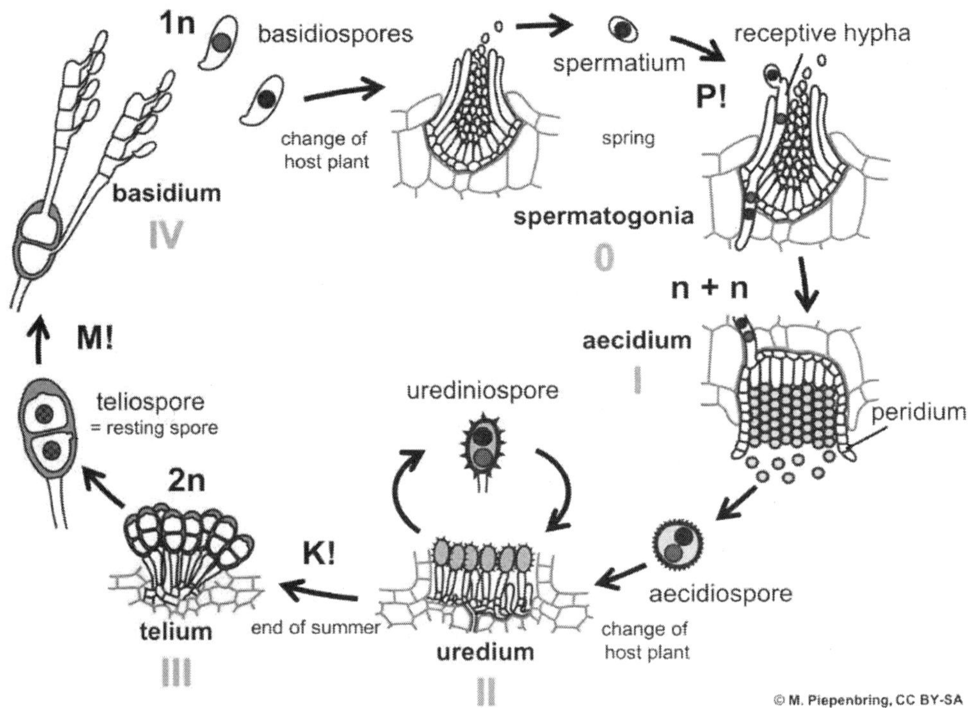

Figure 2.2: Life Cycle of *Puccinia graminis tritici* (*Source*: M. Piepenbring).

Disease Cycle

Puccinia graminis f.sp. *tritici* needs more than one host species to complete its life cycle since it is a heteroecious full-cycle rust. Understanding the many stages in the rust life cycle has been made easier by the thorough investigations conducted by numerous researchers over the past 70 years. There are five spore stages in black rust fungi. Barberry is the alternate host of black rust of wheat. Pycniospores and aeciospores present on barberry. Uredinio spores and telio spores present on wheat. Black rust fungus enters in host through stomata.

Management

1. Growing resistant variety like Lerma rojo, Chhoti lerma, Sonalika, HD- 2733, PBW 343 *etc.* is the greatest option among the numerous controls for wheat rust.
2. Certain agricultural techniques, including altering the sowing date to allow for disease escape or lessening the intensity of infection, may be beneficial.
3. Growing crops with a short or extended lifespan is helpful for these goals, and applying balanced fertilizers to the crop lessens the severity of disease.
4. Eradication of alternate host (Barberry).
5. Release hyperparasite like *Endarluca carisis* or *Darluca filum.*
6. Seed treatment with oxycarboxin @ 2 gm/kg of seed.
7. Spray of oxycarboxin 2.5 gm/liter of water or dusting of sulphur @ 7-8 kg/ha.

B. Leaf/Brown/Orange Rust

Economic Importance and Occurrence

The distribution of this rust is global. The Northern and Eastern regions of India experience the most rust. It does greater harm than stem rust in Punjab, Bihar and Uttar Pradesh. Sometimes the disease has a severe impact when other favourable environmental factors are combined with substantial winter rainfall. 1.5 million tonnes of grain were lost in the epiphytotics of 1972–1973. It can be found on crops growing in the plains and hills of South India.

Figure 2.3: Brown/Leaf Rust of Wheat (*Puccinia recondite*).

Symptoms

The disease's initial sign is the emergence of tiny, rounded, orange sori that are sporadically dispersed on the leaves and infrequently on the stem or leaf sheath. With age, the sori turn brown. The telial stage may develop in the same pustule as the disease worsens. Small, black, two-celled, oval to linear, and covered by the epidermis, the telia are. The leaf sheath is where the telia are also created.

Causal Organism

The fungus *Puccinia recondita* appears in uredial and telial stages on wheat and several other grasses, and in aecial and pycnidial stages on species of Thalictrum. It is unclear exactly what function *Thalictrum flavum* and *Thalictrum javanicum* play as substitute hosts in India. *Isopyrum fumarioides* is well-known for serving as a natural alternative host in Russia. The uredospores are 20–35 in diameter, brown, spherical, minutely echinulate, and contain 7-10 germ pores. They begin to grow by means of germ tubes that protrude from the pores and infect the wheat leaf via the stomata. This rust appears to penetrate by a similar method to *P. graminis tritici*. The uredospores are made visible by the rupture of the epidermal wall covering the uredosori. Telia are uncommon, but when they do form, they are mainly found on the underside of leaves and do not rupture. The smooth, rectangular, thick-walled, brown teleutospores with ounded, conspicuous thickening tip and long paraphyses that split the sori into compartments. They are 35-56 x 12-23 and moderately compressed at the septum. Eleven species of Thalictrum are known to host the fungus pycnidial and accial stages, which resemble

Disease Cycle

The foothills of the Himalayas, as well as the plains of Tamil Nadu and Karnataka in the south, are where the rust begins to take hold in early January, and occasionally much earlier. In the plains of Karnataka, inoculum initially begins to accumulate. It goes towards the north. The northern foothills' inoculum spreads to the plains. This explains why the brown rust shows up in the Western highlands of North India a little later. The rust populations in the north and the south migrate in different ways before uniting and spreading severe disease throughout the states that farm wheat.

Management

1. This disease is also subject to the stem rust control techniques.
2. Numerous wheat cultivars have been found to be resistant to different races of *P. recondite* like Sonalika, Sonara- 64, HD- 2967, Arjun *etc.*
3. Eradication of alternate host and weed.
4. Seed treatment with Plantvax (Oxycarboxin) @ 2 gm/kg of seed.
5. Spray of Triazole fungicide like Propiconazole @ 2.5 ml/liter of water.

C. Yellow/Stripe Rust

Economic Importance and Occurrence

Puccinia striformis, which causes yellow rust, is found all over the world. In the Gangetic plains of India, it affects wheat and barley frequently, while it rarely affects wheat in the western and peninsular regions. Despite being present every year, it is rarely severe and causes significant crop loss.

Symptoms

Bright yellow pustules called uredosori are typically found on leaves, but in cases of severe illness, they can also form on spikelets, glumes, spikelets, and leaf sheaths as well as on grains. Stripe rust gets its name from the elongated sori that are organised in rows between the leaf veins. The majority of the sori are sub-epidermal, remain hidden under the epidermal layer, and only become visible when the crop is fully mature. Late in the season, the teleutosori also appear and are organised in rows. Compared to the leaf blade, they are more numerous on the leaf sheaths. The uredosori may frequently be the site of teleutospore formation. They are subepidermal, compressed, and long and black. Unknown are the pycnidial and accial stages of the fungus.

Figure 2.4: Yellow/Stripe Rust of Wheat (*Puccinia striiformis*).

Causal Organism

Wheat and a few closely similar hosts, such as *Agropyron semicostatum, Bromus catharticus, Hordeum murinum etc.* are hosts for the heteroecious *Puccinia striiformis,* which can develop in both uredial and telial phases. It is unknown if the fungus completes its life cycle or if it can live on other hosts. The uredospores are 6-16 germ pores thick, hyaline, minutely echinulate, and spherical to oval in shape. They measure 23-35 x 20-35 u. They easily germinate, producing germ tubes with binucleate cells that can infect wheat and barley. The teleutospores measure 35-63 x 12-30, have two cells, thick walls, and are flattened at the top. They are dark brown in colour. A small number of spores are present in the teleutosori, which are also filled with numerous unicellular brown, long paraphyses. The uninucleate sporidia and promycelia produced by the teleutospores are easily germinated but are unable to infect wheat or any other known host plant. The fungus has over 40 known races, including races 13, 14, 19, 20, 24, and 31, as well as the 'New Race' A, D, E, F, G, and H, as well as three biotypes of races 14A, 20A, and 38A, which are common in India.

Disease Cycle

It is known for the fungus to overwinter in its uredial stage in England and other nations, but it is unknown how it survives in India. It may be able to live on wheat plants raised in the Himalayas at an altitude of roughly 1,500-1,800 metres. After a time of dormancy, the uredospores begin to germinate and can serve as a source of inoculum for wheat that is sown early. The early wheat crop in Uttar Pradesh is more badly affected by the fungus than the late soy crop. *Agropyron semicostatum* is a common weed. Over 90 additional grasses are also susceptible to stripe rust and are most likely the source of early season infection.

Management

1. This rust of wheat is thought to be less significant than the other two, and it only occasionally manifests in a severe form on the northern plains.
2. Growing resistant wheat cultivars is the most effective way to control this rust, as it is with the other two rusts.
3. Removal of overwintering host.
4. Seed treatment with Oxycarboxin @ 2 gm/kg of seed.
5. Spray of Tebuconazole or Oxycarboxin @ 3 ml/liter of water.

2. Loose Smut

Economic Importance and Occurrence

The spread of the disease is global. It can be seen on wheat fields in both the lowlands and the hills of India. Although the prevalence of smut is generally low, the illness frequently causes serious damage, wiping out up to 30 per cent of the cars in specific fields, although the overall loss is only about 2-3 per cent.

Causal Organism

Loose smut of wheat is caused by *Ustilago nuda* var. *tritici,* which belongs to Kingdom- *Fungi,* Phylum- *Basidiomycota,* Order-*Ustilaginales,* Family – *Ustilaginaceae,* Genus –*Ustilago* and Species- *unda.* The chlamydospores of the fungus are a light shade of olive-brown, spherical to oval in shape, and 5 to 9 mm in diameter. They are also covered in tiny echinulations on the wall. They germinate easily in water, each spore generating a promycelium, a germ tube with four cells. The stigma allows the promycelial cells to enter the ovary where they unite to form the germ tubes, which stay dormant until seed germination. When the plant reaches maturity and a smoky car develops, it becomes visible. At least 8 different races of loose smut exist, with the races LSR-LSR, appearing in Punjab.

Symptoms

Only after the ears emerge from the boot leaf, typically a few days before the healthy ones, do loose smut symptoms appear. It is challenging to distinguish between healthy and unhealthy plants up until that point. Typically, all spikelets are afflicted and have turned into a mass of spores that are black and powdery. Each smutted spikelet has a thin, silvery membrane covering it before it emerges; this membrane breaks as the ear does.

Figure 2.5: Loose Smut of Wheat (*Ustilago nuda* var. *tritici*).

Disease Cycle

The fungus lives on in the seed as latent mycelium, as was already mentioned. When a seed germinates, the dormant mycelium expands along with the seedling to reach the inflorescence at the time of earhead formation, where it arrives and the fungal hyphae begin to multiply. When the binucleate hyphal segments become chlamydospores (smut spores), they enlarge, round out, and develop thick walls. Although the smut spores are likewise binucleate, they unite to form a single diploid nucleus when they reach maturity. Reduction division takes place at the time of spore germination, and the haploid nuclei are transferred to the promycelial cells, which are initially non-infectious before fusing to form the dikaryotic germ tubes, which are able to infect through the flower and establish themselves in the ovary. The fungus is internal to seeds and endures for a very long time in preserved seeds. However, the smut spores are only viable for a short period of time. Therefore, embryo infection is the primary means of continuation.

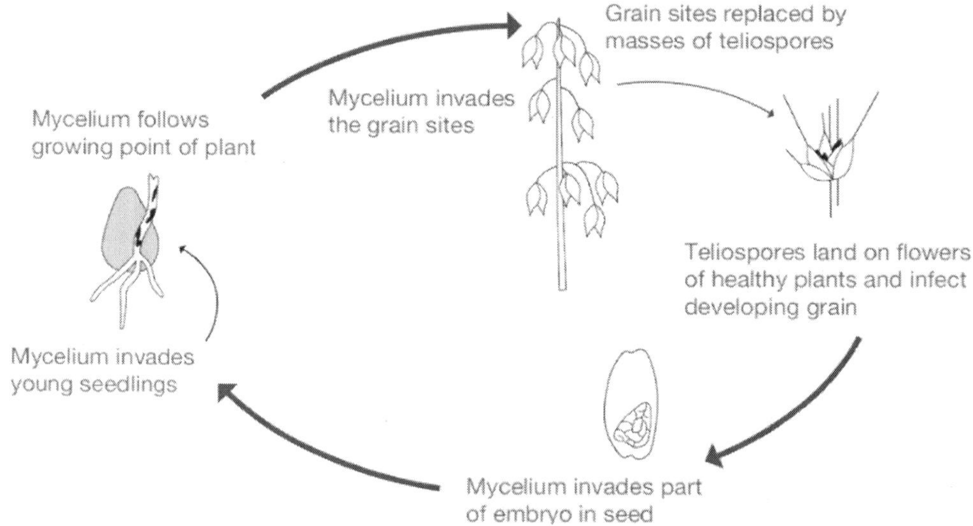

Figure 2.6: Disease Cycle of Loose Smut of Wheat (*Source*: ahdb.org.uk).

Management

1. Hot water treatment of the seeds helps get rid of the interior disease.
2. In order to activate the mycelium and make it more sensitive to heat, the seeds are often steeped in cold water for four hours.
3. Solar heat treatment (given by Luthra and Sattar) should be given for disease management. In Punjab, Luthra created the sun heat treatment, a variant of this technique, in 1934. By this approach, seeds are uniformly soaked in water for approximately four hours, then spread out in thin layers beneath the hot summer sun for at least four hours when the temperature reaches about 50°C. The seeds are then gathered and kept in dry storage until it is time to plant them.
4. Use of disease resistant variety like Kalyansona, WG 307, C 302, PV 18 *etc.* and use of contaminated free seed.
5. Seed treatment with Carboxin @ 2 gm/kg of seed.
6. Spray of carboxin (Vitavax) or Benlate.

3. Flag Smut

Economic Importance and Occurrence

India, Japan, Australia, China, various regions of Europe, and the United States of America (USA) are all familiar with the disease. Even though it was first noted in India in 1906, it is only thought to be of modest significance. It only occurs in Punjab, Uttar Pradesh, Rajasthan, Himachal Pradesh, and Delhi.

Symptoms

Figure 2.7: Flag Smut of Wheat (*Source*: ahdb.org.uk).

Flag smut mostly affects the leaf, although it can sometimes occasionally spread to the stem and vehicle. Grey to greyish black linear sori, which are first hidden by the epidermis before rupturing to reveal the spore mass, develop on the leaf blade and sheath, frequently between the veins. Instead of being at the base, the infection is more prevalent near the top of the leaf blade. In more vulnerable cultivars, the infection may cause the plant to become dwarfed by shortening the internodes and stopping the growth of the inflorescence. The plant is severely stunted, the leaves are bent, and no ears form if the infection occurs when the seedling stage is in progress.

Causal Organism

Flag smut of wheat is caused by *Urocystis agropyri* (Preuss) Schroet, which belongs to Kingdom- *Fungi,* Phylum- *Basidiomycota,* Order-*Ustilaginales,* Family – *Ustilaginaceae,* Genus –*Ustilago* and Species- *agropyri.* The spores are grouped together in balls that have a fertile centre that is black, surrounding by a ring of sterile cells that are lighter in colour. They have an amorphous, sub-spherical, or oblong form. They grow to form three to four septate hyaline basidia or, less frequently, non-septate basidia that, at the tip, create three to four basidiospores. The basidiospores have thin walls, are hyaline, and are rectangular, measuring 12–15 x 3. They develop into slender infection hyphae after germinating. Although there have been reports of different strains of the fungus from other nations, no research has been conducted on this issue in India.

Disease Cycle

Both seed and soil-borne, the spores infect the primary shoot or the developing buds at germination before becoming systemic and causing the recognisable sori.

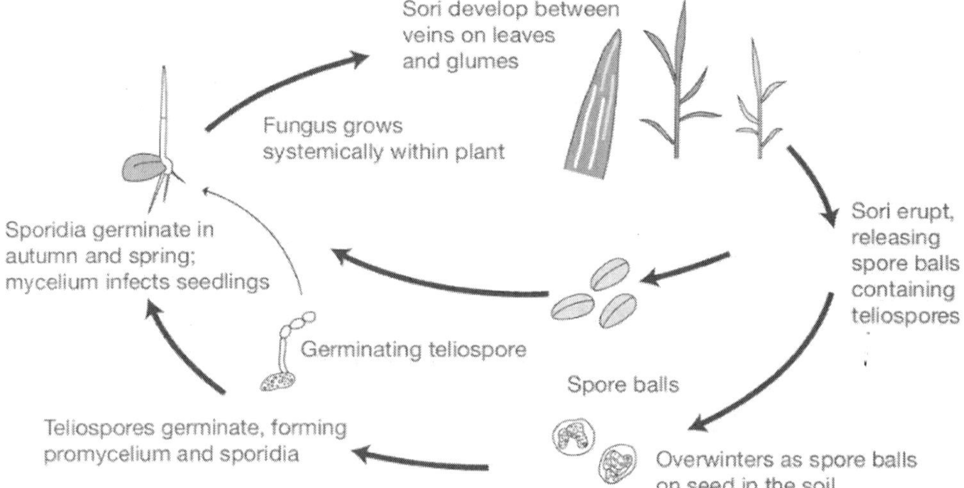

Figure 2.8: Disease Cycle of Flag Smut of Wheat (*Source*: ahdb.org.uk).

Management

1. Pusa-4 is resistant to this smut in India.
2. Crop rotation should be followed.
3. Early crop sowing.
4. Stubble burning and hot weather cultivation are cultural practises that lower the incidence of disease.
5. Given that the organism is both seed and soil-borne, treating the seed with Sulphur or Vitavax @ 2-3 g/kg is beneficial in lowering the initial infection from seed-borne spores.
6. Seed dressing with Evergol extend (Penflufen and Trifloxystrobin) effective against flag smut
7. Vitavax, and Plantvax are effective in suppressing the disease as foliar spray.

4. Bunt Diseases

Economic Importance and Occurrence

Bunt of wheat occurs widely all over the world. It causes huge loss in India of wheat growing states like U.P. Punjab, Haryana and M.P.

Causal Organism

There are three different bunt diseases of wheat in India. They are: Common bunt or Old bunt or Hill bunt or Complete bunt or stinking smut caused by two species of *Tilletia, T. tritici* (= *T. caries*) and *T. laevis* (= *T. foetida*), Karnal bunt or Partial bunt (*Neovossia indica or T. indica*) and Dwarf bunt (*T. controvers*). The three fungi belong to the Phylum- *Basidiomycota*, Order - *Ustilaginales* and Family - *Ustilaginaceae*.

Symptoms

Figure 2.9: Bunt of Wheat (*Source*: Plantix).

No symptoms are present in any of the three bunt disorders until the ear head appears. The plants in the first two ripen before the healthy ones do, and the ears have a dark green colour. The spore balls (sori), which form in place of kernels, push the glumes apart. When the balls are crushed during harvest, the spores are released. Typically, all of the grains in an ear of the first two species are impacted and turn into bunt balls. The heads of some kinds may be slenderer, while those of other varieties may be spreading and more open.

There is a distinctive fishy smell in the air as the spores are dispersed. When it comes to Karnal bunt, the infection is typically, but not always, limited to a few grains in the spike, and the arrangement is a little disorganized. In some instances, the virus could just affect a portion of the grains. The inner glumes expand and the outer glumes spread out as the grains mature, revealing the bunted grains. The pericarp initially encloses the bunt balls, but as it bursts, the masses of spores are exposed and produce the characteristic bunt odour.

Disease Cycle

For the first two bunts, the infection only affects very immature plants; the vulnerable window is between 8 to 10 days following germination. Following infection, the fungus spreads along the shoot's tip. The inflorescence and spikelets are where the fungal hyphae concentrate during flowering, turning the ovary into smutted tissues that enclose masses of chlamydospores or bunt spores. According to reports, the spores can survive in dry conditions and in large quantities for more than 25 years. Although soil-borne spores play a role in some regions, such as the Pacific Northwest of the United States, where soil infestation is the primary factor in the cycle of these species, seed-borne spores are the primary source of infection. In this instance of Karnal bunt, the spores in the soil germinate, resulting in a significant amount of primary sporidia with a needle shape and secondary sporidia with a sickle shape. Soon after the ear develops, the sporidia become airborne and infect the developing kernels or ovaries. The infection only occurs during or just before flowering because the fungus is unable to pass through the established seeds. Due of these unique criteria, the percentage of spikelets infected in an automobile is typically low.

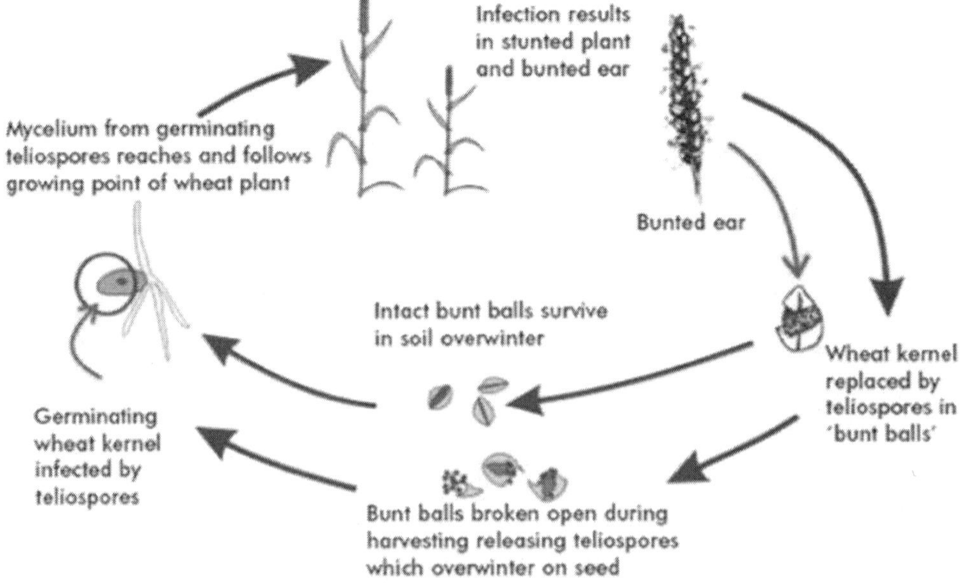

Figure 2.10: Disease Cycle of Bunt of Wheat (*Source*: eagri.org).

Management

1. Use of disease resistant variety like PBW- 343, PBW- 373, HD- 2329, Kalyansona, PV- 18, Panjano- 62, HD- 2012, HI- 7715, WL- 1002, E- 2337, HD- 4513 and HD-4519 are a few examples of resistant varieties.
2. Crop rotation should be followed.
3. Dipping of seeds in 20 per cent brine solution and separate floating seeds.
4. By treating the seeds with substances like Sulphur, $CuSO_4$ or $CuCO_3$ @ 2g/kg seed.
5. Benomyl and other systemic fungicides provide great control. The Karnal bunt cannot be managed with seed treatments because it spreads by air borne spores. There is no known cure for this smut that works well.

5. Powdery Mildew

Economic Importance and Occurrence

Powdery mildew is mostly prevalent in the hills of north and south India. It is of minor importance in northern plains.

Symptoms

The leaf develops a powdery growth that is greyish white. Additionally, leaf sheaths and floral components exhibit same symptoms. The disease-causing powdery growth may eventually turn black, forming distinctive lesions of varied sizes on the affected areas, where tiny dark perithecia are visible, as the disease progresses.

Figure 2.11: Powdery Mildew of Wheat.

Causal Organism

The disease powdery mildew of wheat is caused by *Erysiphe graminis* var *tritici,* which belongs to Phylum- *Ascomycota,* Order- *Erysiphales,* Family- *Erysiphaceae.* The fungus typically spreads in the air in the form of conidia or oidia. The 20-30 by 8-10 µm elliptical, hyaline, single-celled, thin-walled conidia are sized. They land on the surface of the leaf, germinate, and infect the host by haustoria that pierce the epidermis, it continues to spread on the host's surface as an ectoparasite and forms short conidiophores that are perpendicular to the host surface, creating chains of conidia on them. This facilitates the wind's simple dispersion of the conidia. When favourable conditions exist, the fungus creates the perfect or perithecial stage on the host surface as the disease progresses and the infected host tissues perish. The 200 µm diameter black globose cleistothecia feature short appendages, 9-30 asci, and a diameter. The conical to oval asci measure 70-110 x 25-40 µm and feature lengthy pedicels. Eight rectangular, hyaline, thin-walled ascospores with measurements of 20-25 x 10-13 are present in each ascus. In India, the fungus has been documented in five different races.

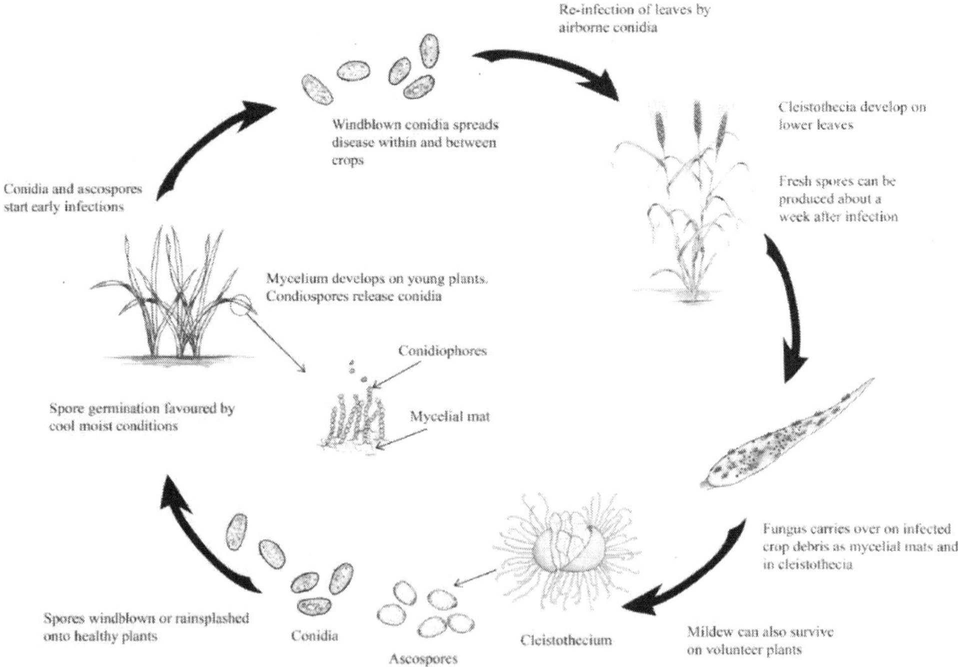

Figure 2.12: Powdery Mildew of Wheat (*Source*: extensionaus.com.au).

Disease Cycle

The pathogen remains in the cleistothecial stage during the off-season. Primary infection is brought on by ascospores discharged during the crop season

(December–January), while secondary infection is brought on by oidia carried by the air: The pathogen generates perithecia as the crop matures, and these perithecia persist on the straw and debris after harvest, serving as the principal source of inoculum for the following season.

Management

1. Extensive efforts have been made in the USA and Canada to develop resistant cultivars, with positive outcomes. In India, NP- 710, and NP- 718, there has been some progress in this direction. The diseases are only mildly resistant to K- 53, E- 750, and C- 591.
2. Hyperparasite *Ampelomyces quisqualis* (AQ-10) should be used for powdery mildew management.
3. Simple water spray can also used for powdery mildew management.
4. Sulphur dusting the plants would lessen the infection, it is expensive and impractical. It has been observed that using 1 per cent Sodium thiosulfate as a foliar spray can effectively manage the disease. Additionally, highly good at preventing the condition is Benlate. Tridemorph (Calixin) is a systemic fungicide widely used for management of powdery mildew.

6. Leaf Blight

Economic Importance and Occurrence

In 1962, Prasada and Prabhu reported this wheat disease in India. In some areas of Maharashtra, it is common. West Bengal, Uttar Pradesh, and Bihar. Certain wheat varieties suffered serious damage from it, and more recently, it is said to have grown to terrible proportions.

Symptoms

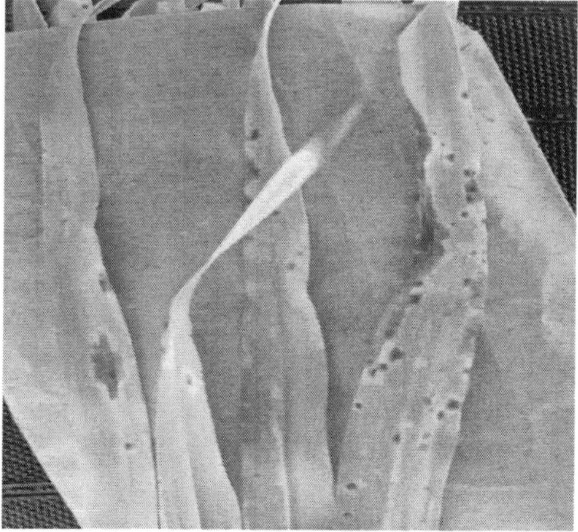

Figure 2.13: Leaf Blight of Wheat.

When the crop is 7-8 weeks old, the disease initially manifests itself in the field. The leaves first exhibit characteristic reddish-brown localized oval patches that grow to become irregular with a brilliant yellow edge during the early stages of infection. The leaf may become blighted when several spots combine to cover sizable portions of it. The youngest leaves, though, typically do not have an infection. Even from a distance, areas that have been severely infected look burned. High humidity and a temperature of about 25 °C encourage the illness. If the infection occurs at or before the boot leaf stage, the drop in grain yield in some cultivars can be up to 99 per cent.

Causal Organism

Late blight of wheat is caused by *Alternaria triticina*, which blongs to Phylum-*Deuteromycouna*, Order- *Monthales*, Family- *Dematiaceae*. There have also been reports of *A. triticola* and *A. tenuissima* producing leaf blight. Hyaline hyphae that change from olive-buff to deep olive-buff and are produced by the fungus cause leaf blight. Being mature through the stomata, conidiophores of the same colour that are 17-28 x 3-6 L in size and containing conidia generated acrogenously, singly or in chains of 2 to 4 are formed. They range in colour from light brown to dark olive-buff, are smooth, cllipsoidal, irregularly oval, and have rounded ends.

Disease Cycle

The conidia that distribute the inoculum, which infects the leaves to cause the blight, are primarily airborne. Additionally, it is externally seed-borne, with seasonally viable spores. Additionally, seed infection may occur during the later stages of kernel development. These infections become deeply ingrained and serve as a source of internal seed-borne illnesses. The only time the fungus sporulates is at night. As a result, wind-borne spores are created, which might cause secondary illness. The fungus has been classified into two races.

Management

1. Use of disease resistant variety like Sonalika, Sonara- 64, HD- 2733, Arjun *etc.*
2. Seed treatment with *Trichoderma viride* @ 4 gm/kg of seed.
3. The seed must be treated before sowing because the pathogen is seed-borne, both internally and externally. To do this, the seed should first pre-soak in water for four hours, followed by a 10-minute dip in hot water at 52°C.
4. Spraying the crop with Zineb @ 2.5 kg/ha can control the disease, starting from the time it first manifests. It is best to avoid using seed from affected fields. By regularly applying Zineb sprays, secondary spread can be somewhat controlled.

7. Tundu or Yellow Ear Rot

Economic Importance and Occurrence

Punjab, Haryana, as well as a few locations in Rajasthan and Uttar Pradesh, have recorded cases of the disease. Hutchinson made the initial observation in 1917. Additionally, reports of the same illness have come from China, Australia, and areas of Egypt.

Symptoms

Yellow slime on the stem and inflorescence, which dries out to form sticky yellow coatings and causes the spikes to bend and twist, is one of the diseases defining signs. The majority of the grains in the earhead are replaced by nematode-created galls that contain the harmful bacterium.

Figure 2.14: Tundu Disease of Wheat (*Source*: Plantix).

Causal Organism

It is caused by the combined action of the wheat nematode *Anguina tritici* and the bacterium *Clavibacter tritici.*

Disease Cycle

Young seedlings are also affected by the nematode, which results in galls, twisted, and distorted leaves. In mature plants, the earhead is twisted and the usual kernel has been replaced with galls. The condition is known as car cockle disease, and it frequently results in very serious damage when the worm and bacterium are together.

Management

1. By soaking the seeds in salt brine at a 20 per cent concentration to separate the galls via floatation, the disease can be effectively managed.
2. Hot water treatment at 52°C for 45 minutes is best for managing this disease.
3. Healthy seed should be used.

Other diseases of wheat which are of minor importance in India but of major concern in other wheat growing areas of the world, are those listed below:

1. Take-all disease caused by *Gaumannomyces graminis*. A disease similar to this was reported to be observed in North India during 1935-36.
2. Seedling blight caused by *Rhizoctonia solani* Kuhn, or *Fusarium* spp. has been reported to occur sporadically in many parts of India.
3. *Sclerotium* rot caused by *Sclerotium rolfsii*, is a pathogen on many crop plants including wheat, causing leaf spot and blight.
4. Leaf blight caused by *Cladosporium herbarum* is present in parts of North India
5. Black-chaff of wheat caused by *Xanthomonas campestris* pv. *undulosum*. Dye is another bacterial disease of wheat occurring in some parts of United Provinces and Punjab.
6. Dilophora leaf spot caused by *Dilophoraalopecuri* (Fr.) is commonly found in Kashmir valley.
7. Molya disease caused by the nematode *Heterodera avenae* is reported from Rajasthan.
8. Mosaic streak of wheat caused by the 'Chirke' virus of large cardamom was reported. Mosaic and striate mosaic are the two other virus diseases.

References

Anonymous (1966). Symposia on Diseases of Wheat, Maize, Sorghum and Millets Bull. Indian Phytopath. Soc. No. 3.

Anonymous (1966). Indian Farming: Special Rice Number. Indian Council of Agr Res., New Delhi.

Anonymous (1967). Proc. Intern. Symp. Plant Pathol, Dec. 27, 1966-Jan. 1, 1967, Indian Phytopath. Soc., New Delhi.

Anonymous (2015). https: //commons.m.wikimedia.orgaccessed on 20 May 2015.

Butler, E.J. and Jones, S.G. (1961). Plant Pathology. Macmillan, London. Gangopadhyay, S. (1983): Current Concepts on Fungal Diseases of Rice. Today and Tomorrow, New Delhi.

Ghosh, R. L. M., Chatge, M. B. and Subramanyam, V. (1960). Rice in India. Indian Council of Agr. Res., New Delhi.

Gnanamanickam, S. S. and Mahadevan, A. (1988). Advances in Research on Plant Pathogenic Bacteria. Today and Tomorrow, New Delhi.

Jacobs, T. H. and Parlevliet, J. E. (1993). Durability of Disease Resistance. Kluwer, Dordrecht, The Netherlands.

Joshi, L. M., Srivastava, K. D., Singh. D. V., Goel, L. B. and Nagarajan, S. (1978). Annotated Compendium of Wheat Diseases in India. Indian Council of Agr. Res., New Delhi.

Kannaiyan, S. (1987). Advances in Wheat Pathology TN. Agr. Univ., Coimbatore, Tamil Nadu.

Manibhushanrao, K. (1996). Sheath Rot Disease of Rice. Daya Publishing House, Delhi.

Mehta, K. C. (1940). Further Studies on Cereal Rusts in India. Part I. Govt. of Indian Publ. Lal Chand, Calcutta.

Mehta, K. C. (1952). Further Studies on Cereal Rusts in India. Part II. Govt of India Publ., Lal Chand, Calcutta.

Mundkur. B. B. and Thirumalachar, M. J. (1952). Ustilaginales of India. Commonwealth Mycol. Inst., Kew, England.

Nagarajan, S. and Muralidharan, K. (1995). Dynamics of Plant Diseases. Allied, New Delhi.

Ou, S. H. (1985). Rice Diseases. Commonwealth Agr. Bureaux, Slough, UK.

Padmanabhan, S. Y. (1977). Fungal Diseases of Wheat in India (A Critical Review), ICAR, New Delhi.

Padwick, G. W. (1950). Manual of Rice Diseases. Commonwealth Mycol. Inst., Kew, Surrey, England.

Pal, B. P. (1966). Wheat. Indian Council of Agr. Res., New Delhi.

Ramakrishnan, T. S. (1963). Disease of Millets. Indian Council Agr. Res., New Delhi.

Rangaswami, G. (1962). Bacterial Plant Diseases in India. Asia Publ. House, Bombay.

Rangaswami, G. and Rajagopalan, S. (1973). Bacterial Plant Pathology. TN. Agr. Univ., Coimbatore, Tamil Nadu.

Raychaudhuri, S. P. (1977). A Manual of Virus Diseases of Tropical Plants. Macmillan, Delhi.

Roelfs, A. P. and Bushnell, W. (1983). Cereal Rusts. Academic Press, New York.

Sprague, R. (1950). Diseases of Cereals and Grasses in North America. Ronald, New York.

Sundararaj, D. D. and Thulasidas, G. (1983). Botany of Field Crops. Macmillan, Delhi.

Suryanarayana, D. (1978). Seed Pathology Vikas, New Delhi. Tarr. S.A.J. (1962): Diseases of Sorghum, Sudan Grass and Broom Corn, Commonwealth Mycol. Inst., Kew, Surrey, England.

Chapter 3

Disease of Barley Crop

Kishan Lal[1], Saurabh Kumar[2], Gajendra Pratap[2], Shivam Kumar[2] & Vishwajit Rai[1]

[1]Assistant Professor, Department of Plant Pathology, National P.G. College, Barahaganj Gorakhpur (U.P.)- 373402
[2]Department of Plant Pathology, Chandra Shekhar Azad University of Agriculture and Technology, Kanpur – 208002.
**Corresponding author: kamalsaurabh8843@gmail.com*

Introduction

Barley (*Hordeum vulgare* L.) is a significant cereal crop their grain used for various purposes, belonging to the Poaceae family. Initially, barley was likely used as a staple food, but it evolved primarily into a grain for Animal feed, Malting and Brewing, partly due to the growing prominence of wheat and rice. Now - a- days, approximately two-thirds of the barley harvest has been designated for livestock feed, 1/3 for malting purposes, and around 2 per cent for direct human consumption. However, over the course of its historical development, barley has consistently served as a vital dietary staple for specific societies, particularly in areas like Asia and northern Africa (Newman and Newman, 2006). Barley can be considered one of the most adaptable cereal grains, thriving in diverse climates, including higher latitudes, elevated altitudes, and even arid desert regions. In extreme environments, like Ethiopia, Himalayan nations, and Morocco it continues to be a primary food source. Early on, barley was recognized for its robust flavor and high-energy content. For instance, Roman fighters were known as "hordearii" or "barley men" due to their consumption of barley for power and stamina (Percival, 1921). In modern times, the incorporation of barley to make different food products and Barley consumption is driven by its capability to provide good health. Barley β-glucans have been widely acknowledged for their effectiveness to reduce cholesterol level in blood and glycemic index in barley-based food products (Behall *et al.*, 2004, Fadel *et al.*, 1987, Newman *et al.*, 1989) and its rich Tocol content, containing Tocopherols and Tocotrienols, contributes to reducing Low-Density Lipoprotein cholesterol through antioxidant action (Qureshi *et al.*, 1986, Qureshi *et al.*, 1991). The recent confirmation of health claimed by the United State Food and Drug Administration for soluble barley β-glucans in minimize cholesterol levels in blood could further stimulate the production of barley-based food products and people having interest in these byproducts. Barley exhibits remarkable genetic diversity among grains of other cereal crop. Barley can be categorised based on factors such as spring types or winter types, 2 - row or 6 – row Barley, hulled or hulless Barley and by end type of use as Malting or for Feed. Additional categorization is rooted in the grain's makeup, encompassing standard, waxy, high amylose starch variations, those high in lysine, rich in β-glucan, and devoid of proanthocyanidins. These diverse classes of barley frequently exhibit unique physical and compositional attributes, leading to a range of processing characteristics and ultimate applications. This wide-ranging genetic variation presents abundant prospects for the discovery and development of barley cultivars customized for particular objectives. Nonetheless, it may present a difficulty for both processors and final consumers when there is a lack of comprehensive guidance for the selection of barley grains suitable for different purposes. Despite considerable endeavors to manage the pathogens accountable, diseases persist as a substantial hindrance to global barley production. These biotic obstacles, coupled with the escalating regulations for environmental preservation, have spurred increased endeavors to explore more sustainable methods of disease management. Such methods must find an equilibrium between shielding crops from biotic risks such as diseases, pests, and weeds while minimizing environmental damage, demanding a comprehensive strategy to crop protection.

A. Fungal Disease	
Leaf blotch, scald or Rhynchosporium	*Rhynchosporium communea*
Powdery mildew of Barley	*Blumeria graminis f.* sp. *hordei*
Net blotch	*Pyrenophora teres*
Brown rust	*Puccinia hordei*
Yellow rust	*Puccinia striiformis*
Take-all	*Gaeumannomyces graminis*
Leaf stripe	*Pyrenophora graminea*

B. Bacterial Disease	
Bacterial stripe	*P. syringae* pv. *striafaciens*
Bacterial leaf blight	*P. syringae* pv. *syringae*
Basal glume rot	*P. syringae* pv. *atrofaciens*
Black chaff and bacterial streak	*Xanthomonas translucens* pv. *translucens*
C. Nematode Disease	
Root knot nematode (Root Knot nematode of Barley)	*Meloidogyne naasi, M. chitwoodi*
Molya disease od Barley	*Heteroderaavenae, H. filipjevi*
Root lesion nematode	*Pratylenchus spp.*

1. Net Blotch

Economic Importance and Occurrence

Net blotch can lead to significant reduction of yield. The Net blotch affects the leaves of the plant, reducing photosynthesis and thus impacting grain production. Quality Impact: Besides yield reduction, net blotch can also affect the quality of barley grains, especially when it is intended for malting and brewing purposes. The presence of disease symptoms on barley grains can lead to lower malt quality, affecting the brewing industry. Farmers may incur costs for disease management, including the application of fungicides or adopting other control measures. These costs can significantly impact the overall economics of barley cultivation. Net blotch disease is found in various barley-growing regions across India. It is more common in some states like Punjab, Haryana, Uttar Pradesh and Himachal Pradesh, which are significant barley-producing states. Climate Influence: The disease occurrence is influenced by environmental conditions. The disease thrives due to the conducive conditions of elevated humidity and mild temperatures during the spring and summer. The presence of volunteer barley plants, especially in rotations after minimum tillage or continuous barley cropping, can play a part to the persistence of disease. The susceptibility of Barley varieties to net blotch varies, and some varieties may be more resistant than others. Planting disease-resistant varieties is one approach to manage the disease.

Causal Organism

The fungi *Pyrenophora teres* f.sp. *teres* (*Drechslera teres*), cause disease net blotch. The division of this fungi is Ascomycota while Subdivision Pezizomycotina. The Class of fungi is Dothideomycetes while Order-Pleosporales, Family- *Pleosporaceae* and Genus- *Pyrenophora*.

Symptoms

In seedlings and tillering plants, a seed infection can lead to the development of brown stripes that originate from the base of the leaves. These symptoms may resemble those of leaf stripe infection. Initially, they appear as isolated spots, which can be easily confused with ramularia leaf spot. As the condition progresses, these spots cylindrical and transform into brown coloured stripes or spots, characterized by a random pattern of darker lines running along and across the leaf. During the winter, these symptoms can become quite extensive, but affected leaves tend to die back. However, in the spring, new leaves may be free of these symptoms. There is also a less common "spot form" of net blotch, where oval lesions are enclosed by a chlorotic yellow halo. In contrast to the net pattern, these marks do not stretch out; instead, they expand to a diameter of 3-6 mm. It's worth noting that this symptom may also be confused with Ramularia leaf spot, but form of spot lesions do not have a rectangular shape and not confined to the veins of leaf. Net blotch disease may occur across the growing season, whereas the symptoms of Ramularia leaf spot appear later when flowering take place. Additionally, severe symptoms of these conditions often lead to yellowing of the leaves. In late-stage infections, the glumes and awns of the plant can also be affected, resulting in dark brown flecking and striping.

Figure 3.1: Net Blotch (*Source*: ahdb.org.uk).

Favourable Condition

Barley residue and volunteer plants, such as second barley crops following minimal soil disturbance, are vulnerable to infection. This susceptibility is heightened when specific conditions are met, including the use of susceptible barley varieties, elevated humidity, and mild spring and summer temperatures. Additionally, early sowing of barley, a high prevalence of seed infection, and dense crop can contribute to the risk of disease.

Disease Cycle

This disease is carried through both seeds and plant debris. Historically, infected seeds played a significant role in spreading the disease. However, nowadays, the primary source of infection is plant debris. Infected stubble and leftover plant material, along with volunteer barley plants, provide a habitat for the pathogen to survive during the winter. The Spores of fungi can be generated both sexually and asexually on infected crop straw. Ascospores produced by sexual reproduction are dispersed by wind and water splashes, and they can initiate infections. However, the majority of disease spread is driven by asexually produced conidia that originate from the fungal mycelium. These conidia are released into the air by strong air currents and then distributed over short distances within the crop canopy through splashing. The primary (initial) source of the inoculum is straw residues left in fields where a new crop is planted. However, adjacent fields with straw residues can also serve as a source of wind-borne infection that occurs during periods of extended high humidity and temperatures ranging from 10 °C to 25 °C. Warmer and drier weather conditions inhibit the perpetuation of the disease. Under optimal conditions, the disease cycle can complete in as little as 14 days.

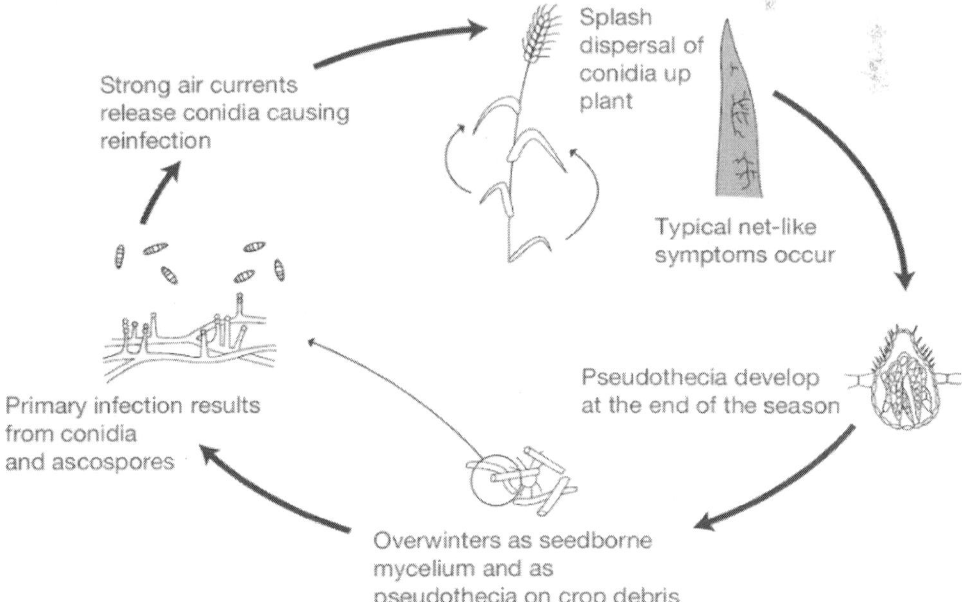

Figure 3.2: Life Cycle of Net Blotch (*Source*: ahdb.org.uk).

Management

1. Grow a resistant variety to minimize crop loss.
2. Avoid second barley crops and choose another non-host crop, remove debris of previous crop and, uproot and destroy volunteer plants.
3. Application of Nitrogen fertilizer to avoid excessive concentrations in plants, apply in 3 split doses.
4. Avoid early sowing and grow timely sown crop.
5. Use healthy seeds for cultivation.
6. Treat infected seed with carbendazim or mancozeb @ 2g/kg.
7. A combination of Pyraclostrobin + Epoxiconazole @ 0.2 per cent or Propiconazole 0.3 per cent as foliar spray 2 time at interval of 15 days most effective.

2. Powdery Mildew

Economic Importance and Occurrence

In India, minimum loss caused by Powdery mildew is 25 per cent in heavily infected crops. The crop affected in both winter and spring season by powdery mildew, causing losses typically ranged from 1 per cent to 14 per cent. When disease onset happens in early stage of crop and there is a high inoculum pressure, it is possible to experience losses that surpass 14 per cent. Beyond impacting crop yield, powdery mildew infection can also lead to a decrease in weight of grain, the quantity of tillers and spikes, as well as hinder root growth (Mathre, 1997).

Causal Organism

The disease Powdery mildew is a commonly appear on barley crop, is caused by the fungus *Blumeria graminis* (DC.) Speer f. sp. *hordei* Marchal. *B. graminis* belongs to Phylum Ascomycota, Class Leotiomycetes while Order Erysiphales, Family- *Erysiphaceae* and Species- *pisi* (Kirk *et al.*, 2008). When crop sown in early stage of crop growth the disease most prevalent on that stage. *B. graminis* is a biotrophic pathogen and special in that it can infect barley if moisture is deficient on host surface.

Symptoms

The symptoms are usually first time can be observed at tillering stage but the ypically, the disease does not continue to exist after the spike emergence stage. The disease can be easily identified by presence of fungal white powder (pathogen signs). At first, they manifest as fuzzy, whitish clusters of fungal mycelium. As time progresses, these mycelial tufts give way to the formation of powdery or fluffy white pustules containing conidial chains. Over time, the mycelium and conidia may adopt a gray or even slightly brownish hue. In times of severe epidemics, not only the leaves and leaf sheaths of plants, but also the entire spikes, can fall victim to powdery mildew infection. As the growing season advances, the fungus will mature into black, spherical cleistothecia, which house the sexual spores, within the fluffy masses of mycelium and conidia. This ailment persists in volunteer plants and grasses, remaining a recurring issue from season to season.

Figure 3.3: Powdery Mildew (*Source*: extensionaus.com.au).

Favourable Condition

Heavy, succulent growth favors mildew development. High dose of nitrogenous fertilizer, high humidity over 80 per cent and low temperatures ambient for development of disease. Cool temperature (15 °C -25 °C) and humid weather favour to disease development but in case of warmer weather and semiarid environments pathogen can also attacks on the crop. Conidia and ascospore produce germ tube after germination. After that germ tube enter into the host cuticle through penetration.

Disease Cycle

Powdery mildew is an obligate fungus that lives on living barley plants. The infection result in the formation of numerous minuscule, white spores, wind is the medium for dissemination of spores. Pathogen infect to plant under conditions of high humidity due to dew, along with temperatures ranging between 15 - 25 °C. Typically, mildew symptoms manifest approximately 5 to 7 days after the initial infection. The pathogen has the capacity to proliferate rapidly, causing crops to become heavily afflicted within a span of 4 to 5 weeks from the onset of the infection. This ailment persists from one season to the next through volunteer plants weeds and grasses. It's important to note that mildews exhibit host specificity, meaning that mildew originating from barley will not affect wheat.

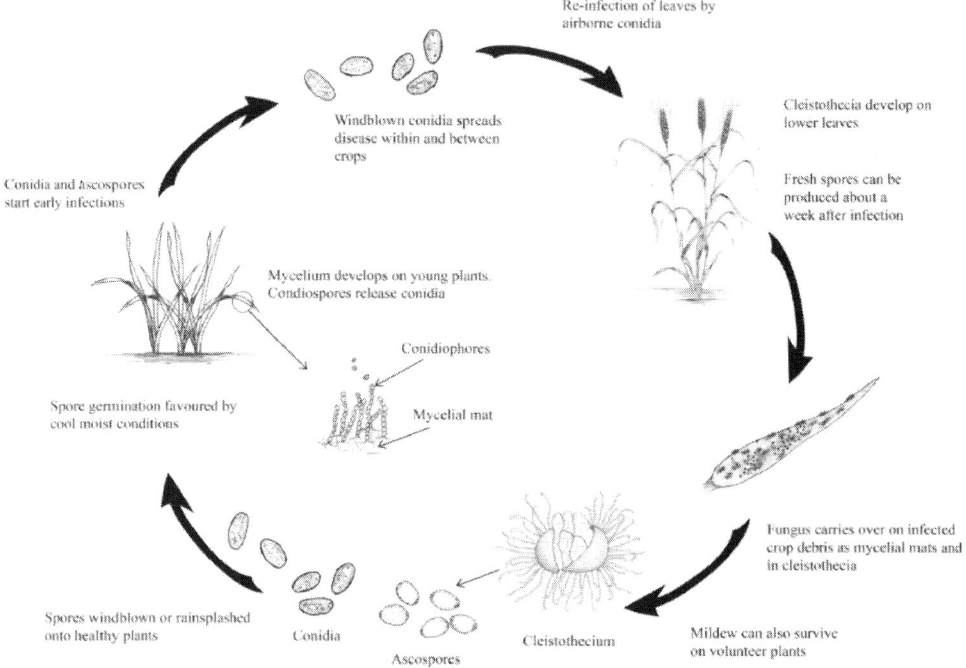

Figure 3.4: Disease Cycle of Powdery Mildew (*Source*: extensionaus.com.au).

Management

1. To minimize the amount of inoculum, eliminate volunteer barley plants through grazing or the use of chemical herbicides.
2. Only rotations of crops are not an effective measure to control powdery mildew as spores can be readily blown from one to another plants.
3. Overhead irrigation might be effective.
4. Foliar spray of Tilt (Propiconazole) or Folicur (Tebuconazole) or Bayleton (Triadimefon) @ 0.1 per cent twice at 15 days of interval.

3. Rust

Barley Rust fungi are highly destructive pathogens of crop plants, as they are obligate biotrophic organisms that rely entirely on living host cells for their nutritional needs. (Cummins and Hiratsuka, 2003; Duplessis *et al.*, 2011). These pathogens have evolved into numerous distinct physiological races or pathotypes. While these pathotypes cannot be distinguished based on their appearance, their identification can be achieved by assessing the host's response to infection using a defined set of Additionally, molecular marker-based techniques are employed to distinguish between these physiological forms (Prashar *et al.*, 2014). There are four distinct barley rust diseases, stem rust, stripe rust, leaf rust, and crown rust, all of which are attributed to pathogens within the genus *Puccinia*, belonging to the family Pucciniaceae. These organisms are classified under the order Pucciniales, within the class Pucciniomycetes, subphylum Pucciniomycotina, Phylum Basidiomycota, and the kingdom Fungi (Bauer *et al.*, 2006).

Yellow (Stripe) Rust

Economic Importance and Occurrence

Yellow rust, a long-standing threat to barley, has caused significant damage in southern Asia, eastern Africa, Western Europe, and the Middle East over the years. In a noteworthy occurrence, a race of this disease primarily affecting barley was identified for the first time in Bogota, Colombia, in 1975 (Dubin and Stubbs, 1986). Subsequently, severe epidemics of barley yellow rust have been documented in countries across northwestern and central Europe, as well as in India, Bangladesh, Nepal, China, and Japan (Chen *et al.*, 1995). Since then, the pathogen has expanded its presence globally. In India, the initial pathotype capable of infecting barley was first discovered in 1939 in the Nilgiri hills of Tamil Nadu, and it was designated as 4S0 (G). In India, the early occurrence of yellow rust can result in substantial crop losses, occasionally even hindering the emergence of ear heads or the development of grains (Prakash and Verma, 2009).

Causal Organism

The disease Yellow rust of barley caused by fungi *Puccinia striformis* Westend. f. sp. *hordei* Eriks. and Henn. (Psh). In northern India pathogen attacks on foliar portion of the plants.

Symptoms

The fungus is partially systemic, so the stripes keep becoming bigger. It can also form on the glumes, necks, and sheaths of leaves. When the right conditions are met (temperature between 10 and 14 degrees Celsius, sporadic rain or dew), pustules appear 8 to 14 days after infection and newly released uredospores become airborne, causing secondary infection and accelerates the progression of the disease (Prashar *et al.*, 2015). As infected barley plants mature, black telia easily form from uredia. *P. striiformis* f. sp. *hordei* uredial and telial spore stages are found on barley and other species of Hordeum (Marshall and Sutton, 1995). Aecial and pycnial spore stages oh *P. hordei* have not yet been reported. Barley yellow rust can be induced by volunteer plants, barley crops seeded in the autumn and wild Hordeum species (Dubin and Stubbs, 1986;; Marshall and Sutton, 1995).

Figure 3.5: Yellow (Stripe) Rust (*Source*: agriculture.gov.au).

Favourble Conditions

In milder temperatures (2–15 °C), where the leaves are wet for extended periods of time (8–10 hours) and create ideal circumstances for infection, yellow rust is primarily a disease of barley. The pustules form narrow stripes on the leaves and hold yellow to orange-yellow uredospores.

Disease Cycle

The obligatory parasite *Puccinia striiformis* f. sp. h*ordei* overwinters on a variety of plants, such as volunteer barley, rye, certain wild barley varieties including *Hordeum jubatum* (foxtail barley), wheat, and several types of perennial grass. A small number of infections that are difficult, even impossible, to identify in the field constitute the initial cause of the disease. This infection may spread quickly, resulting in sudden epidemics and large damages, especially in areas like the Northwest region, where the combination of moisture and cold weather greatly promotes the growth of illness.

Management

1. Late sowing of crop
2. Sanitation surrounding the field and remove all the weed flora from the cultivated land
3. Spray Propiconazole @ 0.1 per cent or Tebuconazole @ 0.1 per cent or Mancozeb @ 0.2 per cent or Triademefon @ 0.05 per cent as foliar application.
5. Brown (leaf) rust:

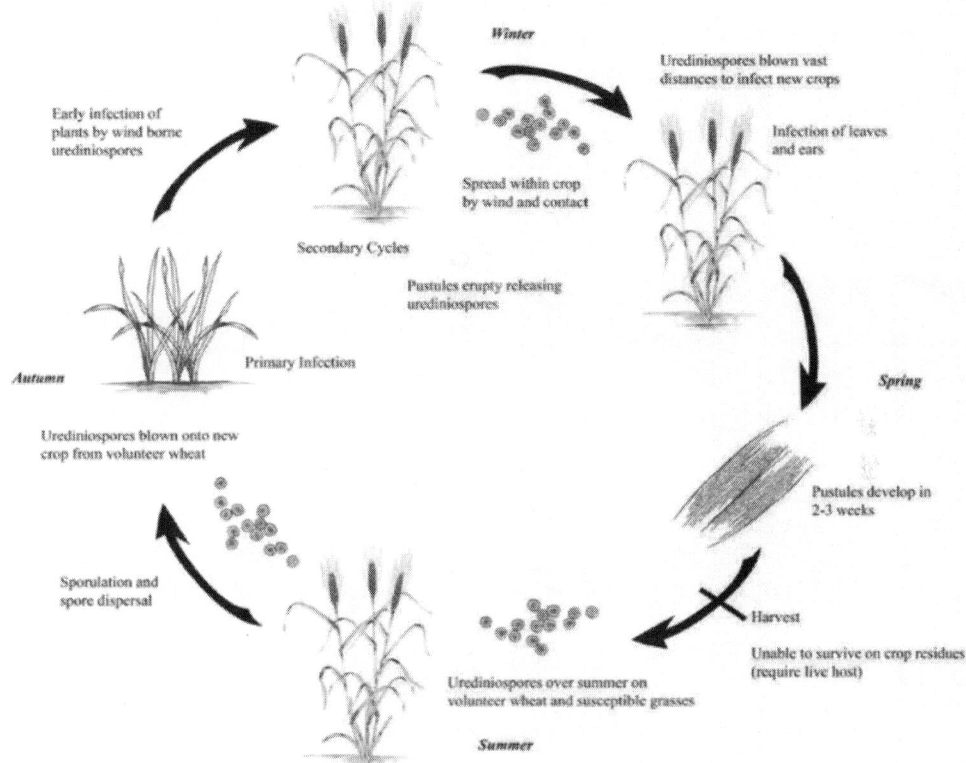

Figure 3.6: Disease Cycle of Yellow (Stripe) Rust (*Source*: ucanr.edu).

Brown Rust

Economic Importance and Occurrence

Although brown rust often occurs in all of India's barley-growing regions, this pathogen hardly seldom spreads widely enough to produce serious outbreaks. However, when the inoculum arrives early and at high levels, vulnerable genotypes may still experience severe yield losses. Highly vulnerable barley genotypes had a loss of production of above 60 per cent under experimental conditions (Das *et al.*, 2007).

Causal Organism

Brown rust caused by *Puccinia hordei* Otth.

Symptoms

P. hordei produces tiny, orange-brown pustules called uredia, which are primarily found on the upper leaf surface of barley leaves but also appear on the lower side the blades of leaves and on the sheaths of the leaves. Green islands or yellow halos may encircle these pustules. Symptoms can also appear on stems, glumes, and awns in instances of severe infection under high inoculum load. Blackish-brown telia develop later in the season, frequently in stripes, and covered by the epidermis, especially on leaf sheaths but also on stems, heads, and leaf blades (Park *et al.*, 2015). The leaf rust pathogen's uredial and telial spore stages are found on barley and a variety of wild Hordeum species, whereas the pycnial and aecial spore stages have been documented on various Liliaceae family hosts, including Ornithogalum, Leopoldia, and Dipcadi (Clifford, 1985).

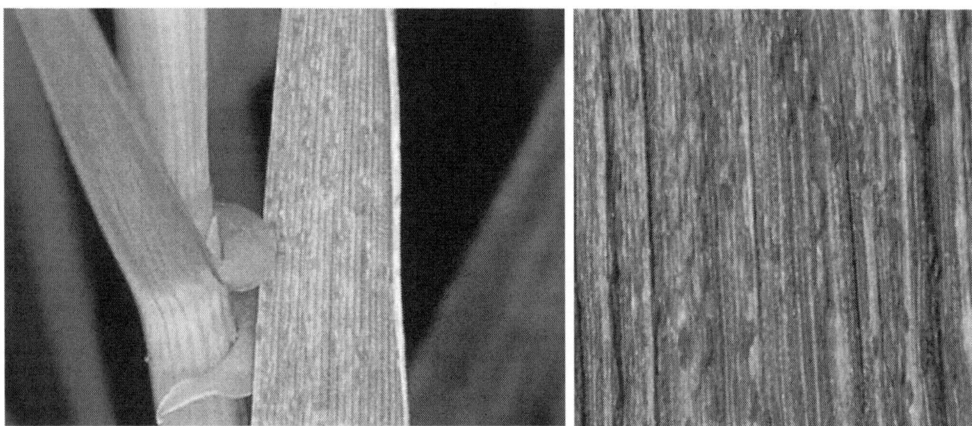

Figure 3.7: Brown Rust (*Source*: cropscience.bayer.co.uk).

Disease Cycle

Only live tissue from leaves will allow the fungus to proliferate and endure. It primarily overwinters on volunteers and crops which were sown early. Spores in the air disperse it. On damaged plant tissue, dark teliospores can appear, signifying a second stage in the fungus's development.

Favourble Conditions

The ideal range of temperatures for sporulation and germination is 15°C to 22°C. Nevertheless, the illness is active from 7°C to 25°C, which is a larger temperature range than that of yellow rust. For spores to germinate, leaves require 100 per cent relative humidity on their surface.

Management

1. Weeds that are left over as a green bridge between crop season and removal.
2. Postpone planting vulnerable cultivars
3. Splitting up the nitrogen application.

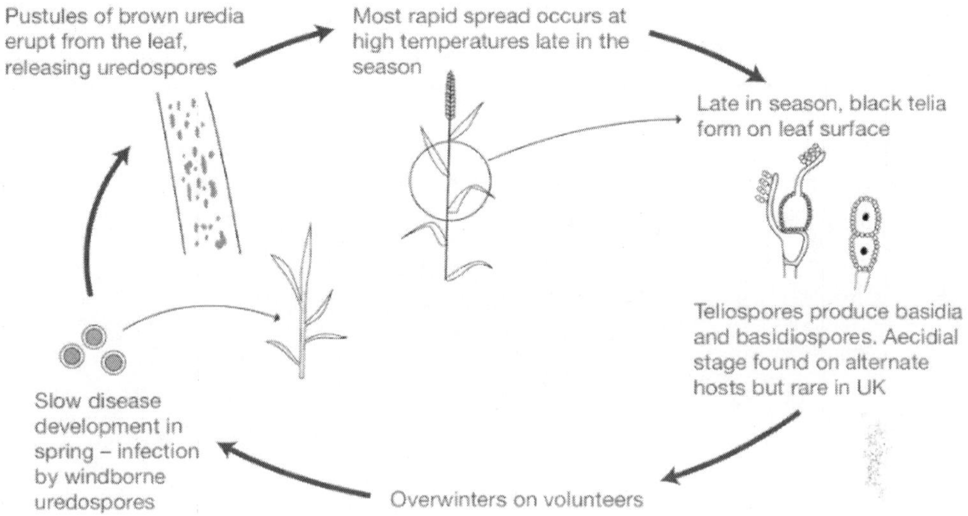

Figure 3.8: Disease Cycle of Brown Rust (*Source*: ahdb.org.uk).

4. Generally speaking, fandango (prothioconazole with fluoxastrobin) will treat brown rust.
5. For crops that have a greater potential for output or that are more susceptible to disease

Stem Rust

Economic Importance and Occurrence

Stem rust has the potential to be the most destructive since it can completely destroy crops. Conditions that are favorable for a significant epidemic are uncommon, though. Leaf rust infections that occur early can cause yield reductions of up to 30 per cent. Although crop stripe rust rarely causes yield loss, in certain years it can affect vulnerable cultivars to the extent of 20 per cent. Since it often infects the crop late in the growing season, the losses are negligible. The epidemiology of barley black rot in India is unrelated to the highlands of North India.

Causal Organism

Mostly affecting Central and Peninsular India, barley stem rust is brought on by *Puccinia graminis* Pers. f. sp. *tritici* Eriks. and Henn. (Pgt) and *Puccinia graminis* Pers. f. sp. *secalis* Eriks. and Henn. (Pgs).

Symptoms

The uredial pustules are elongated, reddish-brown, and much larger. They usually form on the sheath, leaf blade, stem, and occasionally on top of the spike.

The fact that the uredia rip the plant tissue, giving the impacted stem and leaf a noticeably rough look, is one of the most distinctive features of stem rust that helps to distinguish it from the other two rusts. Plant stems that are severely infected and have numerous stem lesions may become weaker and eventually fall off at the severe infection spot. Rust-colored pustules transform into black telia that carry teliospores towards the end of the season (Bhardwaj *et al.*, 2017).

Figure 3.9: Stem Rust (*Source*: ars.usda.gov).

Favourable Condition

A temperature range of 15–28 °C and 6–8 hours of unrestricted wetness on the leaf surface are optimal for infection. If there continues to be precipitation and a temperature between 26 and 30 °C, the disease spreads rapidly. During the flowering season, uredospore generation happens in multiple cycles.

Disease Cycle

This pathogen lives on grass hosts such as barley, wheat, and others in the uredial and telial spore stages. Mahonia spp. and Berberis spp. (barberry) serve as alternating hosts for the pycnial and aecial spore stages. Aeciospores, the principal inoculum, can be obtained from the alternative host species of Mahonia and Berberis, albeit they frequently spread over short distances (Roelfs *et al.*, 1992). Since there are no functional alternate hosts (susceptible Berberis and Mahonia spp.) in India, the Nigiris hills go on to serve as the primary inoculum source. Here, the virus must survive on volunteer plants, summer crops, uredospores, or other grasses in the areas of catchment (Bhardwaj *et al.*, 2016).

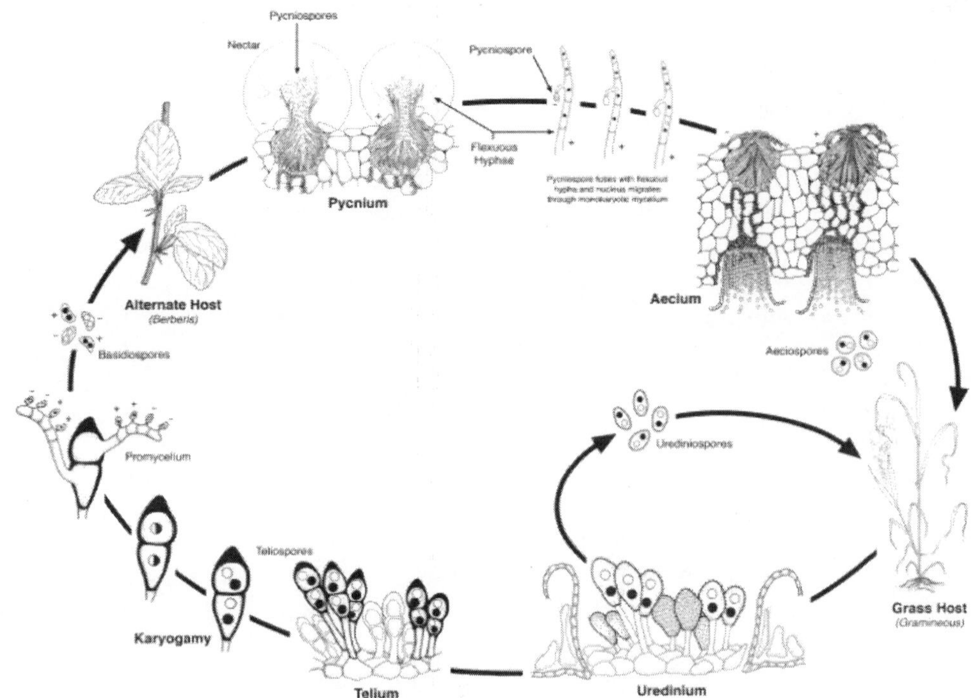

Figure 3.10: Disease Cycle of Stem Rust (*Source*: bsppjournals.onlinelibrary.wiley.com).

Management

1. Barley rusts mostly thrive on self-sown barley during the summer, so eliminating this inoculum source is crucial. Herbicides or intensive grazing should be used to eradicate green growth, especially after.
2. Since self-sown barley is the main source of inoculum for barley rusts during the summer, it is imperative to eliminate this source. It is crucial to use herbicides or vigorous grazing to eradicate green growth, particularly in the approaching season.
3. Combine the following fungicides: 400 litres of water with the following mixtures: trifloxystrobin + propiconazole (Stratego 250 EC), hexaconazole (Cotaf 5 EC), epoxiconazole + carbendazim (Swing 250 EC), cyproconazole at 80 g/liter + propiconazole at 250 g/liter (Artea 330 EC), and epoxiconazole at 125 g/liter + carbendazim at 125 g/liter (Soprano C 250 EC).

Spot Blotch

Economic Importance

In vulnerable barley cultivars, yield losses of 10 per cent - 30 per cent are typical; but, in extremely favourable conditions, yield losses can surpass 30 per cent (Fetch and Steffenson, 1994). When the temperature is between 15 °C and 22 °C in the first two weeks following the development of full ears, spot blotch damage might affect grain yield by 10 per cent to 20 per cent (Steffenson, 1997). A disease's incubation period might range from three to 6 days, depending on the surroundings. Grain yield losses are highest when the infection on the flag leaf occurs early and heavily. In warm, humid climates, such as those found in Uttar Pradesh, Bihar, Jharkhand, West Bengal, Assam, and the plains of northeastern provinces of India, the disease causes a considerable yield loss. Moisture and prolonged warm spells (> 16 hours) are favourable for epiphytotic growth.

Causal Organism

An major foliar disease of barley in India is spot blotch, which is caused by *Bipolaris sorokiniana* (Sacc.) Shoemaker {*Cochliobolus sativus* (Ito and Kuribayashi) Drechs. ex Dastur}. The most significant fungal pathogen of barley has been identified as the pathogen (Arabi and Jawhar, 2004; Valjavec-Gratian and Steffenson, 1997). The pathogen's host range is broad, and its level of pathogenicity varies. Its physiologic races vary widely in terms of virulence and capacity to target particular grasses and cereals.

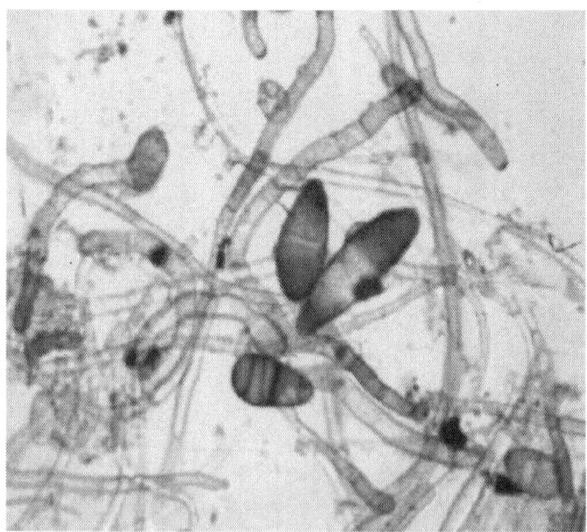

Figure 3.11: Spot Blotch
(*Source*: Dutbayev *et al.*, 2022).

Symptoms

The earliest signs and symptoms are little, dark brown lesions without a chlorotic border, measuring 1-2 mm in length. Typically, the lesions are round to oblong, 2–5 x 15-20 mm, with dark brown edges that are chlorotic, and their width is constrained by leaf veins. These lesions spread swiftly as oval to elongated blotches (2–20 mm) in genotypes that are vulnerable, and they may combine to form larger irregular patches. Excessively diseased leaves prematurely dry out and perish. If the inoculum is accessible and the surrounding conditions are favourable for infection, the disease may progress to its kernel blight phase, often known as the "Black point." Additionally, this fungus produces toxins, primarily prehelminthosporol, which can induce symptoms of many diseases (Kumar *et al.*, 2002). In nature, sexuality is uncommon.

Figure 3.12: Spot Blotch
(*Source*: Dutbayev *et al.*, 2022).

Favourable Conditions

Spot blotch growth is generally facilitated by temperatures between 16 and 32 °C (Acharya *et al.*, 2011). In the Indian subcontinent, the disease primarily expands when temperatures rise above 26 °C since this temperature favours heavy sporulation (Chaurasia *et al.*, 2000). According to Duveiller and Sharma (2009), the anamorph can withstand a sufficiently wide range of temperatures from 4-36 °C, however the teleomorph can only survive up to seven months in natural conditions and develops in a range of 16–24°C, with an ideal temperature of 20°C. Disease severity is increased by high temperatures and high relative humidity. It is thought that in EGP, leaf wetness periods longer than 12 hours caused by dew or rainfall that coincide with high temperatures and humidity favor the start of infection (Duveiller *et al.*, 2005). Furthermore, the rice-wheat cropping system's delayed sowing of wheat results in yield loss from terminal heat stress (Hasan *et al.*, 2021; Narendra *et al.*, 2021). In addition, high residual soil moisture and prolonged leaf wetness due to foggy weather can also exacerbate disease severity (Duveiller, 2004; Duveiller *et al.*, 2005).

Disease Cycle

The fungus hibernates on the seed, in the soil, or in barley straw and stubble. In the spring, spores are formed on barley trash and are carried by the wind and rain. Barley seedlings can contract an infection from an inoculum either in the soil or on the seed.

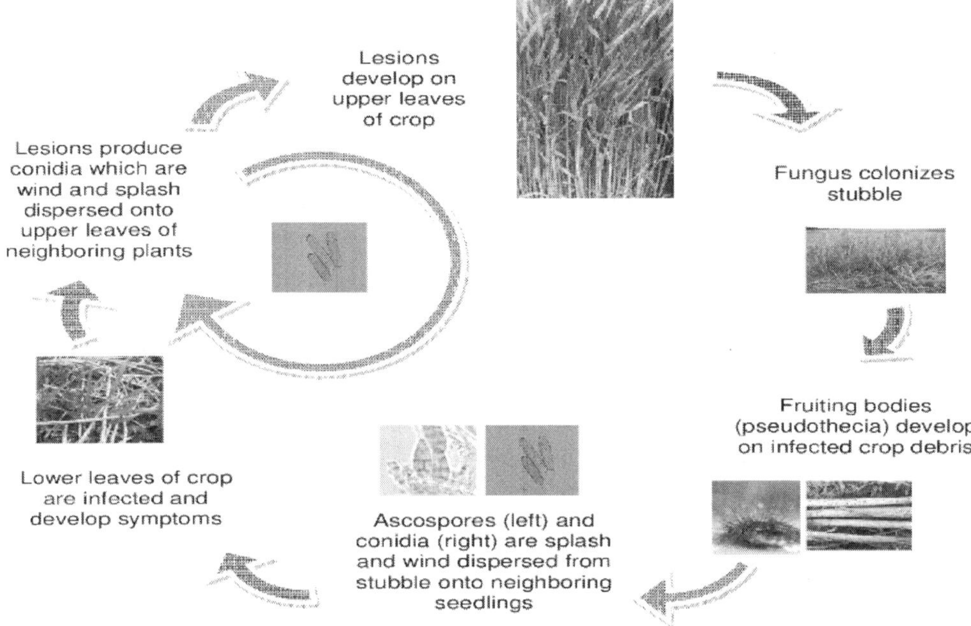

Figure 3.13: Disease Cycle of Spot Blotch (*Source*: M. McLean, 2011).

Management

1. Grow a resistant variety to minimize crop loss.
2. Avoid second barley crops and choose another non-host crop, to reduce infected crop residue and control volunteers plants.
3. Appropriate nitrogen applications follow to avoid excessive concentrations in plants and apply in 3 split doses.
4. Avoid early sowing and grow timely sown crop.
5. Use healthy seeds for cultivation.
6. Treat infected seed with carbendazim or mancozeb @ 2g/kg.
7. A combination of pyraclostrobin + epoxiconazole @ 0.2 per cent or propiconazole 0.3 per cent as foliar spray 2 time at interval of 15 days most effective.

Loose Smut

Economic Importance and Occurrence

Causal Organism

As with wheat, the fungus *Ustilago tritici* (Pers.) Rostr. {*U. nuda* (Jens.) Rostr} is the cause of loose smut in barley. But the specific loose smut pathogen strain that targets wheat cannot affect barley, and *vice versa*.

Symptoms

Until the ear (spike) emerges, afflicted plants frequently show no signs or symptoms. Usually, the affected ears appear before the healthy ones, and a mass of dark brown teliospores replaces all grains. A thick membrane which initially holds the teliopores loosely dissolves quickly, releasing the teliospores on neighboring ears (spikes).

Favourable Condition

Disease is more destructive under favourable condition. At between 16 and 22°C, wet conditions are conducive to infection.

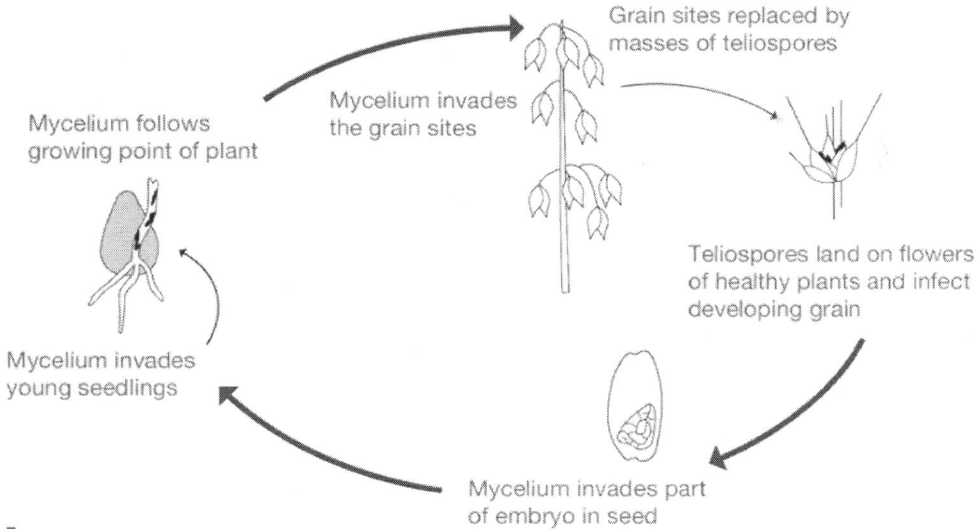

Figure 3.14: Disease Cycle of Loose Smut (*Source*: ahdb.org.uk).

Disease Cycle

The pathogen is internally seedborne and persists from season to season as a latent mycelium inside the barley seed embryo. When untreated and contaminated seeds germinate, the pathogen breaks dormancy and starts to spread throughout the growing barley plant. When the plant's ears emerge, all of its floral parts become infected and are replaced with large smut spores, or teliospores. Teliospores carried by the wind from a smutted ear of barley plant spread the disease to nearby, healthy flowering ears. After germinating, the teliospores spread to the female portions of barley flowers and ultimately colonize the growing embryo. The pathogen remains dormant until the barley seed germinates and the cycle is repeated once the contaminated seed reaches maturity.

Management

1. Prior to harvest, farm-saved seed crops are typically inspected for indications of illness.
2. Avoid farming seed (both crops and volunteers) from highly contaminated areas.
3. Invest in certified seed or check home-saved seed for the pathogen.
4. Apply a 30-minute hot water treatment to the seeds at 50 °C.
5. Before seeding, treat the seeds with the fungicide Carbendazim 2 g/kg (foliar fungicide applications do not effective).

Covered Smut

Causal Organism

Ustilago hordei (Pers.) Lagerh is the causal agent of Covered smut of barley.

Symptoms

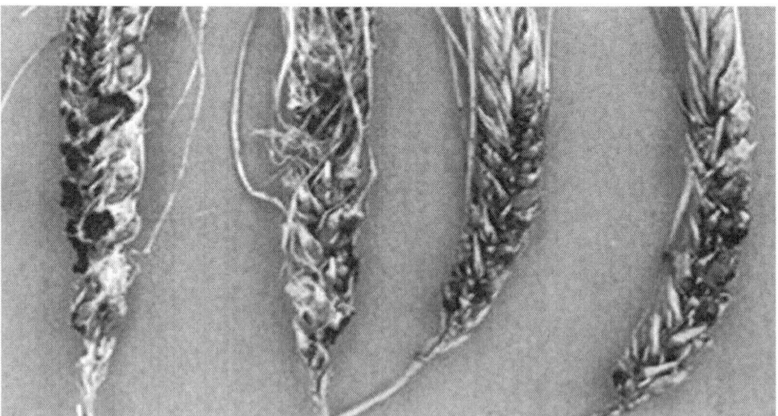

Figure 3.15: Covered Smut (*Source*: ahdb.org.uk).

Smut does not become evident on infected plants until they reach the heading stage. At this stage, smuttered heads start to emerge either alongside healthy ones or a little later. The flag leaf sheath may occasionally capture these diseased heads. The diseased heads on the field are immediately identifiable.

The glume, which is usually a grey membrane, covers the dark black smut sori that replaces the kernels inside these heads. This glume typically doesn't break down until the threshing or harvesting phase. Additionally, the flower bracts and, rarely, the awns remain whole. Clusters of teliospores may even be visible in the harvested grain if the illness is extremely severe. Federal grain inspectors label such seed lots as "smutty," which lowers their market value. Smut sori can also sporadically emerge as lengthy streaks on leaf blades and in nodal tissue.

Favourable Condition

Their ears stand out from the leaf sheaths, giving them a blackened appearance. Seedling infection is more likely in warm, wet, acidic soil. In the range of 10 °C to 21 °C, the majority of seedlings become infected.

Disease Cycle

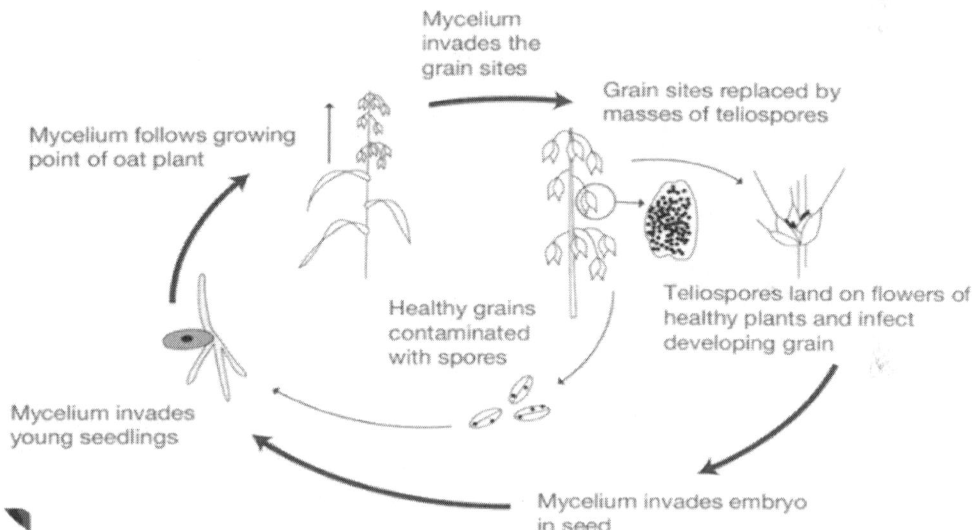

Figure 3.16: Disease Cycle of Covered Smut (*Source*: ahdb.org.uk).

A sick plant has all of its ears, and a sick plant has all of its grains infected. A continuous, tight, greyish-white membrane holds the masses of teliospores that are generated from all of the infected grains in a sick ear in place. Unless the membrane is accidentally torn, the covered masses of teliopores remain enclosed in their enclosing membranes until threshing time. Numerous teliospores are expelled during threshing when the diseased ears are split open. When the seed is spread, many of them lodge on firm kernels and stay dormant (Externally seedborne). Teliospores germinate easily in moist soil or water thanks to the development of a septate promycelium that facilitates the meiosis of the diploid nucleus. Numerous sporidia, also known as basidiospores, are generated, and these can proliferate through budding, germinate through germ tubes, or fuse to form dikaryotic conditions. By means of dikaryotic infection hyphae, the teliopores germinate concurrently with the barley seed and infect the seedling along the epicotyl. The pathogen's hyphae continue to grow with the shoot after it has infected the seedling, eventually replacing the grains with millions of teliopores.

Management

1. Use resistant variety for cultivation.
2. Use disease free seed material for sowing.
3. Depth of sowing keep at 2.5 cm deep.
4. Choose neutral or alkaline field for cultivation of barley crop rather than acidic soil.
5. Rogue affected plants, uproot and burn them.
6. Treatment of seeds with bioagent like *Trichoderma harzianum* 4g/kg of seeds.
7. Seed treatment with Carbendazim @ 1 g/kg or Carboxin 1.5 g/kg or Tebuconazole 1.5 g/kg of seeds.

Molya Disease

Economic Importance and Occurrence

Nematodes that feed on parasites harm barley much like they do most other cereal crops. Especially in northeastern Rajasthan and the neighboring state of Haryana, cereal cyst nematode (CCN) can result in significant output losses.In regions of the world where wheat, barley, and oats are grown, molya disease is common and is especially significant in Europe, Canada, Australia, and India. In 1957, the first reports of molya disease on wheat and barley came from Rajasthan. As of the present, states like Punjab, Delhi, Rajasthan, and Haryana are reported to be affected. Up to 50 per cent of crops in Rajasthan's afflicted areas typically sustain damage. It is projected that Rajasthan loses Rs. 30 million in barley and Rs. 40 million in wheat annually. Other names for the illness include "cereal cyst nematode" and "cereal cyst eelworm." The pathogen (*Heterodera avenae*) can only infect plants that belong to the Poaceae (= Graminae) family, which includes maize, wheat, barley, oats, rye, bajra, and other grasses. In contrast to other nations where oats and barley are more vulnerable than wheat, wheat is a more effective host in India.

Causal Organism

The sedentary endoparasite known as the cereal cyst nematode (Heterodera avenae Woll.) infects the roots of numerous Poaceae crops, including barley. This nematode is the source of molya disease.

Symptoms

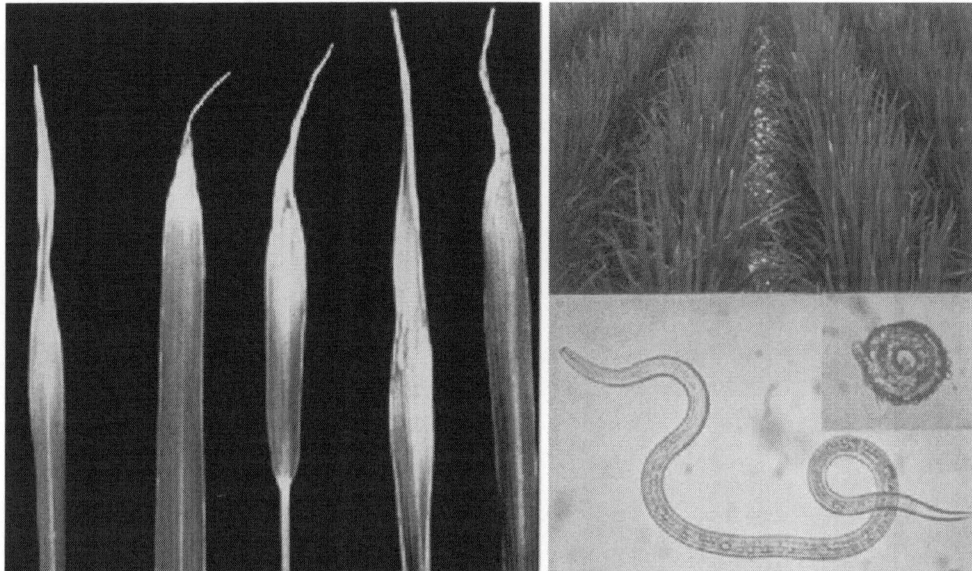

Figure 3.17: Molya Disease (*Source*: Khan, 2015).

The only distinguishing root symptom on barley caused by cereal cyst nematodes is a bushier root system. On the other hand, they induce short branching and swelling (knots) on the roots of wheat seedlings. The plants exhibit indications of nutrient deprivation and are severely stunted above ground, typically with a spotty distribution.

Disease Cycle

The vascular system of the root allows juveniles to enter, and females establish a feeding location. After swelling up and releasing their eggs, the females change into cysts that emerge from the roots. Initially white, these cysts eventually turn brown. The cysts can live for extended periods of time and overwinter, and within them the eggs are created. The next season is when the eggs hatch. The plant H. avenae is found all over the world. Furthermore, Smiley *et al.* (2008) reported the discovery of Heteroderafilipjevi, a new species, in Oregon. Cysts can proliferate due to wind-driven soil movement, machinery, harvesters, shoes, transplants, tubers, and other factors. Nematodes with cereal cysts go through a single life cycle annually. On the other hand, populations on sensitive barley cultivars can grow quickly since each cyst carries several hundred eggs. By rotating with a non-host for a year or two and managing grassy weeds, the illness is decreased. Compared to wheat or oats, barley can withstand cyst nematodes better overall.

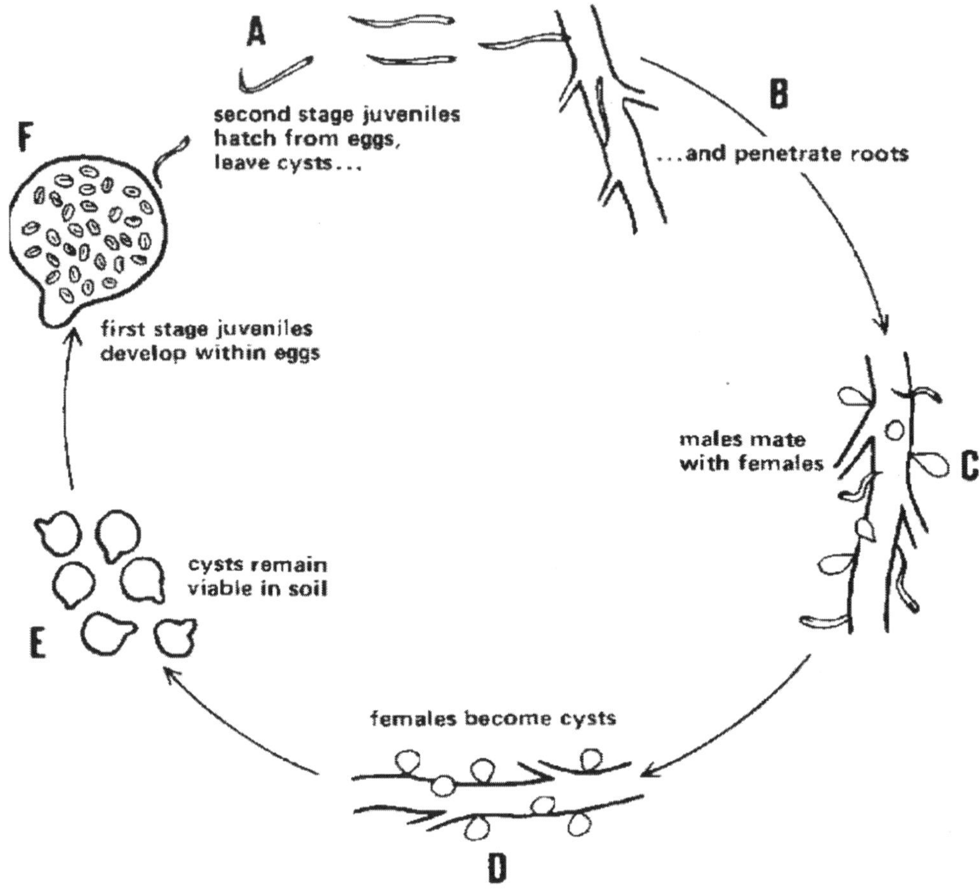

Figure 3.18: Disease Cycle of Molya Disease (*Source*: Khan, 2015).

Management

1. Crop rotation should be follow with non host crops like mustard, pulses, fenugreek or carrot for one or two years.
2. Grow cyst nematode resistant wheat Raj MR-1 or barley RD 2035 or RD 2052.
3. Plough two to three times during summer.
4. Apply carbofuran @ 45 kg/ha.

References

Acharya, K., Dutta, A. K. and Pradhan, P. (2011). Bipolarissorokiniana (Sacc.) shoem.: The most destructive wheat fungal pathogen in the warmer areas. *Aust. J. Crop Sci.***5**: 1064–1071.

Baik, B. K. and Ullrich, S. E. (2008). Barley for food: Characteristics, improvement, and renewed interest. *Journal of cereal science*, **48**(2): 233-242.

Behall, K.M., Scholfield, D.J. and Hallfrisch, J. (2004). Diets containing barley significantly reduce lipids in mildly hypercholesterolemic men and women. *American Journal of Clinical Nutrition*, **80:** 1185–1193

Brown, W.M., Velasco, V.R. and Johnson, R. (2002). Chemical suppression of barley yellow rust - Meeting the challenge of yellow rust in cereal crops. Proceedings of the First Regional Conference on Yellow Rust in the Central and West Asia and North Africa Region, Karaj, Iran, pp. 261-264.

Chaurasia, S., Chand, R. and Joshi, A. K. (2000). Relative dominance of Alternaria triticinapras. et prab. and Bipolarissorokiniana (Sacc.) shoemaker in different growth stages of wheat (*T. aestivum*). *J. Plant Dis. Protect.*, **107:** 176–181.

Dutbayev, Y., Kuldybayev, N., Daugaliyeva, S., Ismailova, E., Sultanova, N., Özer, G. and Yessimbekova, M. (2022). Occurrence of spot blotch in spring barley caused by BipolarissorokinianaShoem. in South-Eastern Kazakhstan. *The Scientific World Journal*, pp. 15-43.

Duveiller, E. (2004). Controlling foliar blights of wheat in the rice-wheat systems of Asia. *Plant Dis.* **88:** 552–556.

Duveiller, E.M. and Sharma, R.C. (2009). Genetic improvement and crop management strategies to minimize yield losses in warm non-traditional wheat growing areas due to spot blotch pathogen Cochliobolus sativus. *J. Phytopathol.* **157:** 521–534.

Duveiller, E., Kandel, Y. R., Sharma, R. C. and Shrestha, S. M. (2005). Epidemiology of foliar blights (spot blotch and tan spot) of wheat in the plains bordering the Himalayas. *Phytopathology*, **95**: 248–256.

Fadel, J.G., Newman, R.K., Newman, C.W. and Barnes, A.E. (1987). Hypocholesterolemic effects of b-glucans in different barley diets fed to broiler chicks. *Nutrition Reports International*, **35**: 1049–1058.

Govindasamy, R., Italia, J. andAdelaja, A. (2001). Willingness to Pay a premium for Integrated Pest Management Produce: A Logistic Approach. *Agricultural and Resource Economics Review*, **30**(2): 151–159.

Hardwick, N., Slough, J. E. and Gladders, P. (2002).Winter Barley: a survey of disease 2002. CSL, ADAS, UK: DEFRA.

Hasan, U.W., Roy, C., Chattopadhyay, T., Ranjan, R. D. and De, N. (2021). Effects of heat and drought stress on yield and physiological traits in wheat (*Triticum aestivum*L.). *J. Crop Weed.* **17** (1): 203–210.

Havis, N. D., Oxley, S. J. P., Piper, S. R. andLangrell, S.R.H. (2006). Rapid nested PCR-based detection of Ramularia collo-cygni direct from barley. *FEMS Microbiology Letters*, **256:** 217–223.

Havis, N. D., Pastok, M., Pyzalski, S. and Oxley, S. J. P. (2006). Investigating the life cycle of Ramularia collo-cygni. *Proceedings Crop Protection in Northern Britain*, pp. 219–223.

Hovmøller, M.A., Yahyaoui, E., Milus and Justesen, A. (2008). Rapid global spread of two aggressive strains of a wheat rust fungus. *Mol. Ecol.*, **17:** 3818-3826.

Narendra, M. C., Roy, C., Kumar, S., Virk, P. and De, N. (2021). Effect of terminal heat stress on physiological traits, grain zinc and iron content in wheat (*TriticumaestivumL.*). *Czech J. Genet. Plant Breed*, **57**: 43–50.

Newman, C.W. and Newman, R.K. (2006). A brief history of barley foods. *Cereal Foods World*, **51**: 4–7

Newman, R.K., Lewis, S.E., Newman, C.W., Boik, R.J. and Ramage, R.T. (1989). Hypocholesterolemic effect of barley foods on healthy men. *NutritionReportsInternational*, **39**: 749–760.

Pins, J.J. and Kaur, H. (2006). A review of the effects of barley b-glucan on cardiovascular and diabetic risk. *Cereal Foods World*, **51:** 8–11.

Qureshi, A.A., Burger, W.C., Peterson, D.M. and Elson, C.E. (1986). The structure of an inhibitor of cholesterol biosynthesis isolated from barley. *Journal of Biological Chemistry*, **261:** 10544–10550.

Qureshi, A.A., Qureshi, N., Wright, J.J.K., Shen, Z., Kramer, G., Gapor, A., Chong, Y.H., Dewitt, G., Ong, A.S.H., Peterson, D.M. and Bradlow, B.A. (1991). Lowering serum cholesterol in hypercholesterolemic humans by tocotrienols (palmvitee). *American Journal of Clinical Nutrition*, **53:** 1021–1026.

SELVAKUMAR, R.Meeta, M., Shekhawat, P.S., Verma, R.P.S. and Sharma, I. (2014). Management of stripe rust of barley using fungicides. Indian Phytopath. 67 (2) : 138-142 (2014).

Reis, E.M. (1991). Integrated disease management: the changing concept of controlling head blight and spot blotch. In: Saunders, DA, Hettel, GP eds. Wheat in Heat Stressed Environments: Irrigated, Dry Areas and Rice Wheat Systems Mexico, *DF*: *CIMMYT*, pp 165–170.

Singh, D.P., Babu, K.S., Mann, S.K., Madhu Meeta, Karwasra, S.S, Kalappanavar, I.K., Singh, R.N., Singh, A.K. and Singh, S.P. (2010). Integrated pest management in barley (*Hordeum vulgare*). *Ind. J. Agrl. Sci.*, **80:** 437-442

Stepanoviæ, M., Rekanoviæ, E., Milijaševiæ-Marèiæ, S., Potoènik, I., Todoroviæ, B andStepanoviæ, J. (2016). Field efficacy of different fungicide mixtures in control of net blotch on barley. *PesticPhytomed*, **31**(1-2): 51–57.

Chapter 4

Diseases of Oat Crop

Anoop Kumar[1], *Puskar Shukla*[2], *Arun Kumar*[3]*, *Mukul Kumar*[4] *and Madan Mohan Bajpeyi*[3]

[1]*Department of Plant Pathology, Sam Higginbottom University of Agriculture, Technology and Sciences, Prayagraj – 211 007, Uttar Pradesh*
[2]*Department of Plant Pathology,* [3]*Department of Entomology,*
Chandra Shekhar Azad University of Agriculture and Technology, Kanpur – 208 002, Uttar Pradesh
[4]*Subject Matter Specialist, Krishi Vigyan Kendra, Jamui – 811 307, Bihar*
**e-mail: arunkumarbujhansi@gmail.com*

Introduction

Oat (*Avena sativa* L.) is an important cereal grain cultivated worldwide for its edible seeds (Forsberg and Reeves, 1995). However, like all crops, oats are susceptible to various diseases that can significantly impact their growth and yield. The oat often known as the common oat is a type of cereal grain produced for its seed. Unlike other cereals and pseudo cereals, the name of this seed is typically used in the plural. Although rolled oats and muesli are suitable for human consumption, one of the most widely used uses for oats is as animal feed. Oats are a nutrient-rich grain that may lower blood cholesterol levels when consumed frequently. As with gliadin in wheat, oats contain gluten proteins called avenins. A tiny percentage of persons may develop celiac disease as a result of them. Aside from that, other gluten containing grains, primarily wheat and barley, commonly infect oat products. Russia accounted for 17 per cent of the total 22.5 million tonnes of oats produced globally in

2021. Stem rust, leaf rust, barley yellow dwarf virus, and Septoria blotch are the four main leaf diseases of Oat, the severity of each disease varies with the season. Oat leaf diseases affect grain productivity and quality as well as hay quality traits including colour and digestibility. Losses (Dietz *et al.*, 2019). The disease infects oat plants at early stages of development under conducive The most significant oat fungal diseases in the world are thought to be crown rust, leaf spot, and *Helminthosporium* leaf spot (*Drechslera avenae*), which results in substantial yield and grain quality losses. The adoption of host resistant plants is a key component in the control of foliar diseases. However, the majority of the commercially available oat cultivars behave as moderately resistant to susceptible. In light of this, the use of fungicides is a typical technique for controlling fungal infections in order to minimise crop weather conditions (Soovali and Koppel, 2011) Since the previous research on oat was published by various scientists, significant progress has been made in our understanding of oat diseases and methods for controlling them. The goal of this chapter is to expand on the description of oat illnesses by highlighting recent research that has been published. Additionally, it evaluates some of the prior research on the taxonomy of oat rust and builds a case for utilisation. Barley Yellow Dwarf viral (BYDV) is the primary cause of reported viral disease issues in oats. Despite the fact that BYDV is thought to be the most severe oat viral disease, there have been questions raised regarding BYDV diagnoses that are based purely on the observation of symptoms and transmission by cereal aphids. Powdery mildew (*Blumeria graminis* f. sp. *avena*), leaf blight (*Helminthosporium* spp.), and loose smut (*Ustilago avenae*) are seriously affecting the fodder and seed yield of oats in the north-western Himalaya. These diseases result in qualitative as well as quantitative losses (Banyal *et al.*, 2016). In Asia, oat diseases have a significant role in oat production. Because the majority of the crop is grown in the spring during warm, humid climates that are typically conducive to disease growth. The yield losse in Oat crop by diseases are typically larger than those for other small grains (Burdon *et al.*, 2008). In many regions, the root- and culm-rot diseases of oats serve as productivity barriers. While, commonly prevalent in all oat fields, the leaf-spot diseases caused by *Helminthosporium avenae* Eidam and *Pseudomonas coronafaciens* Stevens have been of lesser significance in comparison to the disease indicated above. The widespread oat pseudo-rosette disease in Siberia can occasionally cause severe harm to oat crops.

Major diseases of Oat classified as: Fungal, Viral and Bacterial given in following table:

Sl.No.	Name of Diseases	Causal Organism
A. Fungal diseases		
1.	Crown rust	*Puccinia coronate*
2.	Stem rust	*Puccinia graminisavenae*
3.	Victoria blight	*Cochiobolus victoria*

Sl.No.	Name of Diseases	Causal Organism
4.	Culm rot	*Helminthosporium* spp.
5.	Covered smut	*Ustilago kolleri*
6.	Loose smut	*Ustilago avenau*
7.	Powdery mildew	*Blumeria graminis* f. sp. *avenae*
B. Viral diseases		
1.	Soil-borne oat mosaic	Virus
2.	Barley yellow dwarf	Virus
C. Bacterial diseases		
1.	Bacterial stripe	*Pseudomonas striafaciens*

A. Rusts

Economic Importance and Occurrence

Rusts cannot survive on decomposing agricultural left overs; they can only grow on the living cells of their host plants. Rust fungi are much specialised in the host plants they select to infect. Oat rusts have up to five distinct spore kinds in their life cycles, each of which serves a unique purpose. The five varieties can all be found on a single oat plant or can infect two completely different hosts, such buckthorn and oats. A single species or variation of rust may have several distinct races, and no two races of particular rust attack the same kind of oats. This makes it more difficult to create oat types resistant to rust.

1. Crown Rust (*Puccinia coronata*)

Crown rust (*Puccinia coronata*) is also known as leaf rust on oats and is similar to leaf rust of wheat, barley, and rye. This fungus attacks several wild grasses related to oats, yet no other cereal crop is impacted than oats. Buckthorn (*Rhamnus* spp.) serves as the secondary host for crown rust, rust that has two hosts. The majority of Texas, ative buckthorn (*R. Coroliniania*) is not a host plant. Although the diseases severity fluctuates from year to year, crown rust is always prevalent throughout Texas. It is possible to lose up to 30 per cent of the potential crop in years with severe crown rust.

Symptoms

The initial infection of crown rust might happen as early as late October. Mexican spores or spores from untamed grasses carried by the wind contaminate the crop. Orange-yellow blisters that are oval in shape and raised are the initial sign of the disease. They emerge as pustules encircled by the leaf epidermis when these blisters rupture. While some pustules can be seen on the sheaths, stem, and culms, leaf blades tend to host the majority of them (Nažareno *et al.*, 2018). During the growth season, they are spore-rich and spread the disease. Black spore rust stage develops as oats reach maturity.

Figure 4.1: Crown Rust of Oat (*Source*: adama.com).

2. Stem Rust (*Puccinia graminis avenae*)

On the majority of oat types in Texas, stem rust, another two-host fungus, manifests itself. Like crown rust, it is more problematic in humid regions and can result in significant damage when circumstances are right. Reduced yields, a decreased bushel weight, and lodging are all effects of stem rust. The stem rust fungus's alternate host is the common barberry (Berberis), but apart from that, its life cycle and spore stages are strikingly identical to those of crown rust. In the Deep South, the stem rust fungus may survive the winter on oats or wild grasses. The crown rust fungus has more races than the stem rust fungus, which has fewer.

Symptoms

Figure 4.2: Stem Rust of Oat (*Source*: ipmimages.org).

Management

1. Early planting and the adoption of resistant cultivars are the main crown rust prevention strategies.

2. Plants with strong root systems are more resistant to attack from crown rust when they are planted early.
3. In general, early maturing types are less susceptible to rust than later maturing. Only the capacity to infect particular types of oats may identify the many races of crown rust.
4. Oat variants that are resistant to certain races have been developed since the early 1930s as a result of a concerted effort by plant breeders and plant pathologists.
5. Spray Propiconazole 60-120 ml/acre at 15 days of interval.

B. Blights

1. Septoria Blight (*Septoria* spp.)

Crop rotation ought to aid in preventing the illness from escalating to dangerous levels.

Symptoms

The Septoria fungus may infect any portion of the oat plant that is aerial. Lesions on leaves are round to elongate, with a pale yellow to dirty white centre surrounded by a dull brown border, which is itself bordered by yellowish tissue merging into green. Small, black, spore-producing entities may be scattered throughout the lesion's centre. Sometimes, severe lodging is caused by infection of the stems.

Survival

The fungus survives the winter on the field's leftover oat straw and stubble. Sporae are created on this material in the spring and are carried to the green grain by the wind. The leaves are where most early infections are found.

2. Victoria Blight

Oat cultivars with the Victoria type of crown rust resistance initially showed signs of this disease in 1945. It soon became clear that it was one of the most devastating oat diseases ever recorded, lowering yields of vulnerable kinds by as much as 50 per cent.

Symptoms

A longitudinal bronze or lead-colored striping can be seen on the leaf blades and sheaths of infected plants, as well as a basal stem and root rot and a profusion of the fungus black spores on the nodes of mature plants. Around the ground line, infected stems frequently snap off. Throughout the plant, a toxin (poison) is produced by the disease-causing fungus.

Survival and Dispersal

The fungus is spread by seeds and lives on in agricultural waste. When the disease has established, vulnerable kinds are affected every season.

Figure 4.3: A. Septoria Blight; B. Victoria Blight (*Source*: agrobaseapp.com).

Management

1. Crop rotation suppress the disease on susceptible varieties.
2. Seed treatment with mancozeb 2g/kg of seed.
3. Spray Propiconazole 100 ml/acre.

3. Bacterial Stripe (*Pseudomonas striafaciens*)

Symptoms

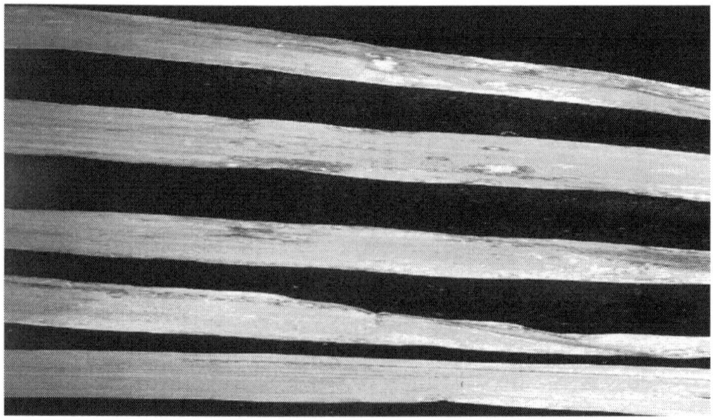

Figure 4.4: Bacterial Stripe of Oat (*Source*: plantdiseases.org).

Small, sunken, water-soaked patches that later combine into lengthy, water-soaked stripes are how the lesions initially manifest. The lesions surface is clearly covered in bacterial exudate. Later, severely diseased leaves become brown and begin to wither away from the tip. When particularly vulnerable types are present, the die-back may extend several inches above the lesions on the leaf, indicating significant extra indirect harm.

Survival

The bacteria overwinter on seed and plant debris in the field, making plant residue the principal source of infection in the spring. Cool, rainy weather encourages the sickness whereas warm, dry weather inhibits it. Normally, no control measures are required.

Management

1. If the issue is extensive, avoid sowing into diseased stubbles and burn or incorporate stubble.
2. The disease can be seed borne.
3. Avoid paddock operations when leaves are wet to prevent disease spread.

4. Culm and Root rots

Culm Rot (*Helminthosporium* spp.)

Symptoms

Culm (stem) rot-infected plants may die or have low yields of poor quality. The grain is light in weight, shrivelled, and frequently difficult to distinguish from sound grain totally. The disease may affect a single plant, a few tiny, symmetrical to asymmetrical patches of plants, or a sizable portion of the field.

Root Rot

Symptoms

Plants are weakened, stunted, and have a pale, purple, yellow, or bleached green appearance. The root structure of affected plants is typically rotten, making them simple to pull out. The roots are frequently dark or black with wet spots. The roots eventually slough off and die. Bleached, brown, or black and rotting crown tissue is a sign of infection. Wind and rain can cause lodging, which is frequent and can be severe. A few shrivelled seed are frequently seen in poorly filled heads. Drought and winter death often cause severe damage to infected plants (Mahapatra *et al.*, 2013).

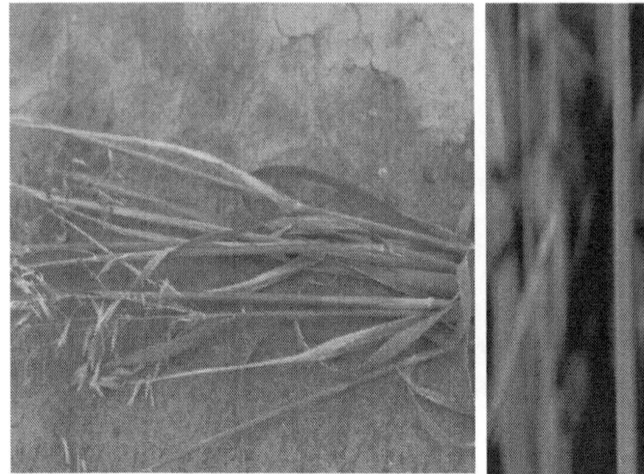

Figure 4.5. A. Culm Rot; B. Covered Smut (*Source*: extensionaus.com.au).

D. Smuts

1. Covered Smut (*Ustilago kolleri*)

Symptoms

According to the type of oat, the thin, greyish membrane that covers the smut mass-which takes the place of the kernel-normally doesn't tear until thrashing or harvest. The spores are dispersed across the grain during harvest and thrashing, and some of them settle beneath the glumes where they germinate and cover the seed coat with an abundance of fungal threads. On the glume's exterior, more spores are still present. These spores germinate alongside the seed when it is planted in warm, wet soil and maintain pace with the plant's growth point. They then infect the young seedling, the glumes' kernels are replaced by a mass of spores when the head develops.

Management

1. Use disease free certified seed.
2. Hot water treatment can eliminate spore and mycelium from contaminated seed, but it must be used carefully to avoid reducing seed vitality.
3. Seed treatment with contact fungicides will control covered smut because the fungus is externally seed borne.

2. Loose Smut (*Ustilago avenau*)

Symptoms

The spore mass has taken the place of the individual blooms in the oat panicle. When smut initially emerges from the boot, loose smut is more noticeable in the field than covered smut because it is not obscured by flowers or chaff. Each spikelet may have a thin white membrane covering the smut mass, but this membrane quickly breaks, allowing the black spores to be transported by the wind to the flowers of healthy heads where they lodge beneath the glumes. Here, they begin to grow and cover the seed coat beneath the hull with a layer of fungal filaments.

Management

1. The most common type of systemic fungicides used for loose smut are from the carboxin group of chemicals.
2. Although carboxins are highly toxic to fish and mildly toxic to other animals, they degrade rapidly in the environment and since there the fungicide is coated on seed and not sprayed there is not much loss to the environment.

3. Powdery Mildew

Symptoms

Powdery mildew may be a major disease on oats, lowering the fodder yields of sensitive cultivars. The fungus first appears as white creamy growth patches on the lower leaves and leaf sheaths since it is an exterior parasite. As the illness worsens, the spots spread to cover larger sections of the leaf and sheath and may finally turn grey or brown.

Figure 4.6: Powdery Mildew of Wheat.

Management

1. *B. subtilis* products should be used for best results. Limit incidence of fungal diseases on foliage with these bio-fungicides.

2. Sulfur and copper fungicides can prevent powdery mildew diseases and control existing fungal problems.
3. Rinse the foliage with Suff Oil-X on a regular spray schedule for continued to control of powdery mildew disease.
4. Spray Karathane @ 1kg/ha.

E. Viruses

Oats may contract a number of viral infections. Major oat diseases should include at least two, soil-borne oat mosaic and barley yellow dwarf, which have both been linked to significant oat losses in Texas.

F. Soil-borne Oat Mosaic

The virus that causes this mosaic disease consists of at least two different strains. The virus can lower yields in vulnerable types by as much as 50 per cent in years when the disease is more prevalent.

Symptoms

The symptoms are more obvious in colder weather. This disease causes infected plants to be stunted and develop in a rosetted state in its most damaging form. Yellow streaks that run parallel to the leaf's axis and a necrotic mottling are two less serious signs that might occasionally be present. The virus overwinters in the soil and is soil-borne.

Management

1. The chemicals that have been reported to provide reproducible control against *P. graminis* in the field are soil fumigants, which are not economically possible for small farmer.
2. Continuous wheat cultivation, *P. graminis* inoculum and SBWM-susceptible cultivars increase the proportion of viruliferous zoospores.

G. Barley Yellow Dwarf

Symptoms

Yellow-green blotches, primarily towards the tips of older leaves, are the initial symptoms of this disease. These regions could then develop a crimson hue, which, depending on the variety and the environment, might range from yellow to vivid scarlet. Plants that are infected are dwarfed, reach maturity sooner, yield seed with a low bushel weight, and have more blast. A variety of aphid species can spread barley yellow dwarf. There are several strains of the virus, and certain species of aphids spread a particular strain more easily than others. Although the barley yellow dwarf virus's whole life cycle is unknown, it is known that wild grasses that are sensitive to the virus may contribute to its overwintering cycle.

Figure 4.7: A. Barley Yellow Dwarf; B. Bacterial Leaf Spot.

Management

1. Control grassy weeds and collateral host.
2. These hosts can harbor barley yellow dwarf, which can then be transferred to your crop by aphids. Small grains planted in mid-summer as companion or cover crops can also become reservoirs of BYDV.
3. Coker-716 is a hexaploid oat cultivar that has resistance to both oat OMV and oat golden stripe virus (OGSV).

H. Seed-borne Diseases

Symptoms

Oats may get a number of diseases via seeds, including Septoria blight, covered and loose smut, and root and crown rots. The pathogens that cause these illnesses typically stick to the seed coat, where they then germinate, develop, and infect the young seedlings when the seed is sown. The traditional seed protection fungicides will aid in preventing these organisms because they are exo-embryo-borne. Fungicides that protect seeds can be used in drill boxes, on farms, at home, and by professional seed treatment facilities. Only clean seed should be used since it typically contains disease organisms and produces less. It's time to plant the seed.

Management

1. Hot water treatment or hot water therapy was first employed by gardeners in Scotland for treating the bulbs of ornamental plants.
2. Solar heat treatment is effective against both seed as well as soil borne diseases.

I. Blast

When the panicle is heading, a type of sterility known as blast manifests as white, empty hulls, generally close to the base of the panicle. Blast appears to be linked to the extremes in temperature, moisture, light, and nutrition that happen during specific crucial times in the growth of the plant. Early-seeded material normally experiences less blast than late-seeded material. Generally speaking, more resistant kinds are those that have been better adapted. Although they won't completely stop the blast, good cultural practises can be helpful.

Management

1. Early planted winter crops and late planted spring crops may be subject to high aphid populations at early growth stages, resulting in high yield losses.
2. The oat varieties have differed in their susceptibility to this disease but there is no effective resistance variety in any of the cereal crops.
3. Control volunteer wheat plants, barley and oats because they may serve as hosts to the aphids and virus.

J. Gray Speck

A shortage in soil manganese is the root cause of grey spots. In alkaline-organic soils and other low manganese soils, it can occasionally be a problem. The initial signs of grey speck are light green to gray-brown streaks and patches, mainly on leaf. Plants can become stunted, and yields might be lower. If the illness is bad enough, the plant could even turn yellow. By adding manganese salts to the soil or spraying the plants with 1 per cent manganese sulphate, Grey Speck can be managed.

Management

1. In general, fluoxastrobin or azoxystrobin, pyraclostrobin, fungicides will control gray leaf spot for 28 days.

Conclusion

Integrated disease management (IDM) approaches that combine various management strategies are highly recommended. IDM involves a combination of cultural practices, biological control, resistant cultivars, and chemical treatments when necessary. Regular monitoring of crops for disease symptoms and timely implementation of appropriate management practices are essential for effective disease control. It is important to note that the effectiveness of management practices may vary depending on the specific oat cultivar, regional climate, and disease prevalence. Therefore, it is advisable to consult with local agricultural extension services or experts to obtain tailored recommendations based on the specific conditions in a particular area. Overall, a comprehensive disease management strategy that incorporates resistant cultivars, cultural practices, and targeted chemical interventions when necessary can help minimize the impact of diseases on oat production and ensure a healthy and productive oat crop.

References

Banyal, D. K., Sood, V. K., Singh, A. and Mawar, R. (2016). Integrated management of oat diseases in north-western Himalaya. *Range Management and Agroforestry*, **37** (1): 84-87.

Bhardwaj, N. R., Banyal, D. K. and Roy, A. K. (2021). Prediction model for assessing powdery mildew disease in common Oat (*Avena sativa* L.). *Crop Protection; 146*, 105677.

Burdon, J. J. and Thrall, P. H. (2008). Pathogen evolution across the agroecological interface: implications for disease management. *EvolutionaryApplications;* **1**(1): 57-65.

Dietz, J. I., Schierenbeck, M. and Simón, M. R. (2019). Impact of foliar diseases and its interaction with nitrogen fertilization and fungicides mixtures on green leaf area dynamics and yield in oat genotypes with different resistance. *Crop Protection*, **121**: 80-88.

Forsberg, R. A. and Reeves, D. L. (1995). Agronomy of oats. In *The oat crop: Production and utilization* (pp. 223-251). Dordrecht: Springer Netherlands.

Mahapatra, S. S. and Sahoo, B. K. (2013). Management of root rot disease of forage oat in Odisha. *Journal of Plant Protection and Environment*, **10** (2): 49-53.

Nazareno, E. S., Li, F., Smith, M., Park, R. F., Kianian, S. F. and Figueroa, M. (2018). *Puccinia coronata f. sp. avenae*: a threat to global oat production. *Molecular plant pathology*, **19** (5): 1047-1060.

Soovali, P. and Koppel, M. (2011). Timing of fungicide application for profitable disease management in oat (*Avena sativa* L.). *Zemdirbyste-Agriculture*, **98** (2): 167-174.

Chapter 5

Diseases of Maize Crop

Sandeep Kumar[1*], Saurabh Kumar[1], Shivam Kumar[1], Dhara Prajapati[2] & Anju Shukla

[1]Department of Plant Pathology, Chandra Shekhar Azad University of Agriculture and Technology, Kanpur – 208 002, Uttar Pradesh
[2]Department of Plant Pathology, B. A. Collee of Agriculture, Anand Agricultural University, Anand, Gujarat – 388110, India
[*]Corresponding author: shivamkumar224229@gmail.com

Introduction

Maize (*Zea mays* L.) is recognized as one of the most vital sources of human food globally and ranks third in production, trailing only wheat and rice (Zhao *et al.*, 2017). It is often referred to as the "queen of cereals" on a global scale due to its great genetic yield potential. Maize is cultivated in most countries for human consumption, animal feed, and various other applications. As one of the most widely cultivated crops worldwide, it stands as the third most sown and produced crop, following rice and wheat (Kaur *et al.*, 2021; Letsoin *et al.*, 2023). According to USDA report 2016, global maize production in 2015-16 was estimated at approximately 96.88 crore tonnes, representing a 4.6 per cent decrease compared to the previous year (101.35 crore tonnes. The nature and extent of losses resulting from plant diseases can vary depending on factors such as the type of plant or plant products, the pathogen involved, the geographic location, environmental conditions, control measures implemented, and their combinations (Agrios, 2005). Projections also indicate that the demand for maize in the developing world will double by 2050. It's crucial to recognize that maize production faces a substantial threat due to maize diseases, which could potentially result in significant economic losses. The global reduction in maize yields due to pathogens and pests has been estimated to range from 19.5 per cent to 41.1 per cent. Despite an annual increment of 2.24 per cent in global maize production from 2015 to 2019, this trend falls short of meeting the projected global demand for 2050. In the case of India, over the past decade, maize consumption has grown at an annual rate of 5.6 per cent, while production has increased by approximately 2.9 per cent. Various biotic and abiotic stresses have led to significant economic losses, impacting both the quality and quantity of maize. Biotic stresses encompass nearly 110 maize diseases caused by fungi, bacteria, and viruses that are prevalent worldwide. Several pathogens responsible for various diseases significantly reduce maize production, posing a threat to global food security and agricultural sustainability. Prominent among the fungal diseases affecting maize are Northern and Southern corn leaf blight, curvularia leaf spot, charcoal rot, downy mildew, smut, rust, ear rot, seedling rot, banded leaf, sheath blight, and stalk rot. This study discusses into the diseases that impact the economically significant maize crop in India, discusses the resultant crop losses, and explores methods for crop management. These diseases have a substantial adverse effect on maize production, but they can be managed through various approaches such as agronomic practices, strategic sowing dates, cultivation of resistant maize varieties, biological control, biopesticides, and, as a last resort, chemical fungicides.

Table 5.1: List of Maize Diseases and their Causal Agent

Sl.No.	*Name of Disease*	*Causal Agent*
A. Physiological disorders		
1	White bud of maize	Zn deficiency
2	Stunted growth and purplish margins of leaves	P deficiency
B. Fungal diseases		
1.	Turcicum leaf blight (TLB)/Northern corn leaf blight (NCLB)	*Exserohilum turcicum*
2.	Maydis leaf blight (MLB)/Southern corn leaf blight (SCLB)	*Biopolaris maydis*
3.	Curvularia leaf spot	*Curvularia lunata*
4.	Banded leaf and sheath blight	*Rhizoctonia solani*
5.	Charcoal rot	*Macrophomina phaseolina*
6.	Fusarium stalk rot	*Fusarium verticillioides*
7.	Common rust	*Puccinia sorghi*
8.	Black bundle disease	*Cephalosporium maydis*

Sl.No.	Name of Disease	Causal Agent
9.	Downy mildew/Crazy top(i) Sorghum downy mildew	*Peronosclerospora sorghi*
	(ii) Phlippine downy mildew	*Peronosclerospora philippinensis*
	(iv) Rajasthan downy mildew	*Pernosclespora heteropogoni*
	(iii) Crazy top	*Sclerophthora macrospora*
10.	Head smut	*Sphacelotheca reiliana*
11.	Anthracnose	*Colletotrichum graminicola*
C. Bacterial diseases		
11.	Bacterial stalk rot	*Dickeya zeae*
12.	Bacterial leaf strip	*Xanthomonas rubrilineans*
D. Viral diseases		
13.	Maize mosaic	Maize mosaic virus (MMV)
E. Nematodes		
14.	Cyst nematode	*Heterodera zeae*

Physiological Diseases

Nitrogen deficiency Development of a pale, light green coloration on the leaves, forming a distinctive V-shaped yellowing pattern that originates at the leaf tips and extends towards the leaf bases.

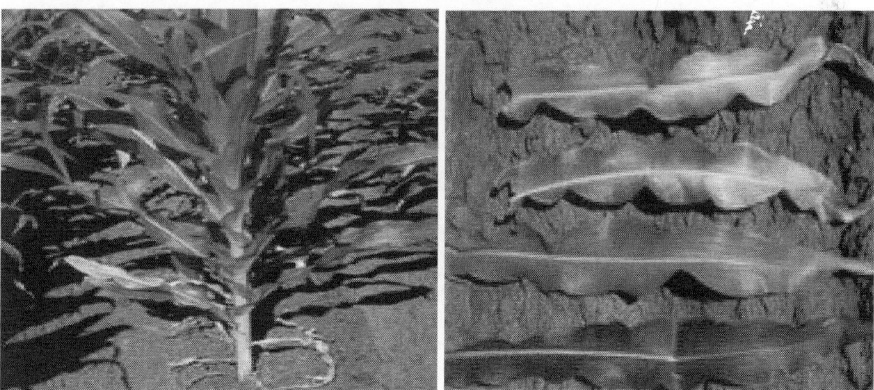

Figure 5.1. Nitrogen Deficiency (*Source*: R.L. Croissant, 2018)

Phosphorous deficiency: a deficiency exhibit a deep green color, with the older leaves displaying a reddish-purple discoloration.

Potassium deficiency: The edges of the leaves exhibit a yellow and brown discoloration, resembling a condition akin to scorching or desiccation.

Sulphur deficiency: The symptoms manifest on newer leaves, displaying a yellow striped pattern between the veins, which is referred to as interveinal chlorosis.

Figure 5.2: Phosphorous Deficiency (*Source*: Mary Ann Hansen, 2018).

Zinc deficiency: The upper leaves exhibit wide bands of yellow discoloration, which eventually transition into a pale brown or grey necrosis with dead spots. These symptoms initially emerge in the center of the leaves and then spread outward.

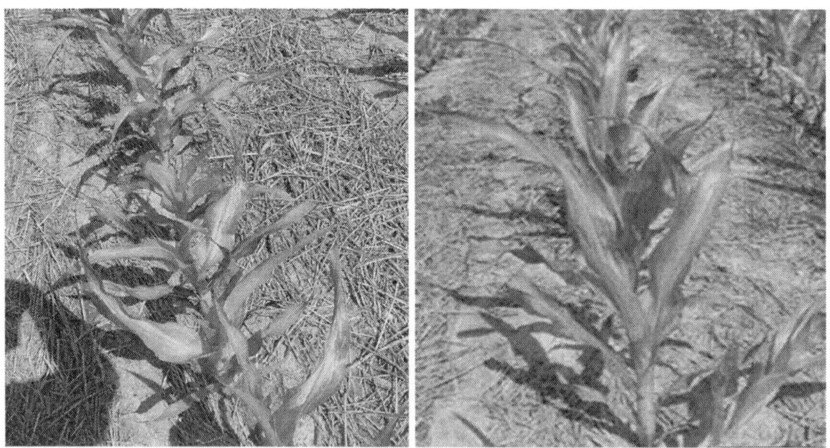

Figure 5.3: Zinc Deficiency (*Source*: Diane Perkins, 2018).

Fungal Diseases

1. Turcicum Leaf Blight (TLB)/Northern Corn Leaf Blight (NCLB)

Occurrence and Economic Importance

Turcicum leaf blight (TLB) represents a prevalent maize foliar disease in Mexico, resulting from the fungal pathogen *Exserohilum turcicum*. This fungus is widely distributed across the globe, primarily inhabiting warm, tropical, and subtropical regions, where it thrives on various plant materials such as grasses, decaying wood, and within the soil. In the Indian state of AP, TLB exhibits its most severe impact in the districts of Warangal, Karimnagar, Medak, Nizamabad, Adilabad, and Ranga Reddy.

Causal Organism

Turcicum leaf blight (TLB), also known as Northern Corn Leaf Blight (NCLB), is caused by the fungus *Exserohilum turcicum*. This fungus falls under the *Fungi* kingdom *Ascomycotina* division, *Dothideomycetes* class, *Pleosporales* order, *Pleosporaceae* family, and *Exserohilum* genus, specifically as the species *turcicum* (Pass.) K.J. Leonard and Suggs (1974). Colonies of *Exserohilum* can have varying colors, ranging from grey to blackish-brown, and their texture can range from suede-like to somewhat floccose. The conidia, which are responsible for reproduction, may come in different shapes, including straight, curved, slightly bent, or ellipsoidal to fusiform. These conidia are produced at the top of the fungal structure through a pore, referred to as poroconidia, located on an elongated and angled structure called a conidiophore. The hilum, a prominent and protruding feature, and the septum just above it typically have a thick and dark appearance. The end cells of the conidia are usually lighter in color, and the walls exhibit a finely roughened texture. Regarding the germination of conidia, *Exserohilum* displays bipolar characteristics.

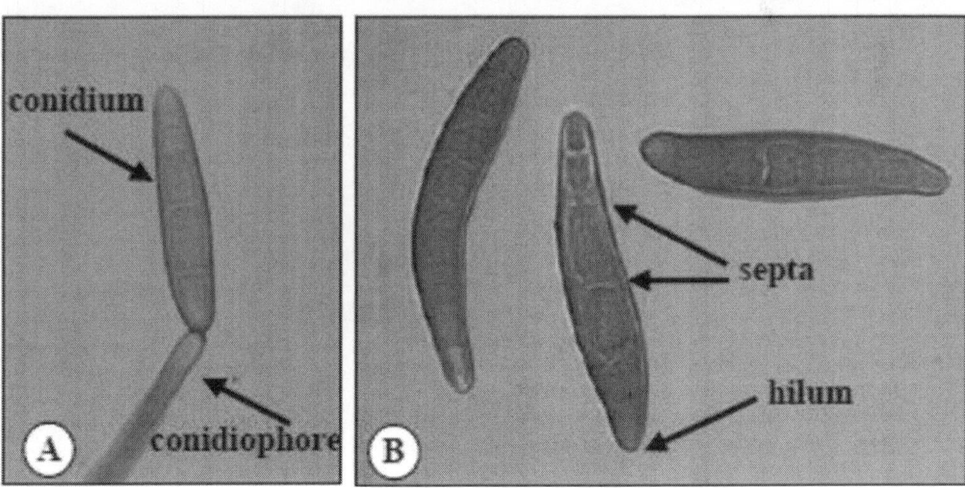

Figure 5.4: A. Conidium on conidiophore; B. Characteristics of Conidium.

Symptoms

The distinctive symptom of turcicum leaf blight on a susceptible maize cultivar manifests as cigar-shaped lesions, typically ranging from one to six inches in length and having a grey to tan coloration. Initially, before these lesions reach their full development, they appear as small light-green to grayish spots, typically emerging about 1 to 2 weeks after infection. As the disease progresses, these lesions extend to cover all leafy structures, including the husks, and give rise to dark gray spores, imparting a dirty appearance to the affected areas. With an increasing number of lesions, the leaves may become so abundant that they eventually succumb, resulting in a significant reduction in yield due to a decrease in available carbohydrates for grain filling. Consequently, the leaves take on a grayish-green and brittle appearance, resembling leaves that have been damaged by frost.

Figure 5.5: Turcicum Leaf Blight (TLB)/Northern Corn Leaf Blight (NCLB).

Favourable Conditions

The onset and progression of the disease are facilitated by specific environmental factors, such as ample dew, frequent rainfall, high humidity, and moderate temperatures. Infection occurs when conidia germinate in the presence of free water on the leaf surface for a duration of 6 to 18 hours, and the temperature remains within the range of 66 to 80°F (18 to 27°C). When these favorable conditions are met, lesions develop and lead to the generation of a new set of spores within a relatively short timeframe, typically occurring within 7 to 12 days in susceptible cultivars. This rapid spore production contributes to the rapid spread of the disease.

Disease Cycle

E. turcicum survives the winter in corn residue that remains infected on the soil surface. As spring temperatures increase, the fungus starts to produce spores on the surface of this residue. These spores are then carried by wind, rain, or irrigation to the lower leaves of a new corn crop, where they can potentially germinate and infect a susceptible hybrid. The presence of consistent moisture on the foliage is a crucial factor in the development of the disease; spores will only germinate when there is a continuous period of 6 to 18 hours of water on the leaf surface. Typically, lesions develop within one to two weeks after the infection has occurred. Spores are produced within these lesions and can be dispersed upward in the plant and to neighboring areas through rainfall or wind. As the growing season ends and plants are harvested, the pathogen survives the winter on infected residue.

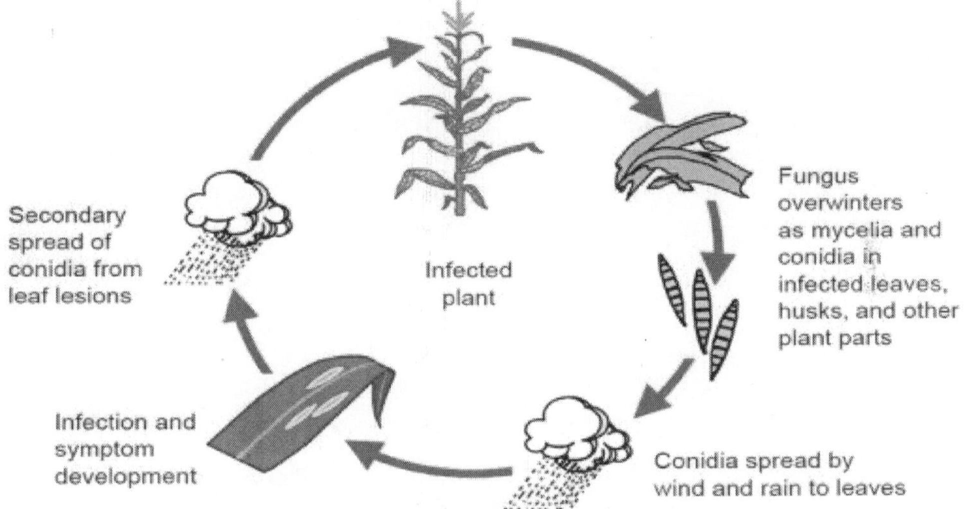

Figure 5.6: Northern Corn Leaf Blight (*Source*: apps.lucidcentral.org).

Management

1. To decrease the severity of Turcicum leaf blight, implement crop rotation by planting non host crops such as legumes. This allows infected residue to decompose over time and prevents the disease from developing in the following corn crop by reducing the overwintering fungus.
2. Eliminate plant debris.
3. Ensure that plants receive adequate nutrients. Avoid excessive nitrogen, but ensure that phosphorus and potassium levels are optimized.
4. Treat seeds with Azospririllum (ACD 15 or ACD 20 Strain) at a rate of 25 grams and Trichoderma at 6 grams per kilogram of seeds, or use Thiram 75 WP at 2 grams or Capatn 50 WP at 2 grams per kilogram of seeds before planting.
5. Apply 2.5 grams of Mancozeb 75 WP or 1 ml of Hexaconazole 5 EC or 1 ml of a combination fungicide like Azoxystrobin 18.2 per cent + Difenconazole 11.4 per cent per liter of water immediately when disease symptoms appear. Repeat the same application after 15 days to also manage Curvularia leaf spot.

6. Administer two Mancozeb sprays (2-3 grams per liter) at 15-day intervals as soon as the disease becomes visible.

2. Maydis Leaf Blight (MLB)/Southern Corn Leaf Blight (SCLB)

Occurrence and Economic Importance

This disease, known as Southern Corn Leaf Blight (SCLB) or Maydis Leaf Blight (MLB), is caused by the fungus *Helminthosporium maydis* (also known as *Bipolaris maydis*, telomorph: Cochliobolus heterostrophus). It poses a significant threat to maize crops worldwide, particularly in regions with warm and humid conditions, as documented by White in 1999. SCLB has emerged as one of the most prevalent and severe diseases affecting maize cultivation in several countries, including Pakistan, India, Nepal, Kampuchea, the Philippines, Indonesia, Vietnam, and China. In India, it is commonly referred to as 'Maydis Leaf Blight' and affects crops such as Corn (Zea mays), Sorghum, and Teosinte. In India, SCLB has become a noteworthy concern, notably in regions like J and K, HP, Sikkim, Meghalaya, PB, HR, RJ, Delhi, UP, Bihar, MP, GJ, MH, AP, KR and TN. The first documented instance of Maydis Leaf Blight in India dates back to 1960, when it was identified by Munjal and Kapoor in Maldah, WB.

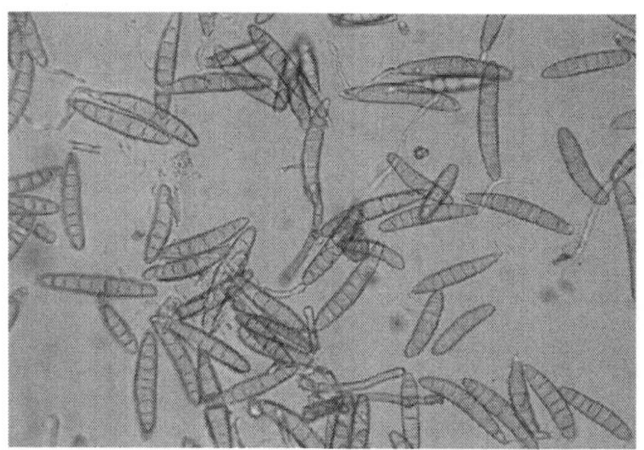
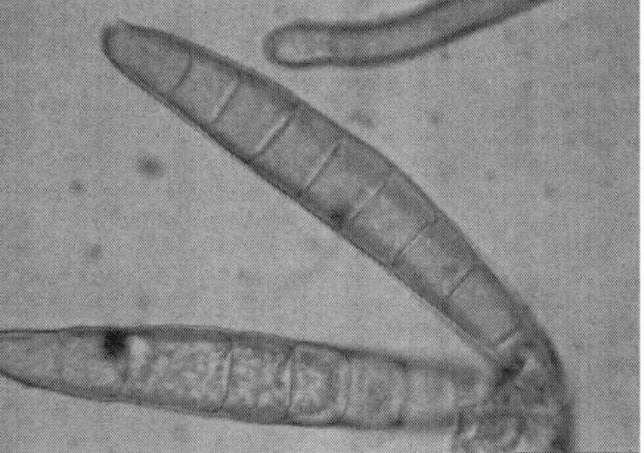

Figure 5.7: Conidia of *Bipolaris maydis.*

Causal Organism

Maydis leaf blight (MLB), also known as Southern Corn Leaf Blight (SCLB), is induced by the fungus Bipolaris maydis. Taxonomically, this fungus is classified as follows: Kingdom: *Fungi* Phylum: *Ascomycotina* Class: *Dothideomycetes* Order: *Pleosporales* Family: *Pleosporaceae* Genus: *Bipolaris* & Species: *maydis*. *Bipolaris maydis* is also recognized by alternative names such as *Drechslera maydis* or *Cochliobolus heterostrophus*. The conidia, which are the reproductive structures, exhibit a curved and spindle-shaped morphology, measuring roughly 70–160 µm in length and 15–20 µm in width. They display a color spectrum from pale to golden brown and are composed of 6-12 cells. However, ascomata, another type of fungal structure, are infrequently observed in natural field conditions.

These structures have a subglobose shape with a diameter of approximately 300-400 µm and appear black. The ascospores found within the ascomata are filiform and measure about 200-300 µm in length, with a width of 7-8 µm.

Symptoms

The symptoms caused by the fungus initially manifest as small, young, and diamond-shaped lesions on the leaves. As these lesions mature, they elongate in shape. Their growth is restricted by nearby veins, leading to the final lesion shape, which is typically rectangular and measures about 2 to 3 cm in length. In some cases, these lesions may merge or coalesce, resulting in the complete browning and damage of substantial leaf areas. This can be particularly detrimental to maize cultivars that have incorporated the Texas source of male sterility.

Figure 5.8: Symptoms of Maydis Leaf Blight (MLB)/ Southern Corn Leaf Blight (SCLB).

Favourable Conditions

The expression "favourable conditions" indicates that there is water present on the leaf surface, and the environmental temperature falls within the range of 60 to 80°F. In such conditions, spores are capable of germinating and penetrating the plant in as little as 6 hours. During the winter season, the fungus persists in corn debris in the form of mycelium and spores, lying in wait for the return of these favorable spring conditions. When these conditions are met, the fungus can generate new inoculum remarkably quickly, with a generation time as short as 51 hours.

Disease Cycle

The disease cycle of *Bipolaris maydis* involves multiple phases and encompasses the release of either asexual conidia or sexual ascospores to initiate infections in corn plants. The asexual cycle is the more prevalent one in natural settings and is of primary concern. When environmental conditions are favorable, marked by moisture and warmth, conidia (the primary source of infection) are discharged from lesions on an infected corn plant. These conidia are then carried to adjacent plants by the wind or rain splashes. Upon landing on the leaves or sheaths of healthy plants, *Bipolaris maydis* initiates germination through polar germ tubes. These germ tubes can either penetrate the leaf or gain entry through natural openings like stomata. The fungus's mycelium subsequently invades the parenchymatous leaf tissue, causing the leaf cells to darken and ultimately collapse. From these lesions, conidiophores emerge, which, under favorable conditions, can either further infect the original host plant (including kernels, husks, stalks, and leaves) or release conidia to infect other nearby plants.

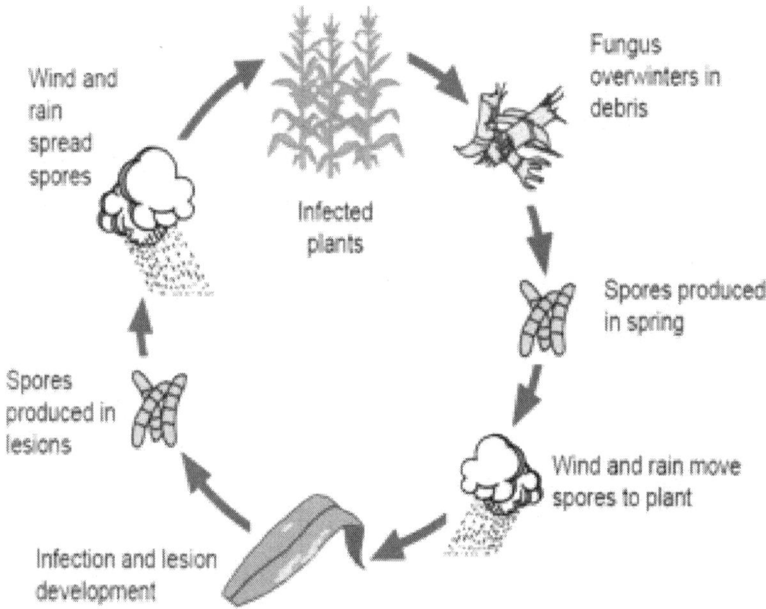

Figure 5.9: MLB/SCLB
(*Source*: Singh *et al.*, 2016).

Management

1. Remove contaminated crop residue from the field.
2. Utilize resistant or tolerant hybrid varieties.
3. Apply a foliar spray of Mancozeb at a rate of 2.5 grams per liter of water approximately 15 days after sowing. Repeat this application two more times at 10-day intervals or immediately upon the appearance of symptoms.
4. Apply a foliar spray of Mancozeb at a rate of 2.5 grams per liter of water approximately 30-35 days after sowing. Repeat this application two

more times at 10-day intervals or immediately upon the appearance of symptoms if necessary.

5. Administer a foliar spray of Azoxystrobin 18.2 per cent + Difenconazole 11.4 per cent w/w SC at a rate of 1 ml per liter of water as soon as symptoms become visible.

3. Curvularia Leaf Spot

Occurrence and Economic Importance

Curvularia leaf spot, which is caused by the fungus *Curvularia lunata,* was officially documented in the United States for the first time in 2017 and has been observed in several states. Typical symptoms of this condition include the presence of small lesions with a tan color and brown margins, often encircled by a yellowish halo. These lesions may be found scattered across the leaves or closely clustered together, occasionally merging into larger necrotic areas. They tend to develop on leaves located in the mid to upper sections of the corn canopy but can appear at any growth stage. It's worth noting that the symptoms of Curvularia leaf spot can be similar to those of eyespot.In general, within the United States, Curvularia leaf spot has been regarded more as an aesthetic concern rather than a disease that significantly affects crop yield.

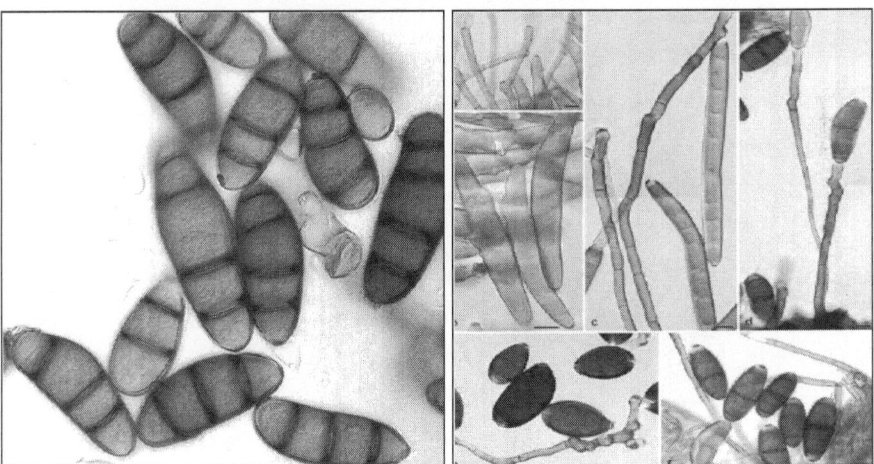

Figure 5.10: Conidia of *Curvularia lunata.*

Causal Organism

Curvularia leaf spot is caused by *Curvularia lunata*. Kingdom *Fungi* Division *Ascomycota* Class *Euascomycetes* Order *Pleosporales* Genus *Curvularia &* Species *lunatus*. The conidia of this fungus have an ellipsoidal shape, and they are often curved or somewhat crescent-shaped, with rounded ends or occasionally tapering slightly toward the base. They exhibit a colour range from pale brown to medium reddish brown to dark brown. Conidia typically contain 3 to 10 septa, although they usually have 3 to 5 septa. The surface of the conidial wall can be smooth or somewhat verrucose (having wart-like projections). In some species, the hilum, which is the point where the spore attaches to its structure, may appear protuberant or prominently raised.

Symptoms

The symptoms of Curvularia leaf spot typically begin as small yellow-brown spots on the leaves, which gradually expand and develop into oblong lesions. Within these lesions, the central portion changes to a brown color, while the margins retain their yellowish appearance. These lesions are more commonly found along the leaf margins. In the upper canopy of the plant, the lesions often exhibit a grey center and maroon margins. Normally, the affected area on a leaf measures about an inch in diameter, and there may be several small lesions present. However, in rare instances, the infection can become so extensive that it covers the majority of one or two leaves. The development of these lesions, attributed to Curvularia leaf spot, is typically associated with conditions characterized by excessive dew or sporadic rainfall.

Figure 5.11: A. Curvularia Leaf Spot Symptoms on a Corn Leaf; B. Curvularia Leaf Spot Lesions.

Favourable Conditions

Curvularia leaf spot is attributed to the fungus *Curvularia lunata*. This fungus survives the winter by residing in corn residue and is typically spread to new corn plants through splashing rain or wind. Fields with continuous corn cultivation or those implementing conservation tillage practices are more likely to experience greater disease severity. The development of this disease is favoured by warm and humid conditions. It's worth noting that the fungus responsible for Curvularia leaf spot can infect a wide range of host plants, and in other countries, various grass species can also contribute to disease development.

Disease Cycle

Common Leaf Spot (CLS) typically occurs in conditions marked by frequent heavy rainfall and high temperatures. The latent period for CLS infection is approximately 7 to 10 days for each infection cycle. *Curvularia lunata*, the causative fungus, persists in the form of mycelia or conidia within infected maize residues and the soil throughout the year, allowing it to infect plants in the next growing season. The presence of moisture on the leaves encourages the germination of conidia and initiates primary infections. Conidia originating from primary lesions can be released and lead to secondary infections, aided by wind and rain, affecting both neighbouring plants and fields. Several factors can exacerbate the spread and severity of the disease, including a high accumulation of the pathogen, dense planting, continuous cropping of susceptible hosts, excessive use of chemical fertilizers, and imbalances in soil nutrient levels. Remarkably, the disease has demonstrated the ability to spread over a distance of 250 meters within a month, primarily through wind dispersal.

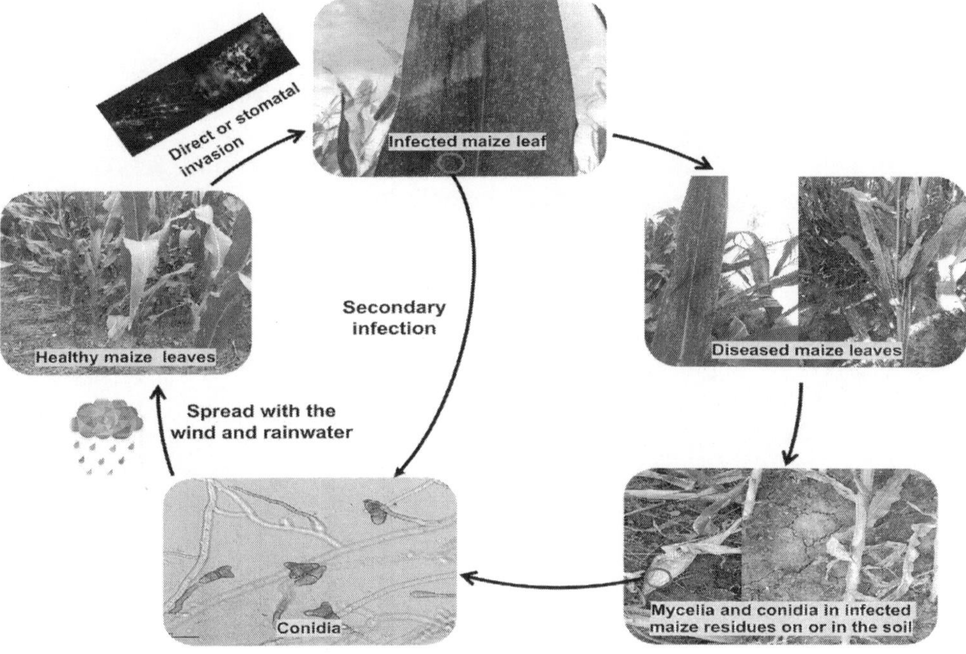

Figure 5.12: Disease Cycle of *Curvularia lunata* (*Source*: Wang *et al.*, 2022).

Management

1. Eliminate weeds both within and around the field to protect the crop from air borne infections.
2. Employ resistant or tolerant hybrid varieties.

3. Treat the seeds with 20 grams of *Trichoderma* chalk formulation in combination with Mancozeb 63 per cent or Thiram 40 F.S. at a rate of 6 grams per kilogram of seed.
4. Apply a foliar spray of Carbendazim 12 per cent + Mancozeb 62 per cent or Zineb 75 per cent at 2 grams per liter of solution at 35 and 55 days after sowing (DAS).
5. Spray Dalero (a mixture of Prothioconazole 16 per cent and Trifloxystrobin 13 per cent) as an overhead application at 20 grams per acre with 15-day intervals.

4. Banded Leaf and Sheath Blight

Occurrence and Economic Importance

This disease was initially documented in Sri Lanka in 1927, originally referred to as "Sclerotial disease" by Bertus (1927). Subsequently, it was reported in Malaysia as "banded sheath rot," in the Philippines as "banded sclerotial disease," and in Japan as "summer sheath blight" (Wiltshire, 1956). The disease has also been identified in various other countries, including Germany, the USA, Nigeria, Venezuela, Sierra Leone, Ivory Coast, and England. Brown Leaf Spot of Maize (BLSB) represents a significant challenge for maize production in regions such as China, South Asia, and Southeast Asia, affecting countries like India, Sri Lanka, Indonesia, Cambodia, Bangladesh, Pakistan, Nepal, Myanmar, Japan, Malaysia, Thailand, Laos, Vietnam, Taiwan, and Korea. Surprisingly, in China, this disease has resulted in yield losses approaching 100 per cent (Singh and Shahi, 2012). In India, the disease has been reported in various states, including Himachal Pradesh, Uttar Pradesh, Haryana, Punjab, Madhya Pradesh, Rajasthan, West Bengal, Meghalaya, Assam, and Orissa (Rani *et al.*, 2013).

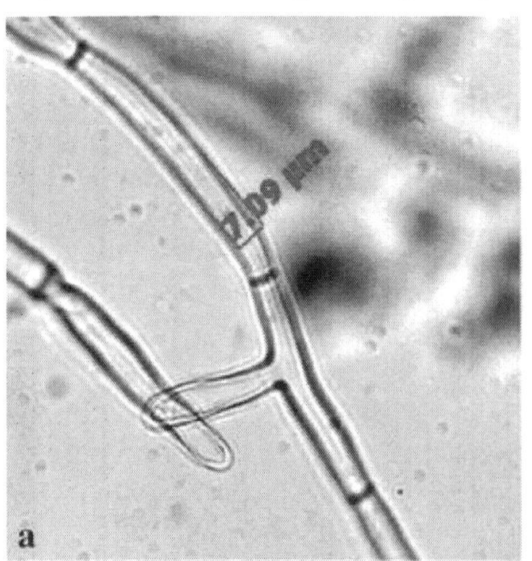

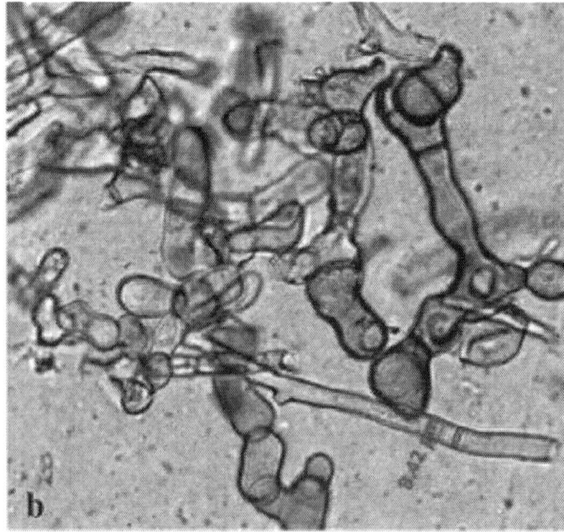

Figure 5.13: Morphological Characteristics of *R. solani*.

Causal Organism

Banded leaf and sheath blight is caused by the fungus *Rhizoctonia solani*. Taxonomically, it belongs to the Kingdom Fungi, Division *Basidiomycota*, Class *Agaricomycetes*, Order *Cantharellales*, Family *Ceratobasidiaceae*, Genus *Rhizoctonia*, and Species *solani* (J.G. Kuhn, 1858).

Symptoms

Figure 5.14: Banded Leaf and Sheath Blight on Lower Maize Leaves and Large Brown Sclerotia Produced by *Rhizoctonia solani* on Infected Leaves and Stem (*Source*: Singh *et al.*, 2019).

The causative agent of Banded Leaf and Sheath Blight (BLSB) affects all parts of the maize plant above the ground, with the exception of the tassel. The disease takes on various forms, including leaf blight, stalk lesions, rind spotting, and stalk breakage. BLSB typically emerges on maize plants during the pre-flowering stage, affecting plants that are approximately 30 to 40 days old. However, infection can also occur in younger plants, leading to severe blighting and potentially causing the apical region of the growing plant to wither (Saxena, 2002). On the leaves, BLSB symptoms manifest as irregularly shaped, either round or elongated, lesions that initially appear as water-soaked areas. These affected areas gradually change to a pale color and then become straw-colored and necrotic (Ahuja and Payak, 1982). The lesions rapidly expand, resulting in discolored areas interspersed with dark bands, which become visible on the lower leaves after approximately 7 to 8 days (Rani *et al.*, 2013). The disease significantly reduces the yield of high-yielding maize varieties, leading to premature plant death, stalk breakage, and ear rot. In India, Singh and Sharma (1976) estimated a 40.5 per cent reduction in maize grain yield, with a BLSB disease index of 71 per cent.

Favourable Conditions

The optimal conditions for the occurrence of the disease are a temperature range of 28-30°C, coupled with a relative humidity of 90-100 percent. Disease development is less likely to occur or is negligible when the relative humidity falls below 70 percent. However, high relative humidity in the range of 88-90 percent during the first two weeks of infection significantly favours the rapid progression of the disease. Disease spread is strongly influenced by high relative humidity and rainfall, as highlighted by Ahuja and Payak in 1981. In particular, rainfall exceeding 100 mm in the first two weeks of infection promotes early infection and the development of the disease in plants.

Disease Cycle

The fungus *R. solani* has the capability to infect maize plants at any stage of their growth, starting from the seedling stage and continuing through to maturity. It persists in the soil and on crop debris that is infected, existing in the form of sclerotia or mycelium, with sclerotia being the primary source of inoculation. The fungus spreads through various means, including irrigation, the transfer of contaminated soil, and infected plant remnants.

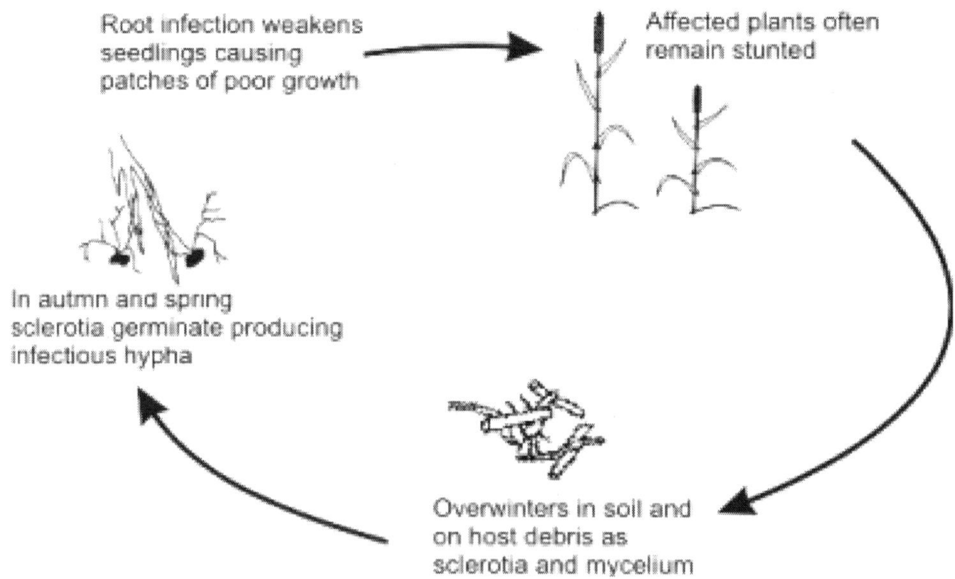

Figure 5.15 Disease Cycle of *Rhizoctonia solani.*

As the growing season commences and favourable conditions prevail, marked by adequate humidity and temperatures ranging from 15 to 35°C, the fungus is attracted to rapidly growing seedlings and water-soaked seed coats.

This attraction is facilitated by chemotropic stimulants that are released by actively growing plant cells and decomposing plant remains. Additionally, the disease can also be transmitted secondarily through contact between diseased leaves or sheaths and healthy plants.

Management

1. Remove the lower leaves and their sheaths by stripping.
2. Implement crop rotation with a non-host crop.
3 Opt for resistant or tolerant hybrid varieties.
4. Apply a foliar spray of Azoxystrobin 18.2 per cent + Difenconazole 11.4 per cent w/w SC (Amistar Top 325 SC) at a rate of 1 ml per liter of water at 50 days after sowing (DAS) or immediately upon the appearance of symptoms. If necessary, repeat the spray at 15-day intervals.

5. Charcoal Rot

Occurrence and Economic Importance

Charcoal rot, caused by *Macrophomina phaseolina* (previously known as *Rhizoctonia bataticola*), is a significant disease of maize. This disease is common in various regions, including J and k, WB, Pb, HR, RJ, Delhi, UP, MP, AP, KR and TN. It tends to be more prevalent in hot and dry weather conditions. Charcoal rot is known for a wide host range, and it is both seed and soil-borne. Controlling it with chemical treatments alone can be challenging, as these treatments may not provide protection throughout the entire crop growth period. Additionally, the disease is favored by soil temperatures ranging from 23-37°C. Therefore, effective management strategies may involve a combination of cultural practices, resistant varieties, and chemical treatments to mitigate its impact on maize crops.

Causal Organism

Disease caused by *Macrophomina phaseolina* belong to Kingdom *Fungi* Division *Ascomycota* Class *Dothideomycetes* Order *Botryosphaeriales* Family *Botryosphaeriaceae* Genus *Macrophomina* Species *M. phaseolina* (Goid 1947). *M. phaseolina* can be identified by its distinct characteristics, including hyaline (colorless) hyphae with thin walls, as well as light brown to dark brown hyphae with septa (cross-walls). When it comes to branching, the branches typically emerge from the main hyphae at right angles, with a noticeable constriction at the point of origin. Another important feature of *M. phaseolina* is the presence of microsclerotia. These microsclerotia are compact masses of hardened fungal mycelium and come in various shapes, including spherical, oval, or oblong. In their early stages, microsclerotia are light brown in colour, but they tend to become darker, ranging from brown to black, as they age.

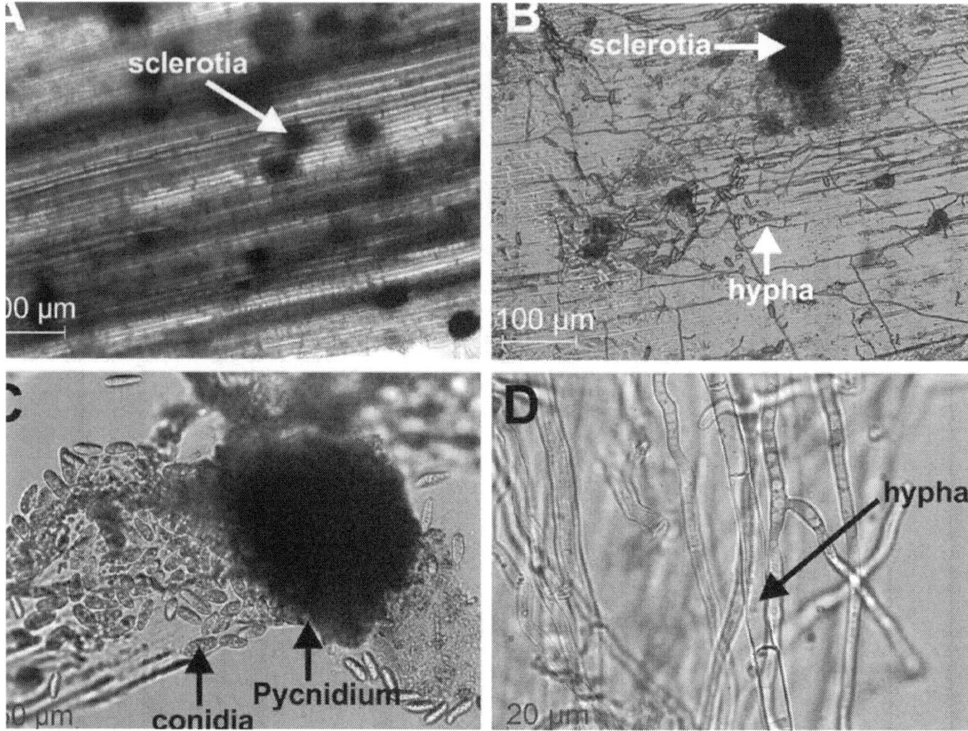

Figure 5.16: Characteristics of *M. phaseolina*: (A) Sclerotia; (B) Hyphal network; (C) conidia from the pycnidium; (D) Aerial hyphae.

Symptoms

Plants affected by *Macrophomina phaseolina*, which is responsible for charcoal rot, typically exhibit symptoms of wilting. Infected plants can be identified by the presence of grayish streaks in their stalks. Within the stalk, the pith becomes frayed, and small grayish-black sclerotia form on the vascular bundles. This fraying of the interior of the stalk often weakens it, making the stalks susceptible to breaking around the crown area. The crown region of the affected plant takes on a dark discoloration. In addition to the symptoms observed in the stalk and crown, the disease can also affect the roots. This is characterized by the fraying of the root bark and the disintegration of the root system. These combined symptoms are common characteristics of charcoal rot in maize plants that have been affected.

Favourable Conditions

High temperature and low soil moisture (drought).

Figure 5.17: Shredding of Stalk.

Disease Cycle

The disease cycle of this fungus involves a broad range of hosts, including sorghum, pearlmillet, fingermillet, and various pulses. The fungus can survive for more than 16 years in infected plant debris. The primary source of infection is the soil-borne sclerotia, which serve as a reservoir of the pathogen. This pathogen's ability to infect multiple hosts contributes to its persistence in the

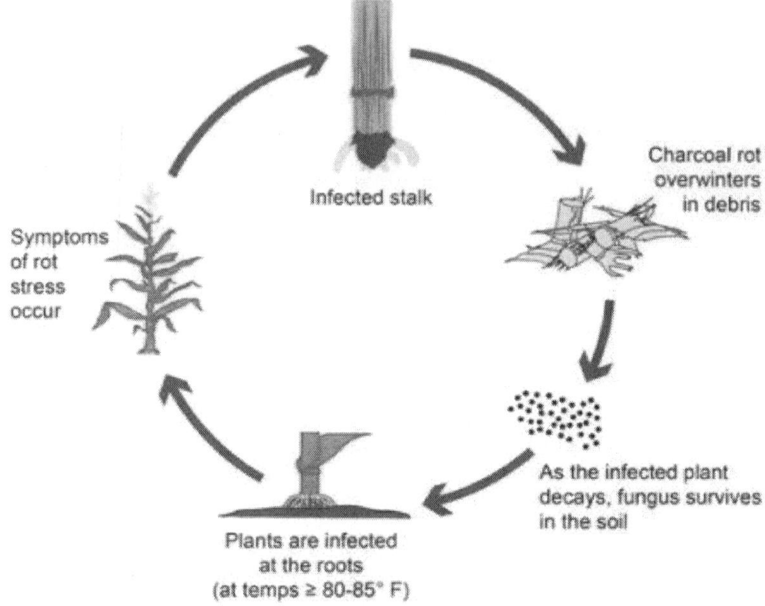

Figure 5.18: Charcoal Rot Disease Cycle.

environment. The fungus is a facultative parasite, meaning it can also exist saprophytically by living on dead organic tissues, especially those produced by many of its natural hosts, leading to the formation of sclerotial bodies. During the winter months, the fungus remains overwinters in the form of sclerotia in the soil. When conditions become favourable and the host crop reaches a susceptible stage, the fungus infects the plant through its roots and subsequently moves towards the stem, continuing its pathogenic cycle

Management

1. Employ crop rotation with non-host crops.
2. Engage in deep summer ploughing.
3. Maintain field cleanliness and sanitation.
4. Prevent water stress during the flowering stage.
5. In the presence of stalk rot, consider early harvesting to reduce ear loss.
6. Use of resistant varieties such as DHM-103 and Ganga Safed - 2, while avoiding susceptible varieties like DHM-105.
7. Treat seeds with 25 grams of PSB (Pseudomonas striata) and 5-6 grams of Trichoderma harzianum/kilogram of seeds before planting.
8. Treat seeds with Carbendazim or Thiram at 3 grams per kilogram, or use Xelora (a mixture of Pyraclostrobin and Thiofanate-methyl) at 450 grams per liter of water to protect against disease infestations.

6. Fusarium Stalk Rot

Occurrence and Economic Importance

Fusarium stalk rot, which is caused by Fusarium verticillioides, was first documented as a significant root and stalk disease in the United States by Pammel in 1914. In India, the initial report of Fusarium stalk rot came from Mount Abu, Rajasthan (Arya and Jain, 1964). This ailment typically affects maize plants aged between 55 and 65 days, coinciding with the tasselling and silking stages, and it tends to manifest immediately after grain formation. In India, Fusarium stalk rot is widespread in numerous maize-growing regions, especially in rainfed areas like J and K, PB, HR, Delhi, RJ, MP, UP, BR, WB, AP, TN, and KR. It is particularly prevalent in regions where water stress occurs after the flowering stage of the crop (Singh *et al.*, 2012).

Causal Organism

Fusarium stalk rot disease caused by *Fusarium verticillioides* fungus belong to Kingdom- *Fungi,* Division- *Ascomycota,* Class- *Sordariomycetes,* Order- *Hypocreales,* Family- *Nectriaceae,* Genus- *Fusarium,* Species- *verticillioides* (Nirenberg, 1976).

Symptoms

A shift in leaf color from a bright green to a more subdued hue and the lower stalks take on a straw-like appearance. Within the infected stalk, a reddish discoloration becomes apparent. The internal pith tissue deteriorates, leaving only the vascular bundles. The fungus gains entry through the roots and progresses upward into the lower stem. If the infection occurs shortly after flowering, the husks exhibit a bleached and straw-like appearance.

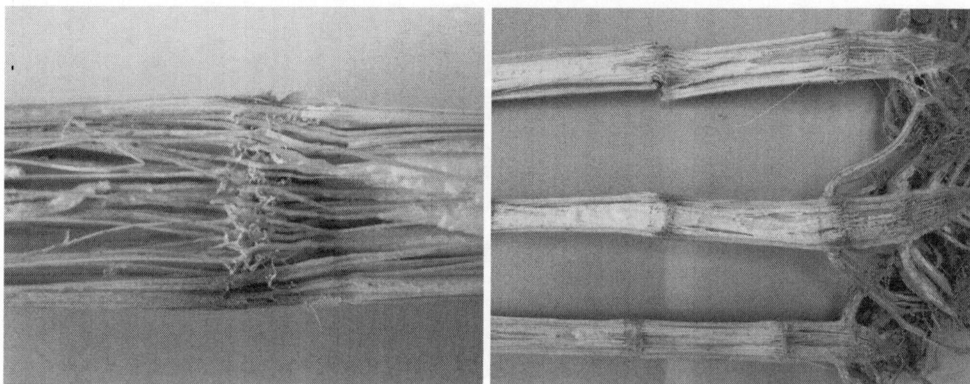

Figure 5.19: Pith Shredding Indicative of Fusarium Stalk Rot.

Favourable Conditions

Fusarium verticillioides exhibits growth at temperatures ranging from approximately 20°C to 37°C, with its optimal growth occurring near 25°C. This species is capable of growing under reduced oxygen tension but not in the complete absence of oxygen. One notable aspect of *Fusarium verticillioides* is its production of fumonisins, which are mycotoxins. This mycotoxin production is possible at a water activity level of up to 0.92. *Fusarium verticillioides* is commonly found in maize and is endemic to this crop, prevalent in all regions, although it is less common in cooler areas. It has been isolated from maize kernels wherever it has been investigated. This fungus is responsible for causing both stalk and cob rots in maize and can be found at all stages of plant development. Therefore, it can be considered a commensal organism in maize. Additionally, it is known to infect other crops such as sorghum, rice, and millet.

Disease Cycle

Fusarium verticillioides can persist on crop residue in the soil or on the soil surface, as depicted in the figure. Under favourable conditions, this fungus has the potential to infect both the roots and stalk of the host plant. *Fusarium verticillioides* may be present throughout the life cycle of the plant, potentially originating from infected seeds, further contributing to its presence and impact in maize plants.

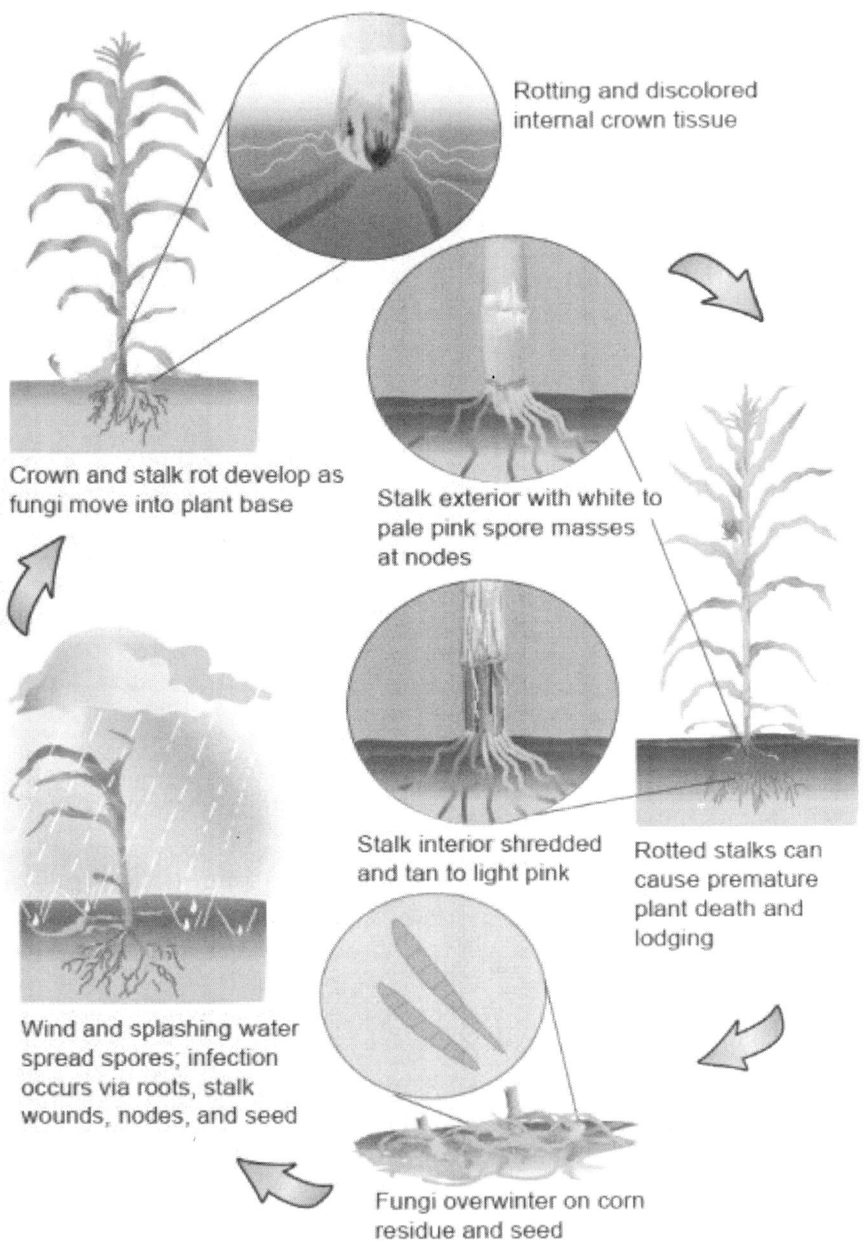

Figure 5.20: Disease Cycle of *Fusarium verticillioides.*

Management

1. Ensure field sanitation by thoroughly clearing and destroying previous plant debris and conduct summer deep ploughing.

2. Implement crop rotation.
3. Application of balanced fertilizer dose, with a reduced amount of nitrogen and a higher dose of potassium fertilizers.
4. When stalk rot is detected, consider early harvesting to minimize ear loss.
5. Use of resistant or tolerant hybrid varieties.
6. Treat seeds with Xelora (a combination of Pyraclostrobin and Thiofanate-methyl) mixed at 450 grams per liter of water to protect against disease infestations.

7. Bacterial Stalk Rot (*Dickeya zeae*)

Maize crop faces a significant challenge in the form of stalk rot, primarily attributed to the presence of *Dickeya zeae*, previously referred to as *Erwinia chrysanthemi* pv. *zeae*. This disease holds considerable economic importance in the context of maize production, leading to substantial yield losses of up to 100 per cent. It is particularly prevalent during the rainy season in India. This bacterium thrives under conditions of elevated temperature and moisture, resulting in the rapid collapse of infected plants within a week.

8. Common Rust (*Puccinia sorghi*)

The ailment gives rise to round to oblong, elongated pustules on the leaves, initially displaying a cinnamon-brown hue. These pustules can be found on both sides of the leaves. As the plant matures, the pustules change from brown to black, mainly due to the substitution of red uredospores with black teliospores. This shift in coloration is a distinctive hallmark of the disease.

9. Crazy Top Downy Mildew (*Sclerophthora macrospora*)

The affected plants exhibit several characteristic symptoms Leaves are narrow, strap-like, and leathery in appearance. Stunted growth and chlorotic (yellowing) striping of leaves. Excessive tillering, which refers to the formation of multiple shoots from a single plant. Rolling and twisting of upper leaves. Partial or complete transformation of normal floral parts into small leaves, a condition known as "crazy top". Plants may produce numerous ear shoots, and there may be phyllody of ears, where the ear's parts take on leaf-like characteristics. These symptoms collectively indicate the presence of a plant disease or disorder affecting the maize plants.

10. Rajasthan Downy Mildew (*Pernosclespora heteropogoni*)

Numerous instances of Rajasthan downy mildew (RDM) have been documented in the primary maize cultivating regions of Rajasthan from 1973 to 1980 (Anon 1973-1980). Recent investigations have shown that this disease occurs naturally to a significant extent, reaching 55-60 per cent on Udaipur district (Anon 2001). Plants that are systemically infected typically do not yield cobs and even those with secondary infections produce smaller cobs.

11. Phlippine Downy Mildew (*Peronosclerospora philippinensis*)

The first documented case of downy mildew on maize in India was reported by Butler in 1913. Eight years later, the causal fungus was identified and described by Weston in 1920 as Sclerospora philippinensis, with initial observations from the Philippines. This disease is known to occur in regions including Bihar, Punjab, and the Union Territory of Delhi. In these areas, both foliar and stalk rot diseases are also prevalent among maize crops.

12. Sorghum Downy Mildew (*Pernosclespora sorghi*)

Affected plants shows symptoms such as yellowing and occasionally stunted growth, leaves with white stripes, and irregular seed development. The yellowed portion of the leaf consistently encompasses the leaf base, with distinct boundaries typically separating the diseased and healthy tissues. A white, fluffy growth can emerge on both sides of the infected leaves. In some cases, the tassels of afflicted plants may demonstrate phyllody. Tolerant plants may exhibit signs of systemic infection but still manage to produce seeds normally.

13. Head Smut (*Sphacelotheca reiliana*)

Head smut, caused by the fungus *Sphacelotheca reiliana*, is characterized by several distinct symptoms: Galls initially form on the tassels (male flowers) and ears (female flowers) of the maize plant. These galls are covered with a thin layer of tissue, which eventually breaks open to reveal black spore masses and the threadlike remnants of the vascular bundles. Leaf-like proliferations develop on the tassel and ears, leading to the growth of abnormal, leafy structures. In some cases, ears may be aborted (fail to develop) and are instead replaced with proliferations of leafy tissue. Infected plants can also experience severe dwarfing, resulting in stunted growth. These symptoms collectively characterize head smut disease in maize-infected plants.

14. Anthracnose

Anthracnose leaf blight, caused by the fungal pathogen *Colletotrichum graminicola*, typically does not lead to significant reductions in yield. The disease tends to display the following characteristics: During the early part of the season, symptoms typically first appear on the lower leaves of the maize plant. As the disease advances, there may be occasional occurrences of symptoms on upper leaves. The lesions on the leaves are oval or spindle-shaped and come in shades of tan to brown, often with dark brown or purple edges. They can reach sizes of up to 1 inch in length and ½ inch in width. As the disease progresses, the fungus generates black, spiky structures on the deceased leaf tissue. In cases of severe infection, these lesions may merge, creating larger dead areas on the leaves, which may eventually turn yellow and wither. While anthracnose leaf blight typically doesn't result in substantial yield losses, it can still affect the overall health and appearance of maize plants.

Reference

Agrios, G.N. (2005). *Plant pathology.* (5th ed.) Burlington, MA.: Elsevier Academic Press.

Ahuja, S.C., Payak, M. M. (1982). Relationship of relative humidity and temperature level with the development of leaf and sheath blight of maze. *Z pflakranth. pflschutz.* **88**: 265-268.

Bertus, L. S. (1927). *Year book.* Department of Agriculture,

Caplik, D., Kusek, M., Kara, S., Seyrek, A., and Celik, Y. (2022). First report of bacterial stalk rot of maize caused by *Dickeya zeae* in Turkey. *New Disease Reports.* **45**(1), 1270-1290.

Diane Perkins, (2018). University of Kentucky, Bugwood.org Licensed under a Creative Commons Attribution-Noncommercial 3.0 http: //www.forestryimages.org/browse/detail.cfm?imgnum=5368762.

Hou, P., Liu, Y., Liu, W., Liu, G., Xie, R., Wang, K., Ming, B., Wang, Y., Zhao, R., Zhang, W. (2020), How to increase maize production without extra nitrogen input. *Resour. Conserv. Recycl.* **160**: 104913.

Kaur, N., Vashist, K. K., Brar, A. (2021). Energy and productivity analysis of maize based crop sequences compared to rice-wheat system under different moisture regimes. *Energy* **216**: 119286.

Letsoin, S. M. A., Purwestri, R. C., Perdana, M. C., Hnizdil, P., Herak, D. (2023). Monitoring of Paddy and Maize Fields Using Sentinel-1 SAR Data and NGB Images: A Case Study in Papua, Indonesia. *Processes,* **11**: 647.

Mary Ann Hansen, (2018). Virginia Polytechnic Institute and State University, Bugwood.org Licensed under a Creative Commons Attribution 3.0 License. http: //www.forestryimages.org/browse/detail.cfm?imgnum=5332037.

R.L. Croissant, (2018). Bugwood.org Licensed under a Creative Commons Attribution 3.0 http: //www.forestryimages.org/browse/detail.cfm?imgnum=5361281.

Rani, V. D., Reddy, P. N. and Devi, G. U. (2013). Banded leaf and sheath blight of maize incited by *Rhizoctonia solani* f. sp. *sasakii* and its management A review. *Int. J. Appl. Biol. Pharma Tech.,* **4**: 52-6.

Samra, A. S., Sabet, K. A., Hingorani, M.K. (1962). A new wilt disease of maize in Egypt. *Plant Dis. Rep.,* **46**: 481–483.

Saxena, S.C. (2002). Bio-Intensive Integrated Disease Man-agement of Banded Leaf and Sheath Blight of Maize. In: Proc. of the 8th Asian Regional Maize Workshop, Bangkok, Thailand, pp. 380-390.

Singh, R. and Srivastava, R. P. (2016). Southern corn leaf blight-an important disease of maize: an extension fact sheet. *Indian Research Journal of Extension Education,* **12**(2): 324-327.

Singh, S. K., Patel, M. B., Thakker, B. N., Hooda, K. S., and Barad, A. K. (2019). *Rhizoctonia solani* f. sp. *Sasakii* inciting banded leaf and sheath blight of maize and their management: an overview.

Therese, K.L., Madhavan, H.N. (2011). *"Exserohilum"*. In Liu D. (ed.). *Molecular Detection of Human Fungal Pathogens. CRC Press*. pp. 83–91.

Wang, S., Lu, Z., Lang, B., Wang, X., Li, Y. and Chen, J. (2022). Curvularia lunata and Curvularia Leaf Spot of Maize in China. *ACS omega,* **7**(51): 47462-47470.

Wiltshire, S. P. (1956). Plant Disease in British Colonial Dependencies; A half-yearly report. *F.A.O. Plant Protection Bulletin,* **5**: 66.

Wise, K. (2011). Northern corn leaf blight. Diseases of Corn. Perdue Extensio.

Wise, K. (2015). Fungicide efficacy for control of corn diseases. *Purdue Extension Publication* **6**: 160-161.

Zhao, C., Liu, B., Piao, S., Wang, X., Lobell, D. B., Huang, Y., Huang, M., Yao, Y., Bassu, S. and Ciais, P. (2017). Temperature increase reduces global yields of major crops in four independent estimates. *Proc. Natl. Acad. Sci. USA,* **114:** 9326–9331.

Chapter 6

Diseases of Sorghum Crop

***Dilip Kumar Reddy*[1]**, *Kamal Khilari*[2], *K.T. Rangaswamy*[3] *and Nishanth, M.*[4]**

[1]Assistant Professor, [3]Head and Professor,
Department of Plant Pathology, JCDR Agricultural College, Tadipatri, Andhra Pradesh
[2]Head and Professor, Department of Plant Pathology, SVPUAT, Meerut, Uttar Pradesh
[4]Research scholar, Division of Plant Pathology, ICAR- Indian Agricultural Research Institute, New Delhi
**Corresponding author: sdkreddy100@gmail.com*

Introduction

Sorghum (*Sorghum bicolor* L.) is a fundamental crop in semi-arid and arid regions of South Asia and Sub-Saharan Africa where faces challenging environmental conditions such as frequent droughts, low and erratic rainfall, and high mean temperatures (Reddy and Reddy, 2019). Globally, sorghum is cultivated on 40.25 million hectares, yielding 58.70 million tonnes with an average productivity of 1458 kg per hectare (FAO, 2021). The top sorghum producing countries include the United States, Sudan, Mexico, Nigeria, India, Niger, Ethiopia, Australia, Brazil, and China, collectively contribute to about 77 per cent of global sorghum production (Aruna and Cheruku, 2019). It is a source of food and fodder for more than 200 million people in these regions, particularly among the traditional, small-holder farming communities (Visarada and Aruna, 2019). Sorghum is a "healthy cereal" that contains 68 per cent carbohydrates, 10 per cent protein, micronutrients, and phytochemicals with nutraceutical properties (Visarada and Aruna, 2019). In major sorghum-growing regions of India, per capita consumption stands at 75 kg per year. Sorghum significantly contributes more than 50 per cent of the iron and zinc requirements of low-income groups (Rao *et al.*, 2006; Ashok kumar *et al.*, 2011; Ashok kumar *et al.*, 2013). Sorghum production is constrained by many biotic and abiotic factors. Among them anthracnose (*Colletotrichum graminicola*), downy mildew (*Peronosclerospora sorghi*), ergot (*Claviceps africana, Claviceps sorghi*), Charcoal rot (*Macrophomina phaseolina*) causes substantial yield losses.

1. Downy Mildew

Causal organism: *Peronosclerospora sorghi*

Economic Importance

Downy mildew is a highly concerning issue for sorghum production, with significant yield losses reported in various parts of the world. It reduces yield by 78 per cent in India (Thakur and Mathur 2002), 11.7 per cent in Africa (Bock *et al.*, 1998), and around 30 per cent in the United States (Craig *et al.*, 1989). Crops such as oat, barley, flax, Sudan grass, and cowpea are also hosts. In practically every state where maize and sorghum are grown, the disease is widespread.

Symptoms

Figure 6.1: A. White downy growth on lower surface of leaf; B. Leaf shredding along the white streaks.

In sorghum, the fungus is the systemic cause of downy mildew. It reaches young plants' growth sites by oospore or conidial infection. Whitish downy growth is visible on the lower surface of the leaves, consisting of sporangiophores and sporangia, and it extends to the upper surface of the leaf in advance. On both sides of the leaf, whitish streaks develop. Leaf shredding results from later tissue rips along the streaks (Pande *et al.*, 1997). The diseased leaves eventually turn necrotic, and many oospores can be observed in the shredded leaves.

Aetiology

The pathogen *Peronosclerospora sorghi* is an obligatory parasite that invades young plants systemically. Mycelium is non-septate and intercellular. Sporangiophores are robust and dichotomously branched at the top, with pointy sterigmata. Sporangia germinate without the generation of zoospores via a germ tube. Sexual spores are known as oospores which are featured by spherical, deep brown coloured and thick walled.

Favourable Conditions

A high relative humidity level is essential for optimum sporulation. During the night, the optimal temperature for sporulation is 21-23 °C. Light drizzle along with chilly temperatures is ideal for infection.

Disease Cycle

The principal mechanism of infection is via oospores found in soil and seeds. The oospores germinate and cause a systemic infection. Oospores can survive in soil for years. The fungus spreads throughout the seeds and becomes an infection source. Secondary spread occurs by airborne sporangia.

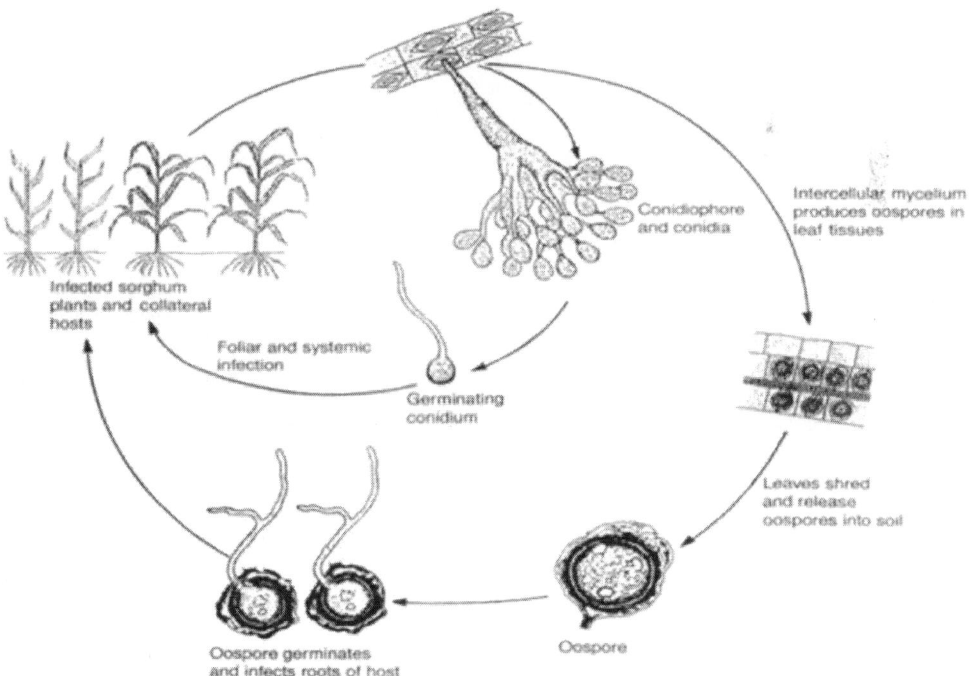

Figure 6.2: Disease Cycle of Sorghum Downy Mildew.

Management

1. Use high-quality, healthy seed and sowing early can help reduce disease spread. Inoculum in the soil can also be reduced by destroying infected plant debris and crop rotation (Tuleen *et al.*, 1980).
2. Seed transmission can control through seed treatment with chemical and by drying.
3. Eliminating sick plants reduces oospore burden in the soil (Janke *et al.*, 1983).
4. Seed treatment using Metalaxyl (Apron 35SD) at a dosage of 4 g per kilogramme of seed can be an effective technique of disease prevention.
5. Application of fungicides like Metalaxyl + Mancozeb (RidomilMZ) @ 0.1 per cent or Ziram or Zineb 1kg/ha can help manage the disease.
6. Growing tolerant varieties of sorghum such as CSH-2, CVS-5.

2. Anthracnose (Red leaf spot)

Causal organism: *Colletotrichum graminicola*

Economic Importance

This is widespread and very severe in all sorghum-growing regions. It is particularly severe in Rajasthan, Madhya Pradesh, Andhra Pradesh, Tamil Nadu, and Delhi. It attacks all parts and all growth stages of plant. The pathogen known to infect grasses of Poaceae family. It reduces yield significantly, generally by 50 per cent to 70 per cent in sorghum crops (Mathur *et al.*, 2002).

Symptoms

Leaf Spot

In sorghum, the pathogen causes leaf spot (anthracnose) as well as stalk rot (red rot). The disease reveals as little red spots on both sides of the leaf. The spots become older and acquire a white or grey core around by a brown or red border. On the white surface of the lesions, there are several small black dots that indicate fungal acervuli. Elongated elliptical, red or purple areas with black acervuli occur in the midrib region. Many lesions join together and produce considerable leaf damage.

Stalk Sot

Symptoms of stalk rot on stem include cankers or circular lesions that have a greyish centre and a crimson to black border. When the infected stem has been cut open, red to white discolouration appears, giving the stem a marbled appearance. Acervuli can also be seen in the stem lesion.

Figure 6.3: Leaf Anthracnose.

Etiology

The fungus mycelium is restricted to the spot. The fungus produces acervuli, which are asexual fruiting structures with long black setae that erupt through the epidermis. While Conidiophores are unicelled, hyaline and short, whereas conidia are unicelled, hyaline, vacuolate, and falcate in form, with an oil globule on the surface.

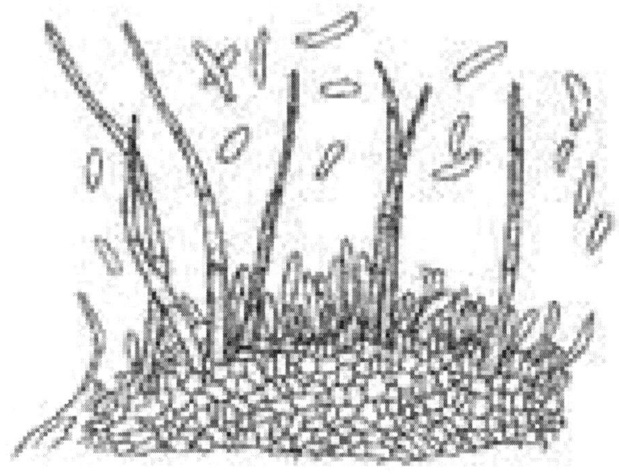

Acervulus

Favourable Conditions

Disease development is enhanced by abundant rainfall and high humidity levels. Additionally, temperatures within the range of 28-30°C encourage disease occurrence.

Disease Cycle

The disease spreads by seed-borne and airborne conidia, as well as infected plant waste.

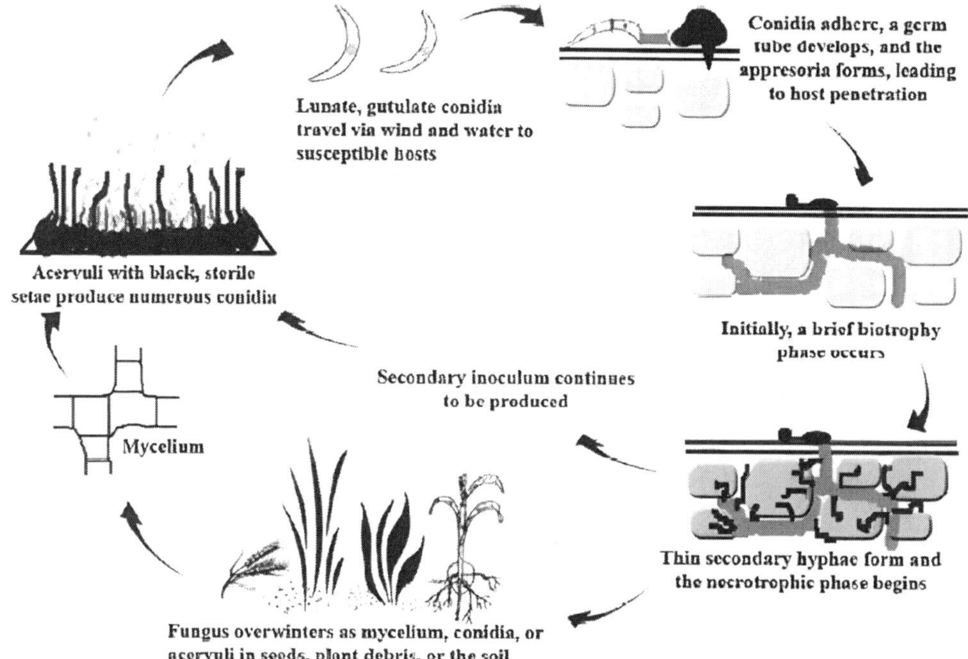

Figure 6.4: Anthracnose or Red Leaf Spot of Sorghum.

Management

1. To prevent anthracnose, it is best not to perform early planting (Tsedaley *et al.*, 2016), choose healthy seeds, clear diseased plant waste, and rotation of crops (Somda *et al.*, 2007).
2. Growing resistant varieties to disease
3. The antagonists *Trichoderma harzianum* decreases seedling mortality, disease incidence (Vasantha Kumari and Shivanna, 2014).
4. An efficient way to control disease is by treating the seeds with fungicides at a rate of 4 g/kg, such as Captan or Thiram.
5. Applying the fungicide Mancozeb at the rate of 2 kg/ha can help in the disease control.

3. Ergot/Sugary

Economic importance Sorghum ergot, caused predominantly by *C. Africana*, has been documented throughout Africa. *Claviceps africana* (sorghum ergot, or sugary disease) was discovered in commercial fodder fields and F1 hybrid production fields in Brazil, making it the first known case of sorghum ergot not native to Asia and Africa. *C. sorghi*, another *Claviceps species*, is now restricted to the Indian subcontinent. A third unnamed Claviceps spp. is exclusively found in Japan. *C. africana* was thought equivalent with *C. sorghi* prior to its designation as a separate species in 1991. Mc Rae first reported ergot disease in Tamil Nadu. The pathogen typically attacks during flowering stage of crop. The disease tends to appears during October – January months when cold weather prevails.

Causal organism: *Claviceps sorghi, C. africana, C. sorghicola*

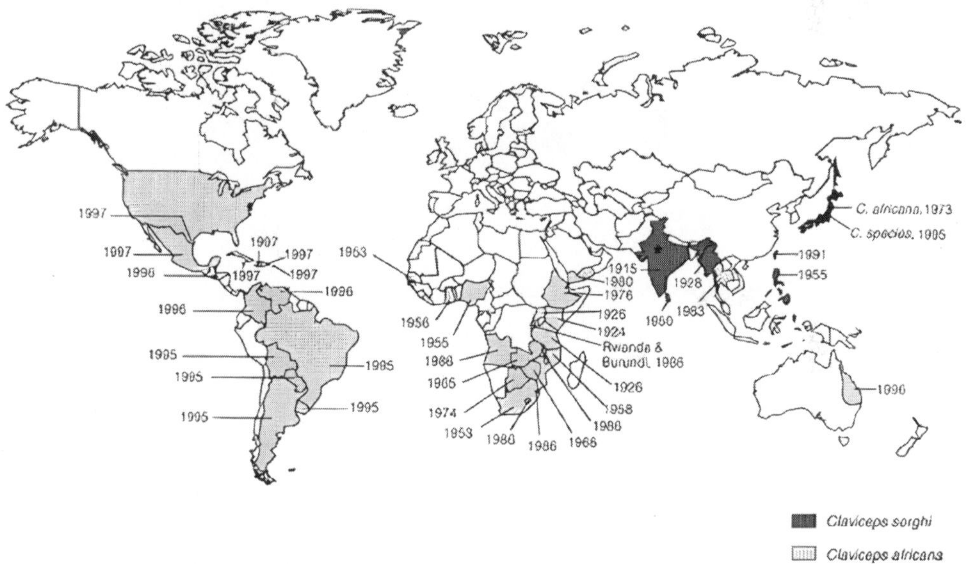

Symptoms

Sorghum ergot is characterized by its confinement to individual spikelets. The fungus attacks flower ovaries, converting them into a white fungal mass. The predominant symptom is honeydew production from affected florets. *Cerebrella sorghivulgaris* frequently colonizes this plant, giving the flower heads a dark colour. Straight or curved, long, rigid, cream to light brown sclerotia form under favourable circumstances.

Etiology

The fungus is categorised as an Ascomycota and generates septate mycelium. The honey dew comprises a concentrated suspension of unicelled, colourless, elliptic or oblong conidia.

Figure 6.5: Ergot of Sorghum (Honeydew stage).

Favourable Conditions

During the flowering season, excessive rains and humidity promotes the disease incidence. The disease spread is aided by cool night temperatures and cloudy weather.

Disease Cycle

The major source of infection is the germination of sclerotia, which releases ascospores that invade the ovaries. Within three to four days, a conidium placed on the stigma of a blooming sorghum spikelet germinates and forms a germ tube that descends the style and enters the unfertilized ovary, where it spreads the infection. The sorghum ovary changes into a white fungal mass called a sphacelium within 5 to 10 days, forcing the glumes apart like an enormous seed. One day later, the sphacelium starts to produce macroconidia, or conidia that are submerged in a virtually transparent sugary liquid that becomes opaque and coloured orange to pink or yellow-brown due to the conidia. It is known as honeydew, the watery combination of conidia.

Management

1. Practise deep ploughing, crop rotation, and removal of diseased panicles at harvest.
2. Change the planting date so that the crop does not blossom during September and October, when excessive rainfall and humidity favour the infection.

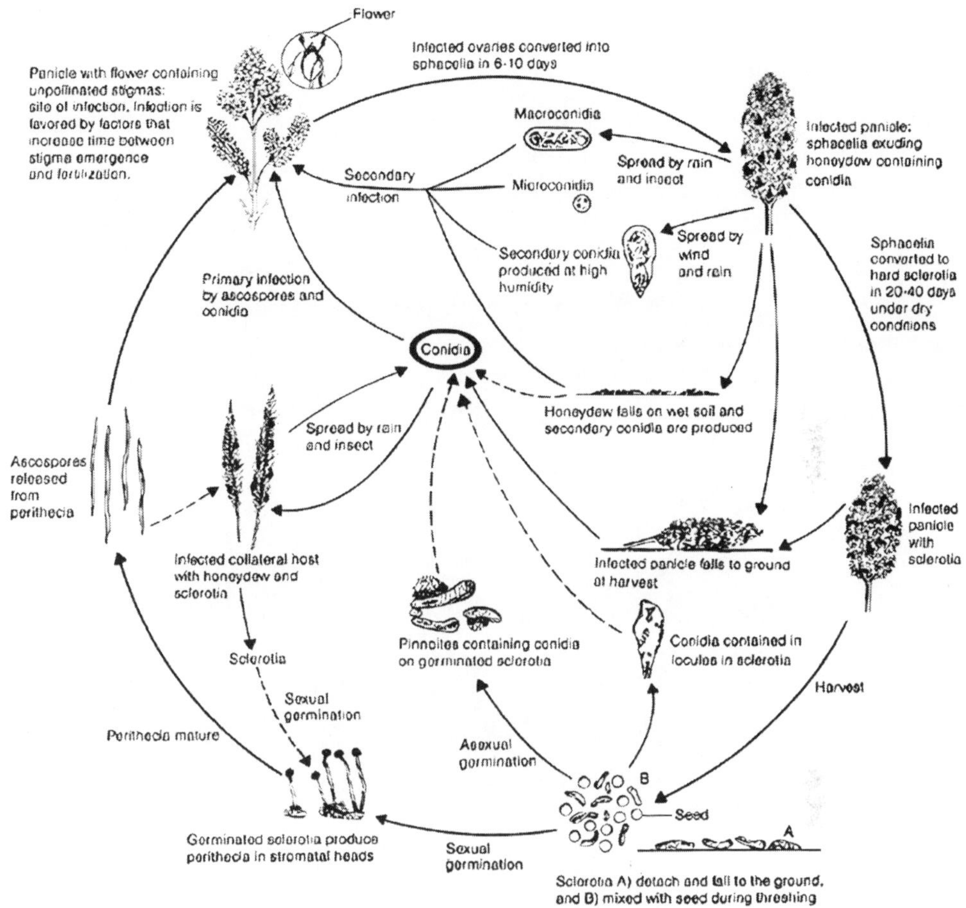

Figure 6.6: Disease Cycle of Sorghum Ergot.

3. Sprays containing carbendazim and tridemorph effectively controlled *C. sorghi* infection and lowers the seed contamination (Lakshmanan and Mohan 1988).
4. Although soaking seeds in a 5 per cent salt solution is seen to be a useful and simple method of eliminating *C. sorghi* sclerotia from seed batches, it is probably not 100 per cent successful (Bandyopadhyay 1992).
5. Triazoles, such as propiconazole and tebuconazole, are efficacious against *C. africana* and, presumably, *C. sorghi* when used topically (Odvody 1997).

4. Loose Kernel Smut

Economic Importance

Pathogen exerts relatively little yield losses, frequently less than 10 per cent (Tarr, 1962).

Causal organism: *Sporisorium cruentum*

Symptoms The diseased plants yield less panicles and grain, which has an impact on the output of grain and fodder. When grain is in the healthy development stage, the disease emerges. Pathogen targets all sorghum groups. The vast majority of grains are transformed into smut sori. Sori has a bigger grain size than typical grain. The sori start as brilliant green but eventually become dark brown. They are protected by a thin membrane, which frequently ruptures at maturity, releasing black/brown spore balls.

Figure 6.7: Kernel Smut of Sorghum.

Etiology

As the flower develops, the fungus attacks it, and its mycelium enters the ovary and transforms into chlamydospores. The sori are balls of spores, and the smut sori does not include columella. After germination, the smut spores develop a promycelium that later becomes sporidia.

Disease Cycle

Spore balls or smut spores found in the soil and seed which act as primary inoculum. During the flowering stage of the crop, spores develop from smuts and produce a large number of sporodia. The ovary becomes infected when sporidia enter the spikelet. Chlamydospores are carried by the wind function as a secondary inoculum.

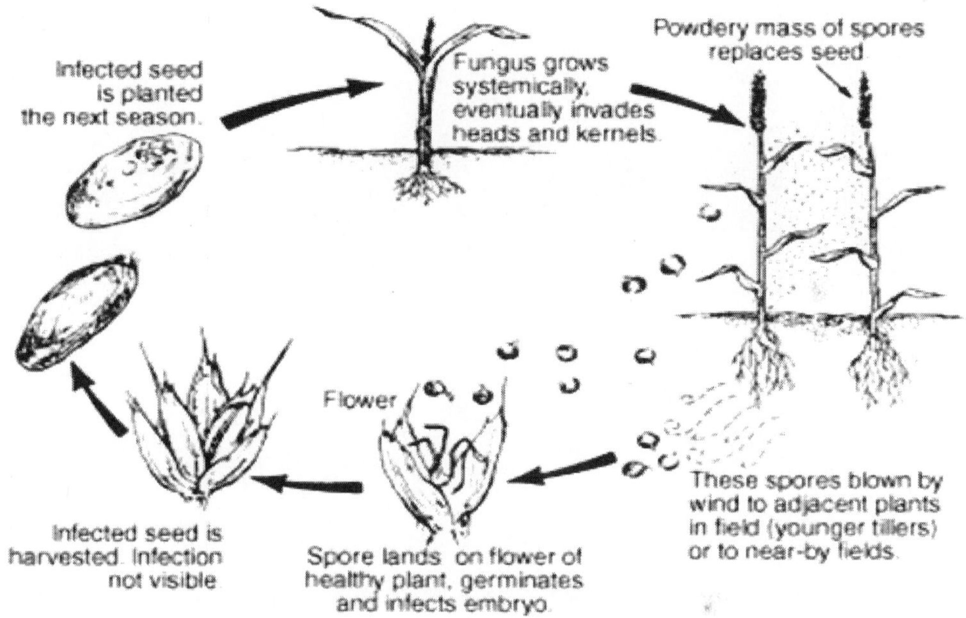

Figure 6.8: Diseases Cycle of Loose Kernel Smut.

Management

1. Use disease free seed.
2. Solar treatment is done in summer.
3. To avoid spreading smutspores, collect grains from panicles free of disease.
4. Seed treatment with chemical thiram @4 g/kg seeds.
5. Use resistant cultivars, such as DC-7 and MPP-7131.

5. Long Smut

Causal organism: *Tolyposporium ehrenbergii*

Economic Importance

Only sorghum-growing regions in Africa and India are affected by the disease. It only occurs in a small number of ear grains and is sporadic in nature.

Symptoms

Individual or few grains are transformed into smut sori. Sorus is surrounded by a pale to dull yellow membrane that is rather thick. Compared to covered and loose smuts, sori are significantly longer. With a comparatively thick, creamy-brown covering membrane known as the peridium, the sori are roughly cylindrical, elongated, and slightly curved. The infected ovary's vascular bundles are represented by the black mass of spores (spore balls) that are released when the peridium breaks at the tip.

Figure 6.9: Long Smut of Sorghum.

Favourable Conditions

During the infection phase, soil temperature should be between 18 and 23 °C, and soil moisture should be between 15-20 per cent.

Management

1. Use healthy seeds that have been certified.
2. Deep ploughing.
3. Rotation of crops aids in disease control.
4. Chemical treatment of seeds with 4 g/kg of thiram.
5. Grow resistant or tolerant varieties to minimize the disease incidence.

6. Head Smut

Causal organism: *Sporisorium holci-sorghi*

Sporisorium holci-sorghi, also known as *Sphacelotheca reiliana* and *S. reilianum*, is the cause of the disease. Aside from Sudan grass, the pathogen affects every group of sorghum (Anitha *et al.*, 2020). All areas that cultivate sorghum are affected by the disease.

Symptom

The disease targets the young panicle that is fully replaced by a large sorus inside the boot leaf. A thick greyish-white membrane covering the sorus bursts even before the head appears, revealing a mass of powdery, dark brown to black teliospores mixed with a dense network of vascular tissue.

Figure 6.10: Head Smut of Sorghum.

Management

1. Use certified disease-free seeds.
2. Deep ploughing.
3. Rotation of crops aids in disease control.
4. Chemical treatment of seeds with 4 g/kg of thiram.
5. Grow resistant or tolerant varieties to minimize the disease incidence.

7 Grain Mould

Grain mould is another important impediment to sorghum yield. Grain mould is caused by a fungus complex that comprises species of *Aspergillus*, Fusarium, Alternaria, Phoma, Bipolaris, and Penicillium. Many nations in Asia, Africa, South America, and North America have a high incidence of the disease.

Symptoms

When severe rains occur, the disease mostly damages the crop during the blooming and grain filling stages. A pink or black mould envelops the infected grains. When *Fusarium* spp. are involved, the discolouration is whitish or pinkish, and when *Curvularia* spp. are involved, the grains are discoloured black. Toxic mycotoxins found in mouldy grains make them unfit for human consumption or use as cow fodder. The type of fungus causing the mouldiness determines its hue (Thakur *et al.*, 2006).

Figure 6.11: Grain Mould of Sorghum.

Favourable Conditions

During the stages of grain development and blooming, warm, temperatures and more moist promotes the growth and spread of mould, whereas dry circumstances might lessen its effects. Ear heads that are small are more susceptible to mould growth.

Management

1. Adjustment in the sowing dates also may reduce in disease incidence.
2. Choose the varieties/hybrids which are fewer compact head to mitigate the disease incidence.
3. When storing grains, keep the moisture content at 10 per cent or below.
4. Grow resistant sorghum cultivars, such as IS 4963 and IS 5726.
5. The pathogen's development is also efficiently inhibited by antagonists such *T. hamatum* and *T. koeningii*. *P. fluorescens*, *T. viride*, and *T.*

harzianum promote germination and vigour in mould-infected seeds (Indira *et al.*, 2004).

6. In the event that frequent rainfall happens during ear-head emergence, one week later, and during the milky stage, use Mancozeb 1 kg/ha or Captan 1 kg + Aureofungisol 100g/ha.

8. Charcoal Rot

Causal Organism

Macrophomina phaseolina (Tassi) *Goidanich* (Perfect state: *Sclerotium bataticola Taub*). Charcoal rot, commonly known as root and stalk rot disease, is a serious limitation in sorghum-growing areas. The pathogen has a wide host range and is transmitted by both seeds and soil (Iqbal *et al.*, 2010; Kaur *et al.*, 2012). The disease can cause significant production losses at all phases of crop growth, up to 100 per cent lodging and a 64 per cent reduction in grain yield (Mughogho and Pande 1984).Crops cultivated in the aftermath of the rainy season are more vulnerable to fungal pathogen invasion.

Symptoms

The infection targets the lowest portion of the stalk, causing the tissues to discolour and become hollow. The pith disintegrates and the tissue shreds longitudinally into fibres when the diseased stem splits apart. There are tiny, asymmetrical black scleortial bodies visible on the tissue's surface. The crop lodges when a plant suddenly withers away and dies.

Figure 6.12: Charcoal Rot (*Source*: extension.okstate.edu).

Favourable Conditions

Extended durations of dry weather, high temperatures (35–38 °C), and low soil moisture are conducive to the disease's growth. In addition to increasing the plant's susceptibility to root and stalk rot, drought stress lowers agricultural production quantity and quality (Reddy *et al.*, 2012; Madhusudhana 2019). High plant populations and excessive nitrogen fertiliser use both increase the risk of disease.

Disease Cycle

Sclerotial bodies are the primary inoculum that thrive in soil. Primary infection develops by sclerotial bodies. Sclerotial bodies that disperse by irrigation water are the source of the secondary mechanism of infection.

Management

1. Deep summer ploughing, destruction of infected plant material, timely and adequate fertilizer and irrigation applications, crop rotation and adjustment of planting times (Rajkumar and Kuruvinashetti 2007).
2. Grow resistant varieties such as PJ-1430 (Kumari *et al.*, 2015)

   The treatment of charcoal rot may be effectively achieved by using bioagents namely *Trichoderma viride* (Jamadar and Desai, 1996), *Pseudomonas chlororaphis* strain SRB127 (Das *et al.*, 2008), and actinomycetes (Gopalakrishna *et al.*, 2011).
4. Applying carbendazim as a seed treatment lowers the crop's lodging percentage while increasing the yield of fodder (AICRP- Sorghum 2017).
5. Foliar fungicides should not be used as they are inefficient in eradicating the infection. (Anitha *et al.*, 2020).

9. Leaf Blight

Causal organism: *Exserohilum turcicum*

Symptoms

The symptoms of leaf blight usually start off as little flecks and develop into narrow, long lesions with a clear border of deep red colour and a brownish core. These lesions darken more during sporulation and finally become necrotic lesions as the disease progresses. Then, if the weather is favourable, numerous of spots combine to form massive necrotic areas on the leaf blade and leaf sheaths, which result in complete leaf wilting. If the strain of *E. turcicum* is extremely virulent, the infection can spread in an epidemic manner under optimum meteorological circumstances (Smith 2017). Although the lesions generated on sorghum and maize may seem identical, the pathogen isolates' pathogenicity on sorghum and maize may differ (Craven, 2016).

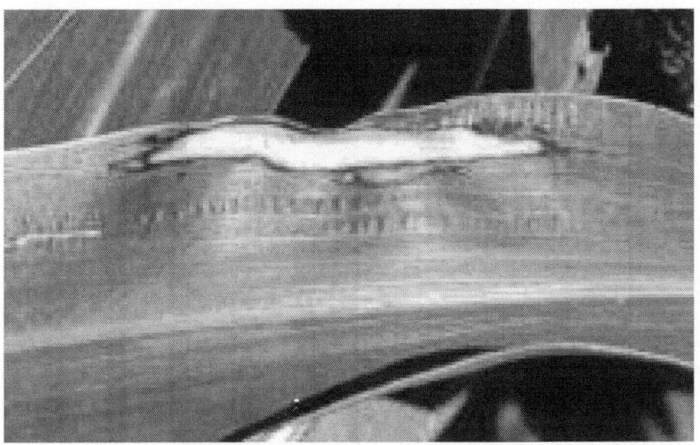

Figure 6.13: Leaf Blight.

Favourable Conditions

In colder, more humid parts of the world, this polycyclic disease is more prominent. The main source of inoculum is the mycelia, sclerotia, or chlamydospores on infected agricultural waste or in the soil (Casela and Frederiksen 1993). There is also evidence of conidia spreading across long distances via wind and seed. (Basu Chaudhary and Mathur 1979; Bergquist 1986; Cardwell *et al.*, 1989).

Management

1. Seasonal tillage, crop rotation, growing non-host crops in between, and clearing away infected plant debris all work to delay the onset of disease.
2. It has been demonstrated that cultivating species combinations lessens the severity of disease (Barrera and Frederiksen 1994).
3. The severity of anthracnose and leaf blight diseases is lessened when susceptible sorghum is interplanted with resistant sorghum cultivars or maize, according to Ngugi *et al.* (2001). 4. Grow resistant strains of sorghum, IS 2683, IS 3490, IS 9303, IS 10775, IS 12466, IS 19163, IS 25400, and IS 26863 (Mathur *et al.*, 2011).
4. It was shown that the bioagent *Trichoderma harzianum* 2×10^8cfu/g at 0.4 per cent was beneficial in raising grain production to 53.60 q/ha and decreasing disease intensity to 5.40 per cent (Wani *et al.*, 2017). 6. In sweet sorghum, it was discovered that a spray schedule using 0.1 per cent hexaconazole was beneficial in lowering disease and raising fodder output (Kiran and Patil 2019).
5. It had been found that strobilurin and DMI fungicides worked well (Shah and Dillard 2010; Testa *et al.*, 2015).

6. Higher yields of grain and fodder were obtained with the use of seed treatment or spray, dithane M-45, bavistin, and neem seed extract (NF1) (Bunker and Mathur 2005;Wani *et al.*, 2017).
7. Amistar Top, a systemic fungicide belonging to the triazole group, works well to prevent sorghum leaf blight.

10. Zonate Leaf Spot

Causal organism: *Gloeocercospora sorghi*

Symptoms

Small lesions emerge first on the lower leaves and progressively grow into huge purple-red or dark brown lesions with 2-8 rings. Semi-oval lesions are visible close to the midribs or along the leaf border. In later stages, lesions on leaves and leaf sheaths that range in colour from dark red to blackish purple or brown eventually combine to blight the entire region. *G. sorghi* sporodochia appear as slimy, salmon-colored lumps on the blotch's top surface. There are black dots of spherical sclerotia grouped in a linear pattern on dead, greyish-tan tissue. When there is a serious infection, the seed surfaces are dark brown or red with black oval dots.

Disease develops when temperatures ranging from moderate to high and high relative humidity conditions. Rain, wind, or water splashing can spread the infection from the soil to the lower leaves.

Management

1. Reducing leaf moisture by avoiding overhead watering.
2. Field cleanliness, crop residue elimination, and crop rotations lasting four years or more.
3. In the Tarai area of India, Purohit *et al.* (2014) found 16 resistant and 36 moderately resistant sorghum varieties. According to Prom *et al.* (2015), 13 sorghum lines shown resistance to the disease.
4. The bio agent *Trichoderma harzianum* was effective against the pathogen (Kharayat and Singh, 2012).
5. Bavistin seed treatment followed by two foliar sprays lowers disease severity and increases plant height and stem diameter (Kharayat and Singh 2012a).

11. Grey Leaf Spot

Causal organism: *Cercospora sorghi*

It is extensively found in all agroclimatic zones of India. The disease spreads more easily in warm, humid weather with frequent rainfall. Plant infection is caused by contaminated seed, closely spaced plants, crowded plants, and improper management of the soil, nutrients, and irrigation.

Symptoms

Small, round to elliptical patches that are usually dark purple or crimson in colour with tan or brownish centres first emerge on leaves and leaf sheaths. Velvety grey spore masses on both sides of the leaf indicate that heavy sporulation and spot enlargement occur gradually. In extreme situations, lesions combine to create a blighted resemblance in the shape of uneven patches or longitudinal stripes (Odvody, 1986).

Figure 6.14: Grey Leaf Spot
(*Source*: pestoscope.com).

Management

1. Crop rotation, the use of disease-free seeds or disease-resistant varieties, and establishing optimum plant populations all contribute to disease prevention.
2. Disposing of infected plant residues and maintaining proper field cleanliness
3. Growing resistant varieties is the best choice to combat against the disease.
4. Identified the pattern of resistance response inheritance by Fredericksen and Franklin (1980).
5. Spray two times 10 per cent Cow dung+ Asafoetida solution in 10 days interval.
6. Spray the fungicides such as Amistar 0.5 ml/lt or chlorothalonil 2g/lt spray at 15 days of interval.

12. Rust

Causal organism: *Puccinia purpurea* Cooke

There is a higher prevalence of the disease in areas that are humid and chilly. If the conditions are appropriate, it could adversely affect the entire crop.

Symptoms

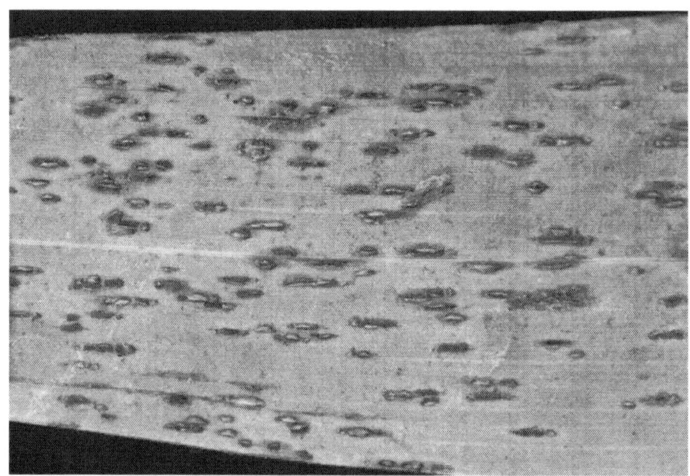

Figure 6.15: Sorghum Rust.

Every stage of the crop might be affected by the pathogen. Initially, the base leaf surfaces develop reddish-brown pustules, and subsequently, the infection spreads to the younger leaves. Rust sori have a diameter of 1.0–2.0 mm, are small and have an elliptical to circular form. Infected leaves wither quickly, and many sori may combine to form a bigger area. Severe circumstances result in lodging. Additionally, the leaf sheaths and inflorescence stalks may develop pustules. When these pustules holding uredospores burst, reddish-powdery masses of uredospores are released. Eventually, in older, longer-looking uredosori or teliosori, teliospores emerge. Temperatures between 10 and 12 °C encourage teliospore germination, whereas 20 to 24 oC are optimal for infection and disease development. The disease is enhanced during the wet season. During crop growth, high rainfall and relative humidity contribute to an increase in disease severity.

Management

1. Use healthy seed for cultural control. Early sowing aids disease resistance.
2. It is best to eliminate any infected debris from weed hosts like *Oxalis corniculata* and other crops.

3. Urediniospore infection can be eliminated by treating seeds with a contact fungicide.
4. Choose hybrids that can withstand rust, such CSH 26 and CSH 32.
5. Spraying oxycarboxin at 10-day intervals from 28 to 35 days after planting is beneficial (White *et al.*, 2012).
6. Hexaconazole (5 EC) at 0.1 per cent or difenconazole (25 EC) at 0.1 per cent are efficient fungicides to apply via foliar sprays (Chavan *et al.*, 2016).

13. Erwinia Stalk Rot

Causal organism: *Dickeya dadantii*

Symptoms

Tropical and sub-tropical climates are the most common locations for Erwinia stalk rot. Early symptoms of the disease are longitudinal spots and early leaf tip withering. At the base of the stalk, water-soaked symptoms first emerge and progressively change to a reddish-brown hue. The diseased stem pith eventually leads to the entire plant to wilt as it disintegrates and exhibits slimy soft-rot symptoms along with an unpleasant odour. Early infection causes early plant mortality, but late infection causes extensive crop lodging. Eventually, the entire stalk dries up and turns into a shredded bundle of fibrous tissue. According to Kharayat and Singh (2013), the rind turns pale-straw in colour and the lower leaves and their sheaths turn chlorotic.

Figure 6.16: Stalk Rot.

The pathogen spreads by irrigation water and rainfall and survives on crop debris in the soil. Disease outbreaks are caused by cloudy weather, relatively high temperatures (>30 °C), and frequent rains (Saxena *et al.*, 1991).

Management

1. Use techniques which includes adding organic amendments to the soil, preventing floods, irrigating excessively during hot weather, and limiting the amount of nitrogen used (Kloppers and Tweer 2009; Kumar *et al.*, 2016).
2. Actinomycetes, *Bacillus subtilis, P. fluorescens*, and VAM fungi have all been shown to be useful biological control agents against *E. chrysanthemi* (Karkouri *et al.*, 2010; Nagaraj *et al.*, 2012; and Singh *et al.*, 2018).
3. It was also discovered that pre-planting P. fluorescens strain Psf-173 and administering one or two foliar sprays thereafter helped to reduce the symptoms of sorghum stalk rot and promote plant development (Singh *et al.*, 2018).
4. In 1987, Hepperly and Davila found that penicillin antibiotics were unsuccessful and that tetracycline-like antibiotics were extremely effective against *E. chrysanthemi.*
5. Plants have a limited ability to withstand soft rot bacteria. Only a small number of resistant/tolerant maize lines have been found, nevertheless (Ahamad *et al.*, 2015; Kumar *et al.*, 2016).

14. Phanerogamic Parasite: *Striga asiatica* and *Striga densiflora*

Figure 6.17: *Striga asiatica.*

The phanerogamic parasite Striga is a semi root parasite that is mostly present in sorghum under rainfed condition. It is a short plant with radiant green leaves that usually grows to a height of 15 to 30 cm. *S. asiatica* blooms in red to pink, whilst *S. densiflora* blooms in white. Groups of plants clusters around their host plants. Every fruit yields a large number of microscopic seeds, which are viable in the soil for a number of years. Sorghum root exudates stimulate the germination of parasite seeds. For a period of one to two months, the parasite develops beneath the soil's surface, inserting haustoria into the host's root and forming stems and roots. The parasite emerges near the root of the plant and develops more quickly. In extreme situations, the host leaves begin to discolour and wilt. Infected plants get stunted and may die before seed set.

Management

1. Adoption of rotation of crops with non-host crops like groundnut, sunflower. Intercropping different legumes with maize and sorghum reduces *Striga*.
2. Collection and destruction of *Striga* before flowering.
3. Even when there is no host, a spray of ethrel mixed with soil causes Striga to germinate and causes suicidal death of striga.
4. Biological control can be done with the help of natural enemies such as herbivorous insects, microorganisms and smother plants. *S. hermonthica* can be controlled biologically by using *F. oxysporum*.
5. Spray Feroxone (2, 4-D) @ 450 g/500 litres of water or Paraquat @ 1 kg/ha.
6. In case of limited water availability, spray 1 per cent Tetrachloro dimethyl phenoxy acetate for immediate striga eradication.

References

Ahamad, S., Lal, B., and Kher, D. (2015). Screening of maize germplasms against stalk rot diseases in the intermediate zone of Jammu region. *Int. J. Innov. Sci. Eng. Technol*, **2**(12): 828-831.

AICRP-Sorghum (2017). Annual progress report of All India Coordinated Research Project on Sorghum, Indian Council of Agricultural Research – Indian Institute of Millets ReserchHyderabad.https: //aicrp.icar.gov.in/ sorghum/achievements-2/crop protection/sorghum-pathology/

Anitha, K., Chakrabarty, S. K., Rao, R. D. V. J. P., Babu, B. S., Babu, A., Varaprasad, K. S., and Khetarpal, R. K. (2005). Quarantine processing of exotic cereals and millets germplasm during 1986-2003. *Indian Journal of Plant Protection*, **33**(1): 105-110.

Anitha, K., Das, I. K., Holajjer, P., Sivaraj, N., Reddy, C. R., and Balijepalli, S. B. (2020). Sorghum diseases: diagnosis and management. *Sorghum in the 21st century: food–fodder–feed–fuel for a rapidly changing world*, 565-619.

Aruna, C., and Cheruku, D. (2019). Genetic improvement of grain sorghum. In *Breeding sorghum for diverse end uses*, pp. 157-173.

Ashok, K. A., Anuradha, K., and Ramaiah, B. (2013). Increasing grain Fe and Zn concentration in sorghum: progress and way forward. *Journal of SAT Agricultural Research*, **11**(12): 1-5.

Ashok K. A., Reddy, B. V., Ramaiah, B., and Sharma, R. (2011). Heterosis in white-grained grain mold resistant sorghum hybrids. *Journal of SAT Agricultural Research*, **9**: 6pp.

Bandyopadhyay, R. (1992). Sorghum ergot. *Sorghum and Millets Diseases: A Second World Review. WAJ de Milliano, RA Frederiksen, and GD Bengston, eds. International Crops Research Institute for the Semi-Arid Tropics, Patancheru, India*, 235-244.

Bashasab, F., and Kuruvinashetti, M. S. (2007). Genetic variability of sorghum charcoal rot pathogen (Macrophomina phaseolina) assessed by random DNA markers. *The Plant Pathology Journal*, **23**(2): 45-50.

Basu Chaudhary, K. C., and Mathur, S. B. (1979). Infection of sorghum seeds by Colletotrichum graminicola 1. Survey, location in seed and transmission of the pathogen. *Seed Science and Technology*, **7**(1): 87-92.

Bergquist, R.R. (1986). Leaf blight. In: Compendium of sorghum diseases. Am PhytopatholSoc**48**: 56–58

Beshir, M. M., Ali, A. M., and Okori, P. (2012). Inheritance of resistance to turcicum leaf blight in sorghum. *African Crop Science Journal*, **20**(1).

Binyam, T., Girma, A., and Fikre, L. (2016). Distribution and importance of sorghum anthracnose (Colletotrichum sublineolum) in Southwestern and Western Ethiopia. *Plant Pathology Journal (Faisalabad)*, **15**(3): 75-85.

Bock, C. H., Jeger, M. J., Mughoho, L. K., Cardwell, K. F., Adenle, V., Mtisi, E., and Blair-Myers, C. (1998). Occurrence and distribution of Peronosclerosporasorghi [Weston and Uppal (Shaw)] in selected countries of West and Southern Africa. *Crop Protection*, **17**(5): 427-439.

Bunker, R. N., and Mathur, K. (2005). *Biology and management of Exserohilum turcicum (Pass.) Leonard and Suggs. inciting sorghum [Sorghum bicolor (L.) Moench] leaf blight* (Doctoral dissertation, Ph. D thesis, MaharanaPratap University of Agriculture and Technology, Udaipur, Rajastana, India).

Canama, A. O., and Hautea, D. M. (2010). Molecular mapping of resistance to bacterial stalk rot (*Pectobacterium chrysanthemi* pv. *zeae*Burk., McFad. and Dim.) in tropical white maize (Zea mays L.). *Philippine Agricultural Scientist*, **93**(4): 429-438.

Cardwell, K. F., Hepperly, P. R., and Frederiksen, R. A. (1989). Pathotypes of Colletotrichum graminicola and seed transmission of sorghum anthracnose. *Plant Disease*, **73**(3): 255-257.

CASELA, C. R., and Frederiksen, R. A. (1994). Pathogenic variability in monoconidial isolates of the sorghum antracnose fungus Colletotrichum graminicola from single lesions and from monoconidial cultures. *FitopatologiaBrasileira*, **19**(2): 149-153.

Craig, J., Odvody, G. N., Wall, G. C., and Meckenstock, D. H. (1989). Sorghum downy mildew loss assessment with near-isogenic sorghum populations. *Phytopathology*, **79**(4): 448-451.

Craven, M. (2016). A look at Exserohilum leaf blight of sorghum. https: //www. grainsa.co.za/a-look-at-exserohilum-leaf-blight-of-sorghum

Das, I. K., Indira, S., Annapurna, A., and Seetharama, N. (2008). Biocontrol of charcoal rot in sorghum by fluorescent pseudomonads associated with the rhizosphere. *Crop Protection*, **27**(11): 1407-1414.

El Karkouri, A., El Hassani, F. Z., El Mzibri, M., Benlemlih, M., and El Hassouni, M. (2010). Isolation and identification of an actinomycete strain with a biocontrol effect on the phytopathogenic *Erwinia chrysanthemi* 3937VIII responsible for soft rot disease. *Annals of microbiology*, **60**(2): 263-268.

FAO (2021). Available at: http: //www.fao.org/faostat/en/#data/QC

Frederiksen, R. A., and Franklin, D. (1980). Sources of resistance to foliar diseases of sorghum in the International Disease and Insect Nursery. *Sorghum Diseases: A World Review. ICRISAT, Patancheru, India*, 265-268.

Freetown, S. L. (1995). Development of leaf anthracnose and its effect on yield and grain weight of sorghum in West Africa. *Plant disease*, **79**(2): 151.

Gopalakrishnan, S., Humayun, P., Kiran, B. K., Kannan, I. G. K., Vidya, M. S., Deepthi, K., and Rupela, O. (2011). Evaluation of bacteria isolated from rice rhizosphere for biological control of charcoal rot of sorghum caused by *Macrophomina phaseolina* (Tassi) Goid. *World Journal of Microbiology and Biotechnology*, **27**: 1313-1321.

Gopalakrishnan, S., Kiran, B. K., Humayun, P., Vidya, M. S., Deepthi, K., Jacob, S. and Rupela, O. (2011). Biocontrol of charcoal-rot of sorghum by actinomycetes isolated from herbal vermicompost. *African Journal of Biotechnology*, *10*(79), 18142-18152.

Hepperly, P. R., and Ramos-Davila, E. (1987). *Erwinia chrysanthemi* Burk., Mc Faddan Dimock: A bacterial whorl and stalk rot pathogen of sorghum [*Sorghum bicolor* (L.) Moench]. *J. Agric. Univ. Puerto Rico*, **71**(3): 265-275.

Indira, S., Muthusubramanian, V., Tonapi, V. A., and Seetharama, N. (2004). Role of Biocontrol agents in the supression of seed borne pathogenic mycoflora and their effect on seed quality in Sorghum (Sorghum bicolor (L.) Moench). In *ISTA seed symposium* (pp. 17-19).

Iqbal, U., Mukhtar, T., Iqbal, S. M., Ul-Haque, I., and Malik, S. R. (2010). Host plant resistance in blackgram against charcoal rot (*Macrophomina phaseolina* (Tassi) Goid). *Pakistan Journal of Phytopathology,* **22**(1): 126-129.

Jamadar, M. M., and Desai, S. A. (1996). Management of charcoal rot caused by *Macrophomina phaseolina* (Tassi.) Goid in sorghum, *Sorghum bicolor* (L.) Moench. *J Biol Control,* **10:** 93-96.

Jyotika, P., Yogendra, S., Holeyachi, P. B., and Nitika, G. (2014). Field reaction of sorghum germplasm to zonate leaf spot disease. *Environment and Ecology,* **32**(1): 174-177.

Kaur, S., Dhillon, G. S., Brar, S. K., Vallad, G. E., Chand, R., and Chauhan, V. B. (2012). Emerging phytopathogen Macrophomina phaseolina: biology, economic importance and current diagnostic trends. *Critical Reviews in Microbiology, 38*(2), 136-151.

Khan, Z.R. Midega, C. A. O. Hassanali, A. Pickett, J. A. and Wadhams, L.J. (2007). Assessment of Different Legumes for the Control of *Striga hermonthica* in Maize and Sorghum, Crop Ecology, Management and Quality, https: // doi.org/10.2135/cropsci2006.07. 0487

Kharayat, B. S. and Singh, Y. (2015). Characterization of *Erwinia chrysanthemi* isolates inciting stalk rot disease of sorghum. *African Journal of Agricultural Research,* **10**(22): 2309-2314.

Kharayat, B. S. and Singh, Y. O. G. E. N. D. R. A. (2013). Unusual occurrence of erwinia stalk rot of sorghum in tarai region of Uttarakhand. *Int. J. Agric. Sci,* **9**(2): 809-813.

Kharayat, B. S. and Yogendra, S. (2012). Biological control of zonate leaf spot of sorghum caused by Gloeocercospora sorghi. *International Journal of Plant Protection,* **5**(2): 401-404.

Kharayat, B. and Singh, Y. (2012). Evaluation of efficacy of fungicides against zonate leaf spot of sorghum caused by *Gloeocercospora sorghi. Vegetos,* **25**(2): 136-142.

Kiran, B. M., andPatil, P. V. (2019). Integrated management of leaf blight of sweet sorghum caused by *Exserohilum turcicum* (Pass.) Leonard and Suggs. *Indian Phytopathology,* **72**(1): 63-69.

Kloppers, R. and Tweer, S. (2009). Northern corn leaf blight fact sheet. Online publication. PANNAR Seed (Pty) Ltd.

Kumar, A., Hunjan, M. S., Kaur, H., Kaur, R. and Singh, P. P. (2016). Evaluation of management of bacterial stalk rot of maize (Dickeyazeae) using some chemicals and bio-agents. *Journal of Applied and Natural Science,* **8**(3): 1146-1151.

Kumar, A., Hunjan, M. S., Kaur, H., Kaur, R., and Singh, P. P. (2016). Evaluation of management of bacterial stalk rot of maize (Dickeyazeae) using some chemicals and bio-agents. *Journal of Applied and Natural Science,* **8**(3): 1146-1151.

Kumari, N., Sharma, I., Alam, A. and Sharma, V. (2015). Screening of sorghum genotypes and biochemical changes for resistance to damage caused by *Macrophomina phaseolina. Archives of Phytopathology and Plant Protection,* **48**(9-12): 760-775.

Lakshmanan, P. and Mohan, S. (1988). Studies on the effect of various fungicides on *Sphaceliasorghi. Pesticides,* **22**(9): 27.

Madhusudhana, R. (2019). Marker-assisted breeding in sorghum. In *Breeding sorghum for diverse end uses* (pp. 93-114). Woodhead Publishing.

Mathur, K., Thakur, R. P., Neya, A., Marley, P. S., and Casela, C. R. (2003). Sorghum AnthracnoseProblem and Management Strategies. *Sorghum and millets diseases,* 211-220.

Mathur, K., Thakur, R. P., Rao, V. P., Jadone, K., Rathore, S., and Velazhahan, R. (2011). Pathogenic variability in Exserohilum turcicum and resistance to leaf blight in sorghum. *Indian Phytopathology, 64*(1), 32.

Mughogho, L. K., andPande, S. (1983). Charcoal rot of sorghum.

Nagaraj, M. S., Umashankar, N., Palanna, K. B., and Khan, A. (2012). Etiology and management of tip-over disease of banana by using biological agents. *International Journal of Advanced Biological Research,* **2**(3): 483-486.

Ngugi, H. K., King, S. B., Holt, J. and Julian, A. M. (2001). Simultaneous temporal progress of sorghum anthracnose and leaf blight in crop mixtures with disparate patterns. *Phytopathology,* **91(**8): 720-729.

Odvody, G. (1997). Continued spread of sorghum ergot in the Western Hemisphere. *SICNA Proceedings.*

Pande, S., Bock, C. H., Bandyopadhyay, R., Narayana, Y. D., Reddy, B. V. S., Lenne, J. M., and Jeger, M. J. (1997). Downy mildew of sorghum. Information bulletin no. 51.

Prom, L. K., Isakeit, T., Cuevas, H., Rooney, W. L., Perumal, R., and Magill, C. (2015). Reaction of sorghum lines to zonate leaf spot and rough leaf spot. *Plant health progress,* **16**(4): 230-234.

Purohit, H., Hampton, A., Bhatt, S., Shalin, V. L., Sheth, A. P., andFlach, J. M. (2014). Identifying seekers and suppliers in social media communities to support crisis coordination. *Computer Supported Cooperative Work (CSCW),* **23**(4): 513-545.

Rao, P. P., Birthal, P. S., Reddy, B. V., Rai, K. N. and Ramesh, S. (2006). Diagnostics of sorghum and pearl millet grains-based nutrition in India. *International sorghum and millets newsletter,* **47:** 93-96.

Reddy, B. V., Reddy, P. S., Sadananda, A. R., Dinakaran, E., Ashok Kumar, A., Deshpande, S. P. andPatil, J. V. (2012). Postrainy season sorghum: Constraints and breeding approaches. *Journal of SAT Agricultural Research,* **10**(1): 1-12.

Reddy, P. S., and Reddy, B. V. (2019). History of sorghum improvement. In *Breeding Sorghum for Diverse End Uses* (pp. 61-75). Woodhead Publishing.

Saxena, S. C., Mughogho, L. K., andPande, S. (1991). Stalk and top rot of sorghum caused by *Erwinia chrysanthemi* in India. *Indian Journal of Microbiology,* **31**(4): 435-441.

Shah, D. A. and Dillard, H. R. (2010). Managing foliar diseases of processing sweet corn in New York with strobilurin fungicides. *Plant disease,* **94**(2): 213-220.

Sharma, R., Upadhyaya, H. D., Manjunatha, S. V., Rao, V. P., and Thakur, R. P. (2012). Resistance to foliar diseases in a mini-core collection of sorghum germplasm. *Plant Disease,* **96**(11): 1629-1633.

Sifuentes Barrera, J. A., andFrederiksen, R. A. (1994). Evaluation of sorghum hybrid mixtures for controlling sorghum leaf blight. *Plant disease,* **78**(5): 499-503.

Singh, P., Singh, Y., Purohit, J., andMaharshi, A. (2018). A comparative evaluation of bioagents and chemicals for the control of stalk rot of Sorghum caused by *Dickey adadantii. Journal of Applied and Natural Science,* **10**(3): 1053-1058.

Smith, K. (2017). *Evaluating the effects of fungicides for the control of Exserohilum turcicum on sorghum in South Africa* (Doctoral dissertation, North-West University (South Africa), Potchefstroom Campus).

Somda, I., Leth, V., and Sereme, P. (2007). Evaluation of lemongrass, eucalyptus and neem aqueous extracts for controlling seed-borne fungi of sorghum grown in Burkina Faso. *World Journal of Agricultural Sciences,* **3**(2): 218-223.

Tarr, S. A. J. (1962). Diseases of sorghum, Sudan grass and broom corn. *Diseases of sorghum, sudan grass and broom corn.* (60s).

Testa, G., Reyneri, A., and Blandino, M. (2015). Foliar fungicide application to maize: yield and grain hardness enhancement in different environmental conditions. *Crop Science,* **55**(4): 1782-1790.

Thakur, R. P. andMathur, K. (2002). Downy mildews of India. *Crop Protection,* **21**(4): 333-345.

Thakur, R. P., Reddy, B. V. S., Indira, S., Rao, V. P., Navi, S. S., Yang, X. B. and Ramesh, S. (2006). Sorghum grain mold information Bulletin No. 72.

Thomas, M. D., Rosewich, U. L. and Frederiksen, R. A. (1995). Preliminary studies on total DNA restriction fragment length polymorphism (RFLP) of West Africa isolates of Colletotrichum graminicola from sorghum. *International sorghum and millets newsletter (USA).*

Tuleen, D. M., Frederiksen, R. A., and Vudhivanich, P. (1980). Cultural practices and the incidence of sorghum downy mildew in grain sorghum. *Phytopathology,* **70**(9): 905-908.

Vasanthakumari, M. M., and Shivanna, M. B. (2014). Biological control of sorghum anthracnose with rhizosphere and rhizoplane fungal isolates from perennial grasses of the Western Ghats of India. *European journal of plant pathology,* **139:** 721-733.

Visarada, K. B. R. S., and Aruna, C. (2019). Sorghum: A bundle of opportunities in the 21st century. In *Breeding Sorghum for diverse end uses* (pp. 1-14). Woodhead Publishing.

Wang, M. L., Dean, R., Erpelding, J., and Pederson, G. (2006). Molecular genetic evaluation of sorghum germplasm differing in response to fungal diseases: Rust (Puccinia purpurea) and anthracnose (*Collectotrichum graminicola*). *Euphytica,* **148:** 319-330.

Wani, T. A., Ahmad, M., and Anwar, A. (2017). Evaluation of fungicides, bioagents and plant extracts against Exserohilum turcicum causing Turcicum leaf blight of Maize. *International Journal of Current Microbiology and Applied Sciences,* **6**(8): 2754-2762.

White, J. A., Ryley, M. J., George, D. L. and Kong, G. A. (2014). Optimal environmental conditions for infection and development of *Puccinia purpurea* on sorghum. *Australasian Plant Pathology,* **43:** 447-457.

White, J. A., Ryley, M. J., George, D. L., Kong, G. A. and White, S. C. (2012). Yield losses in grain sorghum due to rust infection. *Australasian Plant Pathology,* **41:** 85-91.

Chapter 7

Diseases of Pearl Millet Crop

Mayur R. Thesiya[1]*, Jaina Patel[2] and Priyankaben Patel[2]

[1]*Department of Plant Pathology, Punjab Agricultural University, Ludhiana – 141 004, Punjab*

[2]*Department of Plant Pathology, Anand Agricultural University, Anand – 388 110, Gujarat*

**e-mail: mayur-pp@pau.edu*

Introduction

Pearl millet/Bajra (*Pennisetum glaucum* L.) is one of the main *Kharif* crops in the arid and semi-arid farming regions of India. To meet the growing demand for pearl millet, it is crucial to address the lower productivity of the crop in India. According to FAO, millet production in the world is 89.17 million metric tons from an area of 74 million hectares (Anonymous, 2023). India is the largest producer of pearl millet in the world, which has an area of 12.53 million hectares and produces 15.53 metric tons per year, with an average productivity of 1237 kg/ha throughout 2021–2022 (Anonymous, 2022). It comprises around 138 lakh hectares area with 173 lakh tonnes production of the total food grain of the India (FAOSTAT, 2021). It is usually grown under the most adverse agro-climatic conditions where other crops like sorghum and maize fail to produce economic yields. Bajra, being a C_4 plant, has a very high photosynthetic efficiency and dry matter production capacity. The cultivation of bajra is primarily focused on producing food and dry fodder. Grains possess high nutritional value, containing ample amounts of metabolizable energy and protein (Basavaraj *et al.*, 2010). Bajra is often referred to as a "poor man's crop" and known for its abundant content of carbohydrates, protein, fat, fiber, and minerals (Kargwal *et al.*, 2023). Bajra faces various biotic and abiotic challenges that can negatively impact its growth and productivity. When compared to crops like wheat, rice, maize, and sorghum, pearl millet exhibits relatively lower susceptibility to pests and diseases (NRC, 1996). While rice has over 350 reported diseases, wheat has around 200, maize has about 70, and sorghum has 63, pearl millet has approximately 50 reported diseases. This can be attributed to the anatomical structure of pearl millet, which includes cell walls made up of tissues that are resistant to pathogen penetration. This chapter provides detailed information on the significant of fungal, bacterial and viral diseases on pearl millet along with etiology, disease cycle and integrated disease management.

Diseases Classified as Fungal, Bacterial, Viral and Phanerogamic Plant Parasite

A) Fungal Diseases

1. Downy mildew: *Sclerospora graminicola*
2. Ergot: *Claviceps fusiformis*
3. Bajra smut: *Tolyposporium penicillariae/Moesziomyces bullatus*
4. Blast: *Pyricularia grisea*
5. Bajra rust: *Puccinia penniseti, Puccinia substriata* var. *indica*
6. Curvularia leaf blight: *Curvularia penniseti*
7. Zonate leaf spot: *Gloeocercospora sorghi*
8. Cercospora leaf spot: *Cercospora penniseti*
9. Bipolaris leaf spot: *Bipolaris setariae*
10. Phyllosticta leaf blight: *Phyllosticta penicillariae*

B) Bacterial Diseases

1. Bacterial leaf spot: *Pseudomonas syringae*
2. Bacterial leaf stripe: *Pseudomonas avenae*
3. Bacterial leaf streak: *Xanthomonas axonopodis* pv. *pennamericanum*

C) Viral Diseases

1. *Maize dwarf mosaic virus*
2. *Maize streak virus*

D) Nematode Diseases

1. Burrowing nematode: *Radopholus similis*
2. Cyst nematode: *Heterodera gambiensis*
3. Dagger nematode: *Xiphinema americanum*

E) Phanerogamic Plant Parasite

1. *Striga asiatica / S. hermonthica* (Orobanchaceae)

A. Fungal Diseases

1. Ergot Disease

Economic Importance and Occurrence

This disease has been observed in various regions of Africa and India. It was first reported (1956) from 'Satara' area in Maharashtra (India) occurring in an epidemic form. The increased severity of the disease is attributed to the introduction of the hybrid variety HB-l, 2. It has also been detected in Delhi, Rajasthan, Karnataka, Tamil Nadu, and Gujarat. In the years 1967-68, the disease spread extensively, affecting a significant number of newly introduced hybrid bajra varieties in all regions. The damage caused by the disease in hybrid varieties ranged from 58 per cent to 76 per cent. This particular fungus produces a toxic alkaloid called ergotoxin/ergotin, which poses a threat to both animals and humans.

Causal Organism

The ergot disease of pearl millet caused by *Claviceps fusiformis*, which belongs to Kingdom- *Fungi*, Phylum- *Ascomycota*, Class- *Sordariomycetes*, Order- *Hypocreales*, Family- *Clavicipitaceae*, Genus- *Claviceps*, and Species- *fusiformis* (Kirk *et al.*, 2008). Sclerotia elongate to globose, pale to dark brown. Ascospores hyaline, filiform, non-septate. Macroconidia hyaline, fusiform, unicellular. Microconidia hyaline, globular, unicellular (Thakur *et al.*, 1984).

Symptoms

This disease specifically manifests during the flowering stage. Its symptom is characterized by the secretion of small droplets of light pinkish or brownish sticky fluid (referred to as honey dew) from the infected spikelet. Groups of spikelets, ranging from a few to many, exhibit darkening as they age, which attracts various insects. As the disease progresses, small greyish or dark brown sclerotia develop. These sclerotia replace the ovary or grain and have a hard and woody texture, measuring about 0.5-1 cm in length and 1-2 mm in diameter. The disease solely occurs during the flowering period. The fungus responsible for this disease attacks the ovary and undergoes vigorous growth, resulting in the formation of abundant hyphae that ultimately form the sclerotium. After one month, the sclerotia germinate, giving rise to one or two stripes where the asci (sac-like structures) are produced.

Favourable Conditions

The disease exhibits rapid spread in the field under high humidity conditions, particularly when the relative humidity ranges from 60 per cent to 90 per cent in the morning, followed by a decrease to 45-50 per cent in the evening. Rainfall during the flowering stage is conducive to the development of the disease. Factors such as cloudy weather, limited sunshine, frequent showers, moderate temperatures (around 28-30 °C), and air movement during the flowering period promote the progression of the disease.

Figure 7.1: Ergot Symptoms in Pearl Millet (*Source*: Schumann and Uppala, 2000).

Disease Cycle

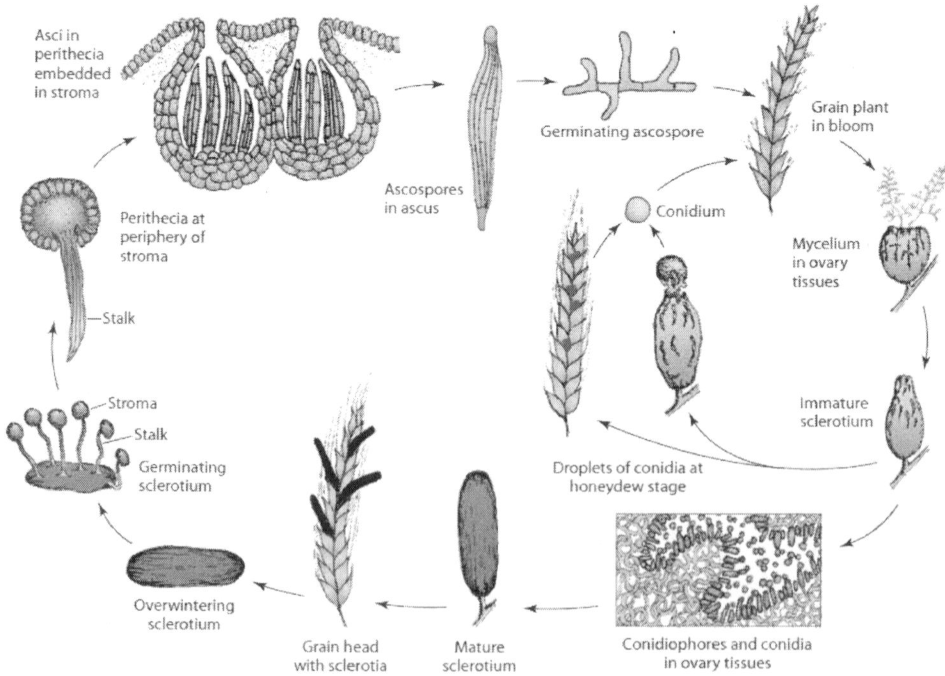

Figure 7.2: Disease Cycle of *Claviceps* spp. (*Source*: Schumann and Uppala, 2000).

The primary source of infection for this disease is the presence of sclerotia in the field, which can be left behind during harvesting or get mixed with seeds, serving as a primary source of infection to establish itself. Secondary infection occurs when honey dew, containing a large number of pathogen conidia, becomes visible. These conidia are then disseminated through various means such as rain splashes, wind, insects, and physical contact facilitating the spread of the disease.

Management

1. It is advisable to sow the seeds immediately after the onset of monsoon.
2. To minimize the presence of sclerotia, practice repeated deep summer ploughing, as it helps reduce the primary source of pathogen infection.
3. Implement a long crop rotation strategy to prevent the recurrence of disease. Avoid sowing bajra in a field that previously severe ergot infestation.
4. Once the disease appears, it is challenging to eliminate it completely. However, you can take the following precautions to prevent its further spread:
5. Immediately burn the ears infested with honey-dew as soon as they are observed in the field.
6. After harvesting the crop, bury the debris using a furrow turning plough. This ensures that the ergot sclerotia rot in the soil.
7. After threshing the ergot-affected crop, burn the leftover ear heads of bajra on the threshing floor.
8. By following these measures, you can minimize the impact of ergot disease and prevent its widespread occurrence.
9. Ensure that the seeds used for sowing are certified or clean. Begin by immersing the seeds in a 20 per cent brine solution or a common salt (NaCl) solution, and remove any floating sclerotia. Proceed to wash the seeds thoroughly with ordinary water and ensure they are completely dried before sowing.
10. Biological control: *Trichoderma viride* (Monitor) found most significant.
11. When 5-10 per cent of the flowers have opened, spray the crop with Carbendazim 50 per cent WP (Bavistin) or Zineb 75 per cent WP (Indofil Z-78). Repeat the spraying process at the 50 per cent flowering stage. These fungicides help manage the spread of disease.

2. Downy Mildew

Economic Importance and Occurrence

E. J. Butler first reported the disease in India in 1907. High-yielding varieties can suffer losses of up to 30 per cent due to this disease. In 1975, Karnataka and

Maharashtra experienced a severe outbreak of downy mildew, resulting in a complete loss of crops in thousands of hectares.

Causal Organism

The Downy mildew of pearl millet disease caused by *Sclerospora graminicola,* which belong to Kingdom- *Chromista,* Phylum- *Oomycota,* Class- *Oomycetes,* Order- *Peronosporales,* Family-*Peronosporaceae,* Genus- *Sclerospora* and Species- *graminicola* (Kirk *et al.,* 2008). The fungus is an obligate parasite. It causes systemic infection in plant and sends haustoria in plant cell. The sporangiophores are short, stout, determinate, dichotomously branched, hypophyllous or amphigenous, clavate and non-septate.

Symptoms

(a) Seedling Stage

The infected plants showed dwarfism due to internode shortening and excessive tillering. The foliage becomes pale, chlorotic, and develops broad streaks that extend from the base to the tip of the leaf. Upon close examination, a downy, dirty white fungal growth can be observed on the lower surface of the leaf, particularly during rainy and humid seasons. The leaves become distorted and eventually dry up.

Figure 7.3: Downy Mildew Symptoms in Pearl Millet (*Source*: PlantVillage plantvillage.psu.edu).

(b) Green Ear Head

In the case of the green ear head, symptoms are seen very distinct. Severely infected plants either fail to form ears or produce malformed green leafy structures, hence the name "green ear disease." Often, the entire ear is transformed into a leafy structure or sometimes partial infected ear seen in adverse condition. As the disease progresses, the green leafy structures turn brown and dry up.

Favourable Condition

The disease tends to be more prevalent in areas with low-lying and poorly drained soil. It thrives in conditions characterized by high humidity (around 90 per cent) and temperatures ranging between 22-25°C. The presence of water on the leaves further facilitates its development.

Disease Cycle

The oospores can maintain their viability in the soil for a period of five years or more. During this time, they have the potential to cause primary infections in host seedlings. These infections occur through direct penetration of the root hairs and coleoptiles by germ tubes. Secondary spread begins when sporangia (zoospores) become active, particularly during the rainy season. The zoospores enter the leaf stomata to initiate infection. The formation of sporangiosphores is favored by specific conditions, including high humidity (around 90 per cent), presence of water on the leaves, and temperatures ranging from 15 to 25 °C.

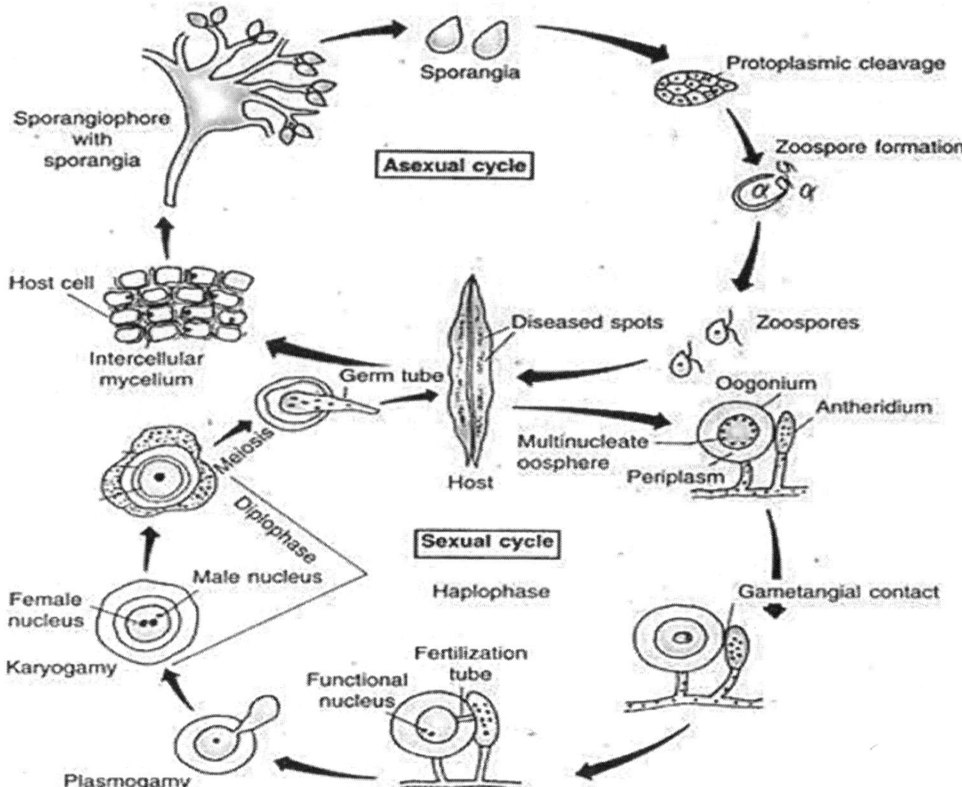

Figure 7.4: Disease Cycle of *Sclerospora graminicola* (*Source*: www.biologydiscussion.com).

Management

1. Sow the crop immediately after the onset of monsoon.
2. Grow resistant variety: Hybrid Bajra (PMH Super-001).
3. GHB-351, GHB-316, GHB-32. Avoidance of low laying and water logging fields for sowing. Crop rotation for more than 5 years should be followed.
4. Rough out infected plants carefully during thinning operation.
5. Late sowing is avoided.
6. Seed treatment with *Trichoderma viride*/*Pseudomonas fluorescens* 5 g/kg seed or spray *Pseudomonas fluorescens* 10g/liter significantly reduced the downy mildew incidence.
7. Seed treatment with Ridomil Gold @ 6g/Kg seed followed by two spray of Metalaxyl 8 per cent +Mancozeb 64 per cent WP (Ridomil Gold) or Fluopicolide 5.56 per cent w/w + Propamocarb Hydrochloride 55.6 per cent SC (Infinito) or Cyazofamid 34.5 per cent SC (Ranman) at 25 and 35 days after sowing.

3. Bajra Smut

Economic Importance and Occurrence

This particular disease is found in Pakistan, Africa, and India, with a higher prevalence observed in Tamil Nadu, Andhra Pradesh, and Maharashtra. While it typically causes minimal harm to the crops, in years of heavy incidence, it can lead to significant reductions in grain yield, resulting in notable losses.

Causal Organism

The smut of Bajra caused by *Tolyposporium penicillariae/Moesziomyces bullatus*, which belongs to Kingdom- *Fungi*, Phylum- *Basidiomycota*, Class- *Ustilaginomycetes*, Order-*Ustilaginales*, Family- *Ustilaginaceae*, Genus- *Tolyposporium* and Species- *penicillariae* (Kirk *et al.*, 2008).

Symptoms

The disease becomes noticeable during the grain setting stage, where only individual grains irregularly distributed in an ear are infected. Infected ears may exhibit occasional replacement of a few grains with oval to top-shaped structures known as sori. These sori are typically two or three times larger than regular grains and vary in color from bright green to dirty black. They often rupture, revealing a black mass inside.

Favorable Conditions

High humidity, Mono-culturing of the susceptible variety is favouring the disease development.

Figure 7.5: Smut Symptoms in Pearl Millet (Source: US Department of Agriculture).

Disease Cycle

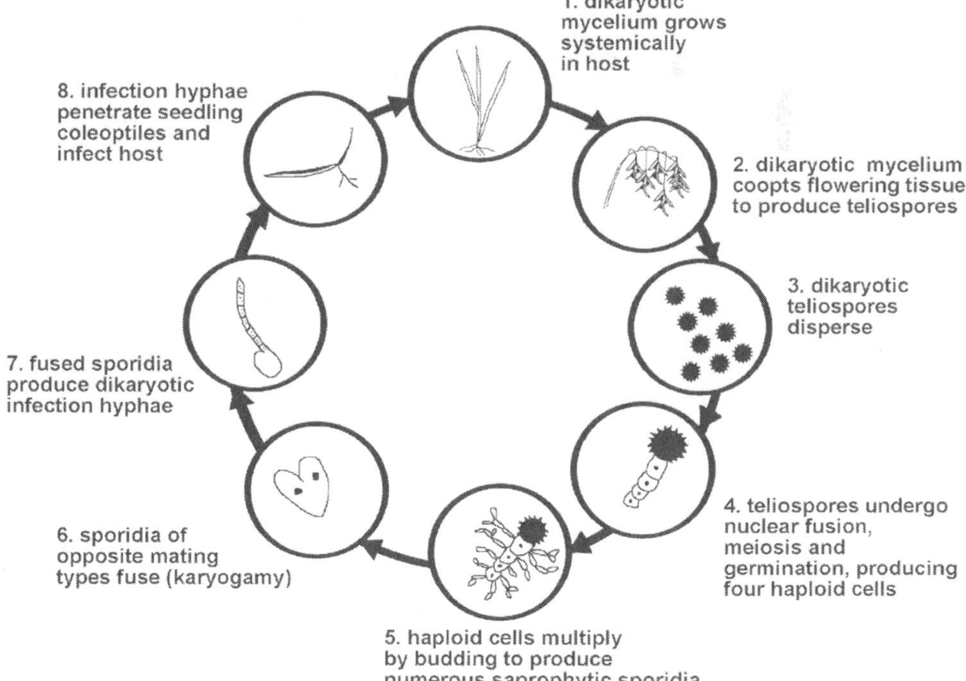

Figure 7.6: Disease Cycle of Smut (*Source*: Meyer *et al.,* 2010).

Primary inoculum consists of smut spore balls which are soil borne or is brought to soil as contaminant with the seed. These smut balls germinate in soil and produce sporidia which are disseminated by wind and reach individual florets where they cause infection. This infection is local and restricted to the individual flowers. The kernel is replaced by smut sorus. The spores that fall on the ground may germinate and produce sporidia which can cause secondary infection of flowers appearing late.

Management

1. Resistant varieties: WC-C75, ICMS 7703, ICTP 8203, and ICMV 155 released in India against smut disease.
2. Infected ear heads should be removed and destroyed.
3. It is important to sow immediately after the onset of monsoon and avoid late sowing.
4. Crop rotation and field sanitation should be practiced to reduce the source of primary infection.
5. Diseased ears should be removed early in the season and destroy them.
6. Heavy doses of nitrogenous fertilizers should be avoided.
7. The crop can be sprayed with Carboxin 37.5 per cent + Thiram 37.5 per cent (Vitavax Power) at 0.1 per cent or Zineb 75 per cent WP (Indofil Z 78) at the boot leaf stage.

4. Blast

Economic Importance and Occurrence

In India, the disease was first reported from Kanpur (Uttar Pradesh). Disease has been sporadically observed on high-yielding cultivars from 1953 onwards in the pearl millet growing regions belonging to northern parts of India. Also newly released hybrids cultivated in Gujarat, Maharashtra, Uttar Pradesh, Rajasthan and Madhya Pradesh.

Causal Organism

The blast disease of pearl millet is caused by *Pyricularia grisea*, which belongs to Kingdom-*Fungi*, Phylum- *Ascomycota*, Class- *Sordariomycetes*, Subclass-*Sordariomycetidae*, Family-*Magnaporthaceae*, Genus- *Magnaporthe* and Species-*grisea* (Kirk *et al.*, 2008). Typically, the conidia are three-celled and formed at the apex of a conidiophore. The ascospores of the pathogen have a hyaline, fusiform shape with three septa. The pathogen produces toxins known as Pyricularin and Picolinic acid, which have inhibitory effects on plant growth.

Symptoms

Circular to diamond-shaped lesions measuring up to 1 cm length observed with lighter colored centers and dark brown margins. These lesions exhibit chlorotic-yellow haloes with extensions of the brown margins into the haloes at opposite ends. Under humid conditions, the lesion centers develop a light grey bloom consisting of conidiophores and conidia.

Figure 7.7: Blast Symptoms in Pearl Millet (*Source*: Nayaka *et al.*, 2017, Rao *et al.*, 2021).

Favourable Conditions

The weather is characterized by moderate temperature (25–30 °C) with high relative humidity (>90 per cent) and cloudy days coupled with intermittent rainfall creating continuous leaf wetness for more than 10 hours are congenial for rapid development and spread of the disease. Continuous rains during heading lead to the occurrence of finger blast, resulting in massive production losses in both finger and foxtail millet. Also, high nitrogen fertilizer application is reported to increase blast disease.

Disease Cycle

Primary source: Mycelium and conidia in infected debris serves as inoculum. The secondary spreads takes place through wind born conidia. The mycelium survives in straw for 1-2 years under dry condition.

Management

1. Use disease free seeds.
2. Grow resistant variety: Hybrid Bajra (PMH Super-001).
3. Seed treatment Carbendazim (Bavistin) or Carboxin+ Thiram (Vitavax Power) @ 2 g/kg.
4. Destruction of weed host from field (*Pennisetum pureum, Pennisetum cenchroides*, and *Cynodon dactylon*).

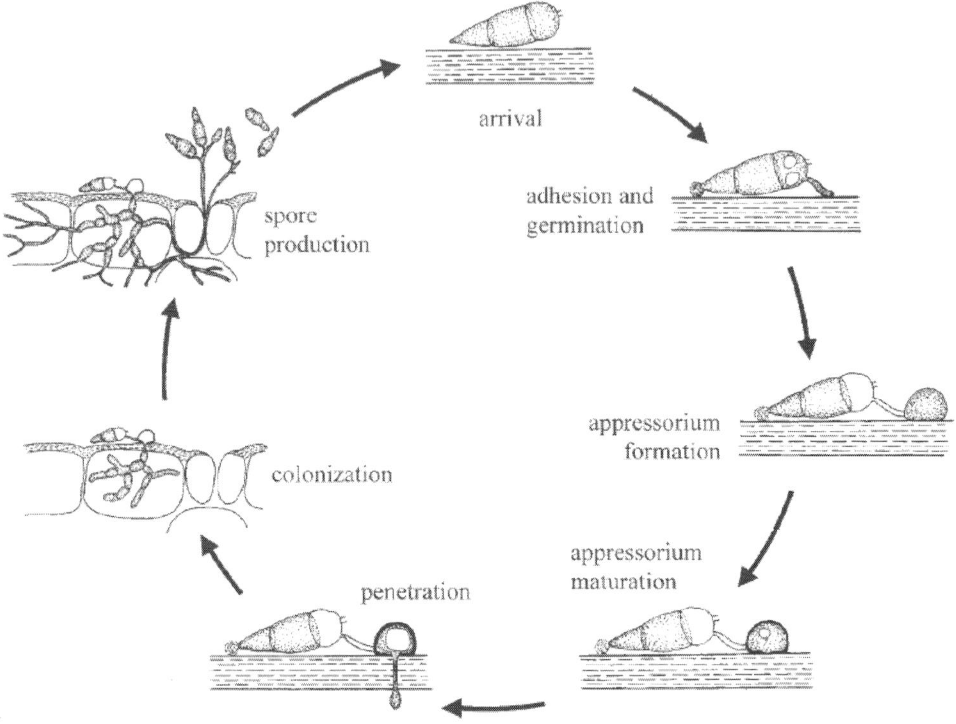

Figure 7.8: Disease Cycle of *Pyricularia grisea* (*Source*: Thines *et al.*, 2004).

5. Adjust planting time. Sow seeds early, when possible, after the onset of the rainy season.
6. Split nitrogen fertilizer application in two or more treatments. Excessive use of fertilizer can increase blast intensity.
7. Silicon fertilizers can be applied to soils to reduce blast.
8. Spray 200 ml Amistar Top 325 SC (Azoxystrobin + Difenoconazole) or 500 g Zineb 75 WP (Indofil Z 78) per acre in 200 liters of water.

5. Bajra Rust

Economic Importance and Occurrence

Rust was first reported in U.S. in 1954 as *P. penniseti*. This disease is prevalent in regions with warmer climates. In cases where the infection occurs during the initial stages, there can be significant and severe losses. The disease has been observed both in dry and irrigated crop conditions.

Causal Organism

The rust of Bajra disease caused by *Puccinia penniseti, Puccinia substriata var. indica,* which belongs to Kingdom- *Fungi,* Phylum- *Basidiomycota,* Class- *Pucciniomycetes,* Order- *Pucciniales,* Family- *Pucciniaceae* and Genus- *Puccinia* (Kirk *et al.*, 2008).

Symptoms

In the early stages, the symptoms manifest as circular pustules called uredinia, displaying reddish-brown to reddish-orange colored uredospores on both sides of the leaves, mainly towards the tip of the blade, as well as on the leaf sheath. As the pustules mature with plant age, they undergo a color change and become darker, black colour elliptical and sub-epidermal teliospores. In severe infection, the plants appear unhealthy and slightly stunted compared with healthy ones.

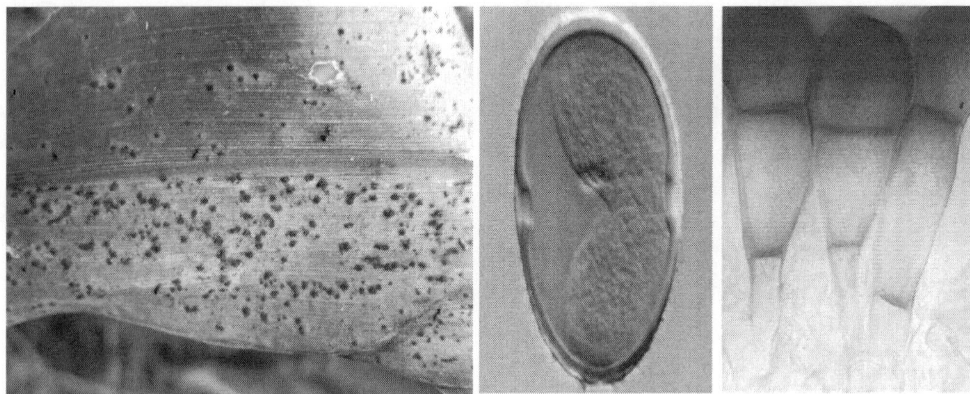

Figure 7.9: A. Rust symptoms in pearl millet (CABI, 2023) with B. Uredospore, and C. Teliospore (de Carvalho *et al.*, 2006).

Disease Cycle

The rust fungus is **heterocious** and alternate host of the rust fungus is brinjal. Uredospores are known as repeating spores and cause infection in the field. The fungus has a long life cycle, producing the uredial and telial stage on bajara and the aecial and pycnidial stage on several species of *Solanum* including brinjal.

Management

1. Grow resistant variety: Hybrid Bajra (PMH Super-001).
2. Immediately sowing after onset of monsoon.
3. Destruction of weed host from field.

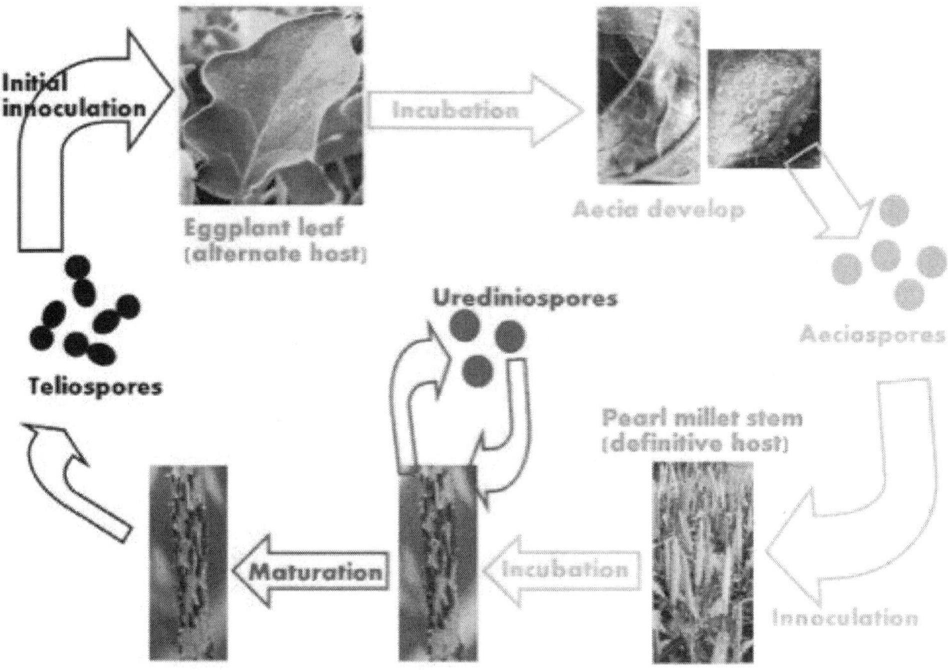

Figure 7.10: Disease Cycle of *Puccinia penniseti* (*Source*: http://rsandss.blogspot.com/).

4. Effective control using bio-control agents (*Chaetomium globossum*, *Trichoderma koningii*).
5. Two sprays Propiconazole 25 per cent EC (Tilt) at 15 days interval starting from the initiation of disease.

Minor Diseases of Pearl Millet

A. Fungal

1. Curvularia Leaf Blight: *Curvularia penniseti*

The initial symptoms observed were irregular yellow spots on the leaf surface, which subsequently turned brown and merged together, resulting in a blight appearance of the leaves. Use healthy, disease free seed. Leaf blight can be managed by spray Chlorothalonil 75 per cent WP (Kavach) or Mancozeb 75 per cent WP (Uthane M-45).

2. Zonate Leaf Spot: *Gloeocercospora sorghi*

The symptoms is marked by the presence of circular, purple bands that alternate with straw-colored areas, resembling the appearance of a bullseye target. Typically, these spots form in semicircular patterns along the edge of the leaf. Selection of resistant and healthy variety. Collect plant residues and destroy them at distance of field. Foliar spray of Mancozeb 75 per cent WP (0.2 per cent) can suppress the pathogen.

3. Cercospora Leaf Spot: *Cercospora penniseti*

Characterized by oval lesions with dark brown borders and pale tan to grey centers. These lesions are marked with rows of black conidiophores and can also be found on the stem. Avoid planting pearl millet varieties that are highly susceptible to *Cercospora* fungi. Spray Carbendazim 500 g/ha or Mancozeb 1000g/ha at initiation of the disease and 10 days later.

4. Bipolaris Leaf Spot: *Bipolaris setariae*

The lesions small, brown specks or oval to oblong or rectangular patches. Over time, these lesions may grow and merge together. Typically, the lesions exhibit a tan or grayish brown color with a clearly defined dark brown border.

Treating seeds with fungicides like Vitavax Power (Carboxin 37.5 per cent + Thiram 37.5 per cent DS) helps protect seeds from infection. Spray Nativo (Tebuconazole 50 per cent + Trifloxystrobin 25 per cent WG) at initiation of the disease and 10 days later.

B. Bacterial Diseases

1. Bacterial Leaf Spot: *Pseudomonas syringae*

Bacterial spots manifest on leaves as small, elliptical, or irregular-shaped marks characterized by a central area of straw color and a darker border. Over time, these spots merge together, forming extensive bands. Initially, the spots exhibit a light yellowish-brown hue, but gradually darken to a deep brown shade. Sprays Copper Oxychloride 50 per cent WP (Blue copper) at 0.25 per cent reduced the incidence of bacterial disease. Field sanitation is important. Also seeds must be obtained from disease free plants.

2. Bacterial Leaf Stripe: *Pseudomonas avenae*

The development of characteristic water-soaked, reddish-brown stripes in the affected leaves. The necrotic tissue showed limited water-soaking, resembling the symptoms caused by the pathogen that affects watermelons. The straw-colored appearance was consistently observed in the midrib region of the affected leaf. Seeds must be obtained from disease free plants. Sprays Copper oxychloride 50 per cent WP (Blue copper) at 0.25 per cent reduced the incidence of bacterial disease.

3. Bacterial Leaf Streak: *Xanthomonas axonopodis* pv. *pennamericanum*

Typical water-soaked streaks are transparent, measuring around 1-4 mm in width and 3-12 mm length. They typically emerge as early as the second leaf stage of the seedling. Over time, the lesions progress, turning red and losing their transparency. They may take on an oval shape with tan centers and narrow red borders, occasionally appearing somewhat irregular. In severe instances, the lesions can merge together, creating lengthy, irregular streaks and patches that span across a significant portion or the entirety of the leaf surface. The affected areas exhibit dead tissue, which is outlined by dark margins between the reddish-brown streaks. Use of resistant variety. Practicing field sanitation is important. Sprays Copper Oxychloride 50 per cent WP (Blue copper) at 0.25 per cent reduced the incidence of bacterial disease.

C. Viral Diseases

1. Maize Dwarf Mosaic Virus

This virus transmitted by sugarcane aphid *Longiunguis sacchari*

Infected plants exhibit mosaic symptoms with stunted growth and malformed ears. The affected ears are pale yellow or brownish-white and produce a limited number of small seeds. From a distance, the affected crop appears yellow in color. Use of tolerant, resistant varieties if available. Viral disease managed by control vector population by spraying insecticide *i.e.* Flonicamid 50 per cent WG (Ulala) or Neem based bio-pesticide.

2. Maize Streak Virus

This geminivirus transmitted by the vector plant hopper (mostly) *Peregrinus maidis* and leaf hopper *Cicadulina mbila*

The disease initially presents itself with thin, pale streaks between the veins of the leaves, and the affected plants show stunted growth. As the disease advances, these streaks turn necrotic. In the early stages, infected plants may die before the emergence of the ear head, while those infected at later stages may produce ear heads with or without grain. Plant early in the season. Eradicate grass weeds. Viral disease managed by control vector population by spraying insecticide *i.e.* Dinotefuran 20 per cent SG (Oshin) or Neem based bio-pesticide.

D. Phanerogamic Plant Parasite (PPP): Partial Root Parasite/Witch Weed

1. *Striga asiatica/S. hermonthica* (Orobanchaceae)

The dormant seeds of this parasitic plant lie in the soil until they are triggered to germinate by the root exudates released by bajra plant. Once germinated, the plant's germ-tubes penetrate the roots of bajra, creating a pathway for the extraction of water and nutrients from the host plant. This enables striga to grow and eventually produce purple flowers. These flowers produce seeds that contribute to the seed bank in the soil, ensuring the parasitic cycle can continue to affect future crops. Hand pulling, uprooting Striga by hand or hand tools is still considered as the cheapest traditional control method. Farmers advised to control it before the weed emerges. Therefore, spray crop with 2, 4-D herbicide (Weedmar Super), two to three weeks after sowing

References

Anonymous (2022). The Associated Chambers of Commerce and Industry of India. *Millets the Super Food for India; The Associated Chambers of Commerce and Industry of India*: New Delhi, India; p. 15.

Anonymous (2023a). International Year of Millets National Conference on *Kharif* Campaign, Ministry of Agriculture and Farmers Welfare. Available online: https: //agricoop.nic.in/Documents/Crops_0.pdf (accessed on 19 June, 2023).

Basavaraj, G., Rao, P. P., Bhagavatula, S. and Ahmed, W. (2010). Availability and utilization of pearl millet in India. *SAT ejournal*, 8.

CABI (2023). CABI Compendium. Wallingford, UK: *CAB International.*

de Carvalho, A. D. O., Soares, D. J., do Carmo, M. G. F., da Costa, A. C. T. and Pimentel, C. (2006). Description of the life-cycle of the pearl millet rust fungus - *Puccinia substriata* var. *penicillariae* with a proposal of reducing var. indica to a synonym. *Mycopathologia*, **161**: 331-336.

FAOSTAT (2021). Food and Agriculture Organization of the United Nations Statistical Database; Statistical Division; FAO: Rome, Italy.

Kargwal, R., Yadvika, Singh, V. K. and Kumar, A. (2023). Energy Use Patterns of Pearl Millet (*Pennisetum glaucum* (L.)) Production in Haryana, India. *World*, **4**(2): 241-258.

National Research Council. (1996). Lost crops of Africa: volume I: grains. *National Academies Press.*

Nayaka, C. S., Srivastava, R. K., Udayashankar, A. C., Lavanya, S. N., Prakash, G., Bishnoi, H. R. and Satyavathi, C. T. (2017). Magnaporthe blast of pearl millet in India-present status and future prospects. *All India Coordinated Research Project on Pearl Millet (Indian Council of Agricultural Research), Mandor, Jodhpur, India.*

Rangaswami, G. and Mahadevan, A. (1998). Diseases of crop plants in India. PHI Learning Pvt. Ltd.

Rao, K. B., Motukuri, S. K., Kumar, K. A., Babu, C. P. and Pathak, V. (2021). Pearl Millet Blast Pathogen Virulence Study and Identification of Resistance Donors on Virulent Isolate. *Journal of Pure and Applied Microbiology*, **5**: 32-45.

Schumann, G.L. and Uppala, S. (2000). Ergot of rye. *The Plant Health Instructor.*

Thakur, R. P., Rao, V. P. and Williams, R. J. (1984). The morphology and disease cycle of ergot caused by *Claviceps fusiformis* in pearl millet. *Phytopathology*, **74**(2): 201-205.

Thines, E., Heidrun, A. N. K. E. and Weber, R. W. (2004). Fungal secondary metabolites as inhibitors of infection-related morphogenesis in phytopathogenic fungi. *Mycological research*, **108**(1): 14-25.

Kirk, P. M., Cannon, P. F., Minter, D. W. and Stalpers, J. A. (2008). *Dictionary of the Fungi* (10th edn). *Wallingford, UK.*

Meyer, S. E., Nelson, D. L., Clement, S. and Ramakrishnan, A. (2010). Ecological genetics of the *Bromus tectorum* (Poaceae) *Ustilago bullata* (Ustilaginaceae) pathosystem: A role for frequencydependent selection?. *American Journal of Botany*, **97**(8): 1304 -1312.

ORCID: https: //orcid.org/0000-0003-2027-5055

Chapter 8

Diseases of Minor Millet Crop

Gajendra Pratap[1]*, Bipin Verma*[1]*, Kishan Lal*[2]*, Naimish Kumar*[1] *and Shivam Kumar*[1*]

[1]*Department of Plant Pathology, Chandra Shekhar Azad University of Agriculture and Technology, Kanpur – 208002, India*
[2]*Assistant Professor, Department of Plant Pathology, National P.G. College, Barhalganj Gorakhpur (U.P.)- 273402*
[*]*e-mail: shivamkumar224229@gmail.com*

Introduction

Millets are small-seeded cereal crops widely known for its nutraceutical importance as well as food and fodder. The earliest evidence of millets has been found in Indus Valley civilization (3,000 B.C.) and was one of the first plants domesticated for food. It includes three major crops *viz.*, Jowar (Sorghum bicolor), Pearl millet/Bajra (Pennisetum typhoides or glaucum), Finger millet/Madua/ Ragi (Eleusine coracana) and six minor crops *viz.*, Foxtail/Italian/German millet (Sateria italic), Sawan/Barnyard millet (Echinochloa frumentaceae), Proso millet/Cheena/Vraiga (Penicum miliaceum), Kodo/Varagu/Haraka/ Arikalu (Paspalum scrobiculatum), Little millet/Kutki/Shavan (Panicum sumatrense). In India, common millets include Ragi, Shavan, Jowar, Cheena, and Bajra. Roughly 60 billion people in Asia and Africa traditionally eat it, and it is grown in about 131 countries. After rice, wheat, and maize, pearl millet is the fourth most extensively grown food crop in India. With more than 55 per cent of the world's millets produced, China, India, and Nigeria are the top three producers worldwide. The major millets-producing states in India are Rajasthan, Karnataka, Maharashtra, MP, UP, and others; they account for 80 per cent of Asia's production and 20 per cent of the world market. India ranks fourth globally in terms of millet exports (the top five are China, the USA, Russia, and the Ukraine). India exported millets for USD 26.97 million in 2020–21 as opposed to USD 28.5 million in 2019–20. In 2020–21, Nepal (22095.7 MT valued at USD 6.09 million), the United Arab Emirates (15556.7 MT valued at USD 4.84 million), and Saudi Arabia (13516.5 MT valued at USD 3.84 million) were the top 3 importers of millets from India. 2018 has been designated as the "National Year of Millets." The United Nations General Assembly has proclaimed 2023 as the International Year of Millets, following the FAO's approval of India's proposal to observe the year as such in 2018. The Agriculture Ministry named millets a "Super food of the world" in 2018 due to their high nutritional value, which includes minerals like calcium, iron, zinc, potassium, and magnesium as well as essential fatty acids, dietary fiber, and B vitamins. In particular for women and children, it can guard against nutritional inadequacy and offer nutritional security. As per the Indian Institute of Millets Research, Hyderabad, ICAR, millets have the following nutritional values: 15-20 per cent dietary fiber, 2-5 percent fat, 65-75 per cent carbohydrates, and 7-12 per cent protein. Millets are grown for grain and fodder in India. Even though millets are thought to be hardy crops, the majority of them are now vulnerable to numerous diseases due to climate change. Certain millets diseases are sporadic in certain climatic situations and have less of an adverse effect on the crop, while many others regularly occur in a severe form under various climatic conditions and cause significant economic loss. There are more types of fungi than bacteria or viruses. The most common and destructive diseases that cause significant losses in grain and forage yield are maize stripe virus, maize mosaic, ergot, smut, anthracnose, downy mildew, blast, rust, charcoal rot, foot rot, sheath rot, banded sheath blight, leaf blight, leaf spot, bacterial leaf spot, and bacterial leaf streak. The main topics of this book chapter are disease occurrence, distribution, symptoms, yield loss, aetiology, pathogen dissemination mechanism, and survival, in addition to integrated disease management options. This information will be very beneficial for a deeper comprehension of the illness. Furthermore, as millets are primarily grown as staple food crops in developing nations on the Asian and African continents, it will be advantageous to boost millet productivity and production as well as food and nutritional security in these regions.

Disease classified as fungal, bacterial and viral diseases:

A. Fungal Disease

1. Sugar Disease/Ergot Disease

Causal Organisms = Ergot of Sorghum (*Claviceps sorghi*) and Ergot of Pearl millet (*C. fusiformis*)

2. Sorghum Smut

Causal Organisms = Head smut (*Sporisorium reilianum*), Covered smut (*Sporisorium sorghi*), Loose smut (*Sporisorium cruenta*) and Long smut (*Tolyposporium ehrenbergii*).

3. Pearl Millet Smut

Causal organisms: *Tolyposporium penicillariae*

4. Small Millet Smut

Grain smut: a). Finger millet (*Melanopsichium eleusinis*) b). Barnyard millet: *Ustilago panici-frumentacei* and c). Foxtail millet (*Ustilago crameri*),

Head smut: d). Barnyard, Kodo and Proso millets (*Sorosporium paspali-thunbergii*)

5. Udbatta Disease: *Ephelis oryzae*

6. Downy Mildew of Small Millet

Finger millet: *Sclerophthora macrospora*

Foxtail millet: *Sclerospora graminicola*

7. Blast

Finger millet blast: *Pyricularia grisea* (Perfect state: *Magnaporthe grisea*)

8. Pearl Millet and Small Millet Blast

Pearl millet, Barnyard, Proso and Little millets: *Pyricularia grisea*

Foxtail millet: *P. setariae*

9. Rust

Sorghum rust: *Puccinia purpurea*

10. Pearl Millet and Small Millet Rust

Kodo millet and Pearl millet: *Puccinia substriata* var. *indica*

Finger millet: *Uromyces eragrostidis*, **Foxtail millet:** *U. setariae-italiae*,

B. Bacterial Diseases

1. Bacterial leaf spot of Sorghum/Pearl millet (*P. syringae* pv. syringae).
2. Bacterial leaf spot of Finger millet (*Xanthomonas eleusinae*).
3. Bacterial leaf stripe of Sorghum (*Pseudomonas andropogoni*).
4. Bacterial leaf stripe of Pearl millet (*P.avenae*).
5. Bacterial leaf stripe of Finger millet (*P. eleusinae*).
6. Bacterial leaf streak of Sorghum (*Xanthomonas axanopodis* pv. avnae).
7. Bacterial leaf streak of Finger millet (*X. a.* pv. *coracana*).

8. Bacterial leaf streak of Foxtail (*P.avenae*).
9. Bacterial stalk rot of Sorghum (*E. chrysanthemi*).

C. Viral Diseases

1. Maize stripe virus (Sorghum).
2. Maize streak virus, which affects Finger, Pearl, and Sorghum millet.
3. The mosaic virus of maize (Sorghum).
4. Maize dwarf mosaic virus (Sorghum, Pearl millet).
5. Sugarcane mosaic virus (Sorghum).
6. Finger millet virus (Ragi mottle streak virus).

1. Small Millet Smut

Causal Organisms

a. *Grain Smut*

Finger millet: *Melanopsichium eleusinis*

Foxtail millet: *Ustilago crameri*

Barnyard millet: *Ustilago panici-frumentacei*

b. *Head Smut*

Barnyard, Kodo and Proso millets: *Sorosporium paspali- thunbergii*

Survival and Spread

Head smut is solely seed borne, whereas grain smut is externally air-seed borne.

Symptoms

Three different smut diseases commonly impact barnyard millet at different stages of growth. Head, grain, and kernel smut all have an impact on the crop; the grain smut is particularly significant. Symptoms of Head Smut: The inflorescence infected with Head Smut is distorted and swollen like a gall on the stem, in the axils of the older leaves, and at the nodes of young shoots. The flag leaf may enclose the infected panicle and prevent it from fully emerging. Symptoms of grain smut: Affected seeds enlarge to two or three times their normal size and develop hair on their surface. When temperatures are between 20 and 25 degrees, the disease first manifests itself during the grain-forming stage. Figure 8.1 shows the development of grain smut sori occurs at random in the grain, main rachis, or peduncle of the following millets: foxtail (d), little (c), barnyard (b), and finger (a). Ovaries change into larger-than-normal grain sori, which resemble velvety galls. When sorus dries, its original greenish hue eventually changes to a pinkish green and then a dirty black. The fungus on foxtail millet affects the majority of the grains in an ear. In proso, barnyard, and kodo millets, head smuts are frequent. In kodo millet, it results in a significant yield reduction. A long sorus is formed out of the entire panicle (d). The flag leaf may enclose the infected panicle, preventing it from fully emerging. The surrounding membrane ruptures, revealing the black mass of spores.

Figure 8.1: Grain and Head Smut of Small Millets *viz.,* (a) Finger millets; (b) Barnyard millets; (c) Little millets; (d) Foxtail millets.

Management

Seed treatment: Chlorothalonil or Carbendazim @ 2g per Kg seed reduces the risk for infestation of Head and Grain smut. Benefits include Aspergillus awamouri (phosphate solubilizing fungus) or Azospirullum brasilense (bacterium that fixes nitrogen) at a rate of 25g/kg.

2. Udbatta

Causal organism: *Ephelis oryzae*

Millet host: Foxtail millet, Kodo millet and Little millet

Survival and Spread

The pathogen can infect grasses like *Cynadon dactylon, Pennisetum* spp. and *Ergostis tenufolia* for survival.

It is seed-borne disease.

Symptoms

Figure 8.2: Udbatta Disease of Millets.

In India, udbatta, a panicle disease, is frequently seen on a variety of small millets. Usually stunted, infected plants occasionally develop narrow stripes along the veins of the flag leaves caused by white mycelia and conidia before panicles emerge. The name "Udbatta disease" comes from the transformation of the affected panicle into a compact, silver-colored, cylindrical spike that resembles an incense stick, much like an agarbatti or udbatta. A panicle that has contracted an infection stops producing grain normally and turns sterile.

Management

1. When sowing, use healthy, disease-free seeds.
2. Eliminating and discarding diseased panicles in the field.
3. Applying 2 grams of Captan or Thiram per kilogram of seed.
4. A hot water treatment applied to the seeds for ten minutes at 50–54°C prior to sowing effectively controls the disease.
5. The pathogen that the seeds contain can also be killed by subjecting them to solar radiation.

3. Downy Mildew of Small Millet

Causal Organisms

Finger millet: *Sclerophthora macrospora*

Foxtail millet: *Sclerospora graminicola*

Survival and Spread

Survive as oospores in host tissues, soil and alternate host.

Primary infection in soil.

Secondary infection through conidia or sporangia.

Symptoms

Downy mildew in finger millet is known as crazy top downy mildew because of appearance of bushy and leafy growth in place of fingers or grains. Usually absent is the white, cottony growth that distinguishes many other downy mildews. This disease results in partial or total sterility by transforming the heads into green, narrow, leafy structures (Figure 8.3a). The "green ear" symptom is the result of a proliferation that begins at the basal spikelet and eventually gives the entire ear a bushy appearance. The host range of the pathogen is broad and includes Eleusine indica, maize, wheat, oats, and Digitaria marginata. As seen in many other downy mildews, Foxtail millet downy mildew is characterized by downy growth of the fungal structures (Figure 8.3b). In the later stages, it is typical for the infected leaves to shred (Figure 8.3c). It is typical to see malformation of the floral organs and the transformation of a few or many spikelets into leafy structures (Figure 8.3d).

Management

1. Use disease free seed materials.
2. Use disease resistant varieties

Figure 8.3: Downy Mildew of Small Millet.

3. Seed treatments with Bioagents like *Trichoderma harzianum* (4 g per kg seeds), *Pesudomonas fluorescens* and *Bacillus species* @10 g per kg seeds also help manage the disease.
4. Seed treatment with fungicides like Captan, Fludioxonil, Metalaxyl.
5. It is also possible to directly manage downy mildew with metalaxyl.

4. Finger Millet Blast

Causal organism: *Pyricularia grisea* (**Perfect state:** *Magnaporthe grisea*)

Favorable Condition

1. The over application of fertilizers high in nitrogen.
2. Rainy, cloudy weather or dew that keeps leaves wet for longer than ten hours.
3. A temperature of 15 to 24°C at night with a relative humidity of more than 90 per cent.

Survival and Spread

Survives in the crop residues and on other cereals.

Initial inoculum comes from weeds or collateral hosts.

Spread by air-borne conidia.

Symptoms

Depending on the crop's stage, blast symptoms can be seen on the finger, peduncle, leaf, or seedling. Characteristic symptoms include water-soaking and a chlorotic halo surrounding the lesions, as well as elliptical or diamond-shaped lesions on leaves with grey centers (Figure 8.4a). In favorable circumstances, leaf blades appear blasted and spots enlarge and coalesce. The elongated black lesion that appears one to two inches below the ear is the primary sign of neck blast (Figure 8.4b). The finger blast symptom begins at the tip and spreads to the brown base of the finger (Figure 8.4c). The most dangerous stage of the illness is a neck infection, which results in significant losses in grain weight and quantity as well as an increase in spikelet sterility.

Management

1. Disease is transmitted by seeds, so it is advised to treat the seeds with a fungicide solution prior to sowing.
2. It is necessary to prevent high plant densities since they promote the development of the blast disease. To prevent seedling crowding and to guarantee adequate aeration, the recommended spacing for finger millet (30 x 15 cm) should be observed.
3. Excessive use nitrogenous fertilizers not recommended.

Figure 8.4: Finger Millet Blast.

4. Inter-cultivation as well as crop rotation of ragi with legumes is not only economical but also helps in soil nutrient enrichment and in reducing the fungal inoculum.
5. Cultivation of blast resistant finger millet varieties include GPU28, GPU48, GPU45, VL Mandua 348, KMR 340, BM2 and PRM.
6. Use of biocontrol agents such as *Pseudomonas* @ 10g/kg seed, *Bacillus* @ 10g/kg seed and *Trichoderma* @ 4g/kg seed as field spray as well

as in seed treatment has been found to be effective against controlling the fungus.

7. In the chemical control measures, application of fungicides such as Tricyclazole (0.05 per cent), Carbendazim 50 WP @1 g/litre or Ediphenphos 50 EC @1g/Litre of water or combination product of Carbendazim + Mancozeb at the rate 1 g/litre water.

5. Small Millet Rust

Causal Organisms

Kodo millet: *P. substriata*

Finger millet: *Uromyces eragrostidis*

Foxtail millet: *U. setariae-italiae*

Little millet: U. linearis

Pathogen *U. setariae-italiae*

The teliospores have smooth, thick walls that are thicker at the apex than the base, are pedicellate, oblong, globose, yellowish brown, and have only one cell. We only know about the uerdial and telial stages.

Favourable Condition

Low temperature and high relative humidity.

Survival and Spread

1. The pathogen survives on alternate host (Brinjal).
2. It spread by air-borne spore.

Symptoms

Small millet rust is frequently observed on finger millet, foxtail, kodo, and small millets. The symptoms appear as tiny to tiny, linearly arranged, dark brown, broken pustules on the upper surface of the upper leaves. As seen in Figure 8.5, the disease is more prevalent on the upper leaf than on the lower or middle leaves. *U. linearis* yields black teleutospores, while *U. setariae-italicae* produces light yellow ones. On the upper surfaces of the leaves, P. substriata develops tiny, oval, brown spots. Telia are produced on the leaf's underside.

Management

1. Rotate with non-host crops like legumes and sorghum.
2. Remove weeds and volunteer plants during crop growth, paying special attention to grasses.
3. Avoid planting close to eggplant or any of its relatives, such as *Solanum torvum* (turkey berry), *P. substriata* hosts these plants.

Figure 8.5: Small Millet Rust.

4. To control rust on millet, fungicides such as Copper compounds, Mancozeb, Sulfur, or Chlorothalonil can be used.
5. A 2.5 g/litre dose of Mancozeb applied topically effectively suppresses rust.

6. Leaf Blight of Small Millets

Causal Organism

Kodo millet: *Alternaria* species

Survival and Spread

It persists as mycelia and conidia in the infected crop residues or in the soil.

It spread by air-borne conidia.

Symptoms

The appearance of long, elliptical, and necrotic lesions on the leaf is a characteristic of small millet leaf blight (Plate 15.6 a). The lesion's center is straw-colored, and its margin is typically dark brown. The degree of host resistance affects the lesions' size and shape. When a susceptible genotype is present, the lesions enlarge and combine to form necrotic areas on the leaf that are purplish gray or tan in color. The lower, older leaves experience the symptoms first, and then the upper, younger leaves experience them as well. Because the pathogen produces spores, especially in damp weather, the surface of necrotic lesions appears dark-gray or black. A serious disease gives the crop a characteristically burned look.

Figure 8.6: Leaf Blight of Small Millet.

Management

1. Use 4 grams of Captan or Thiram per kilogram of seeds.
2. Spray 1.25 kg/ha of Mancozeb.
3. Apply a 1 per cent Bordeaux mixture (2g/lit. water) or Copper oxy chloride, or Dithane Z-78 as a spray.

7. Helminthosporium Leaf Spot

Causal Organisms

Proso millet: *B. panici-miliacei*

Finger and Little millets: *Drechslera nodulosum* (P.S. *Cochliobolus nodulosus*)

Foxtail millet: *Cochliobolus setariae*

Symptoms

Favourable Conditions

Temperature of 25-30°C

Relative humidity above 80 per cent.

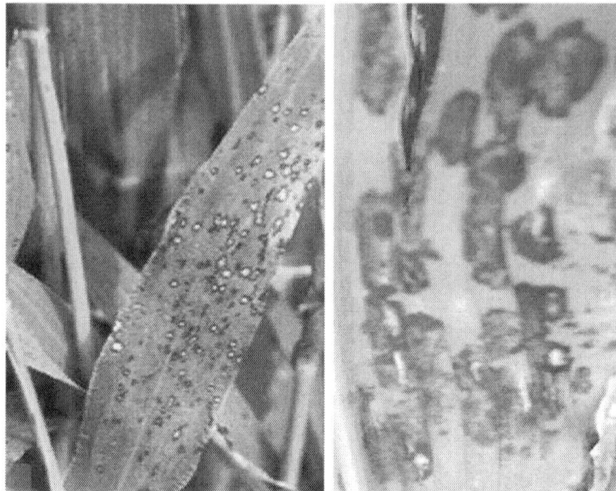

Figure 8.7: Helminthosporium Leaf Spot.

Excess of nitrogen aggravates the disease incidence.

Leaves must be wet for 8-24 hours for infection to occur.

Survival and Spread

The pathogen survive on crop residues, stray crops, and collateral hosts.

It spread by air-borne conidia.

Management

1. Use seeds free of disease.
2. Elimination of collateral and alternative hosts.
3. Seed treatment with 10g/kg of *Pseudomonas fluorescens*, followed by a 30-minute seedling dip with 2.5 kg of products/ha dissolved in 100 liters.
4. Soaking or treating seeds with 2.0g/kg of Thiram or Captan
5. Mist Edifenphos (1ml/lit) or Mancozeb (2.0g/lit) 2 to 3 times at 10- to 15-day intervals.
6. Tricyclazole seed treatment was followed by a mancozeb + tricyclazole spray application during the tillering and late booting stages, which effectively controlled the disease.

8. Banded Sheath Blight

Causal organism: *Rhizoctonia solani* (P.S. *Thanatephorus cucumeris*)

Survival and Spread

The sclerotia of pathogen attached and survive with host tissue in soil.

Symptoms

Figure 8.8: Banded Sheath Blight

A common disease among small millets is banded sheath blight. On the lower leaf and leaf sheath, it is characterized by oval to irregular, light grey to dark brown lesions. On finger (Figure 8.8a), kodo (Figure 8.8b), proso (Figure 8.8c), foxtail (Figure 8.8d), barnyard (Figure 8.8e), and little (Figure 8.8f) millets, symptoms on millet-stalk are easily observed. The lesion then turns white in the center, with thin reddish-brown margins. Later, the spots on the leaf lamina become erratically distributed. In the presence of favorable conditions, lesions grow quickly and unite to cover a significant area of the leaf lamina and sheath. At this stage, the leaves have a banded appearance due to a series of copper color bands across them. When the condition is severe, peduncles show symptoms.

9. Sheath Rot of Kodo Millet

Causal organism: *Sarocladium oryzae*

Symptoms

Figure 8.9: Sheath Rot.

The appearance of large brown discolored patches on the stem-sheath is one of the symptoms (Figure 8.9). When kodo millet is planted after kharif rice, the disease is usually noticed. It was initially noticed Cuddalore district in Tamil Nadu's. The fungus that causes rice to become infected also causes kodo millet.

Management

1. After harvest, the diseased stubbles are eliminated.
2. Potash application at the stage of tilling is also advised.
3. Seed treatment @10g/kg of Bacillus subtilis, followed by a 30-minute seedling dip with 2.5 kg of products/ha dissolved in 100 liters.

4. Soil application of *P.fluorescens* @ of 2.5 kg/ha after 30 days of transplanting. This product should be mixed with 50 kg of FYM/Sand and then applied.
5. Depending on the severity of the disease, apply a foliar spray containing 0.2 per cent Bacillus subtilis three times every ten days starting 45 days after transplanting.
6. Sprinkle 250 g of carbendazim, 1 kg of chlorothalonil, or 1 lit of Edifenphos per hectare.

10. Foot Rot of Finger Millet

Causal organisms: *Sclerotium rolfsii*

Survival and Spread

The sclerotia of pathogen attached and survive with host tissue in soil.

Symptoms

Figure 8.10a-b: Foot Rot of Finger Millet.

Tamil Nadu, Karnataka, Odisha, and Gujarat are the states where finger millet foot rot is most common. The collar region is the site of the infection, and the affected area is only two to three inches above the ground. The base of the affected plant has a wet appearance, which subsequently turns brown and finally dark brown. The affected area of the stem also reduces concurrently (Figure 8.10a). This region is characterized by abundant white cottony fungal growth and tiny, round, white, velvety sclerotia bodies that resemble mustard seeds (Figure 8.10b). In the end, the plant dries off too rapidly as the leaves start to lose their shine, droop, and dry.

11. Bacterial Leaf Spot

Millet host: Finger millet

Causal Organisms

Finger millet: *X. eleusinae*

Survival and Spread

The pathogen survives on residues of crop in soil.

Its dissemination by rain-splash

Symptoms

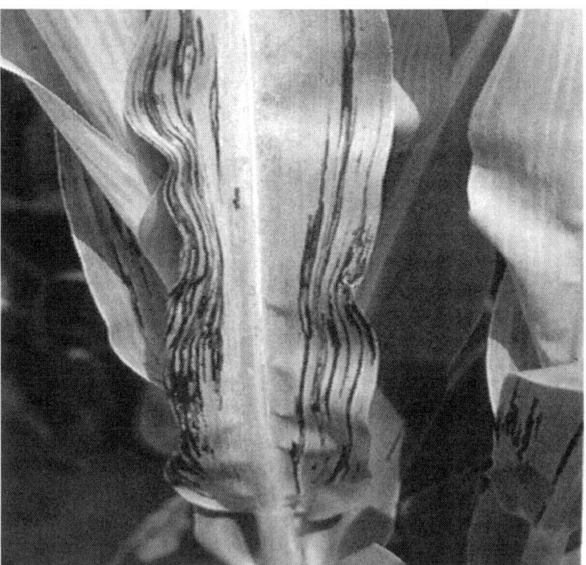

Figure 8.11: Bacterial Leaf Spot.

Finger millet leaves develop bacterial patches. On sorghum, the first signs appear as tiny, irregularly shaped or elliptical spots with a dark border and a straw color center; these spots eventually combine to form broad bands (Figure

8.11). Linear spots that spread along the veins on both leaf surfaces can be observed on finger millet. Spots are initially light yellowish brown, but they quickly turn dark brown. When the leaf reaches an advanced stage, it splits along the streak, appearing shredded. A plant could drop some or all of its leaves. The bacterium primarily damages leaves, but it can also occasionally cause distinctive streaks on the peduncle.

12. Bacterial Leaf Streak (BLS)

Causal Organisms

Finger millet: *Xanthomonas axonopodis* pv. *coracanae*

Foxtail and Proso millets: *P. avenae*

Survival and Spread

The pathogens survives on residues of crop in soil.

It disseminate by rain-splash.

Symptoms

Figure 8.11: Bacterial Leaf Streak.

Small millets, such as finger, foxtail, barnyard, and proso millets, can have streaks on their leaves. Millets first show symptoms as narrow, intraveinal, water-soaked streaks with red borders and tan centers (Figure 8.12). On the lesions, there may be an abundance of bacterial exudates. The first signs of finger millet are translucent, water-soaked, linear streaks that range in color from pale yellow to dark greenish-brown and run parallel to the lamina's midrib. In the end, the hyaline streak becomes a wide, yellowish lesion that eventually turns brown. A leaf that is severely infected turns completely brown and withers out of sight. Infections that occur in the early stages of growth cause the plants to turn yellow and wilt rapidly.

13. Maize Stripe Virus-S

Transmission: By *Peregrinus maidis* (Plant hopper)

Millets exhibit distinctive symptoms in the form of persistent chlorotic bands or stripes that develop between the veins of the diseased leaf. These bands or stripes turn yellow and continue from the base of the leaf to the tip (Figure 8.12a). The height of affected plants varies according to the stage of infection. Sometimes, the infected plant doesn't have the ability to exert its earhead/arrow (Figure 8.12b). It is also known as chlorotic stripe stunt disease.

Figure 8.12: Maize Stripe Virus (MStV-S).

Colour Plates

Figure 1.1: BLB Symptoms in Rice Leaves (*pg.* 3)

Figure 1.3: Bacterial Leaf Streak Symptoms in Rice Leaves (*pg.* 6)

Figure 1.5: Sheath Rot Symptoms in Rice Panicles (*pg.* 8)

Figure 1.7: Symptoms of Stem Rot of Rice (*pg.* 11)

Figure 1.8: Foot Rot or Bakane Disease of Rice (*pg.* 12)

Figure 1.9: Blast of Rice (*pg.* 15)

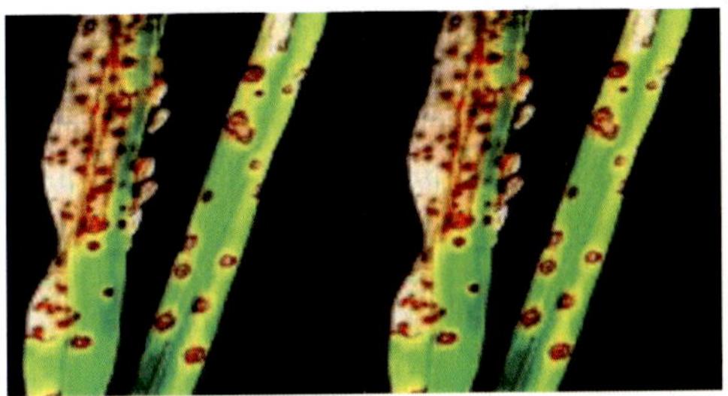

Figure 1.10: A. Irregular lesions on the rice sheaths during tillering (*pg.* 16)

Figure 1.11: A. Irregular lesions on the rice sheaths during tillering (*pg.* 18)

Figure 1.12: False Smut Symptom
(*pg.* 20)

Figure 1.14: Udbatta Disease of Rice
(*pg.* 22)

Figure 2.1: Black Rust of Wheat
(*pg.* 27)

Figure 2.3: Brown/Leaf Rust of Wheat
(Puccinia recondite). (*pg.* 29)

Figure 2.4: Yellow/Stripe Rust of Wheat
(Puccinia striiformis). (*pg.* 31)

Figure 2.5: Loose Smut of Wheat
(Ustilago nuda var. tritici). (*pg.* 33)

Figure 2.7: Flag Smut of Wheat
(*pg.* 35)

Figure 2.7: Flag Smut of Wheat
(*pg.* 37)

Figure 2.11: Powdery Mildew of Wheat. (*pg.* 39)

Figure 2.13: Leaf Blight of Wheat. (*pg.* 41)

Figure 2.13: Tundu Disease of Wheat (*pg.* 43)

Figure 3.1: Net Blotch (*pg.* 50)

Figure 3.3 Powdery Mildew of Wheat. (*pg.* 39)

Figure 3.5: Yellow (Stripe) Rust. (*pg.* 56)

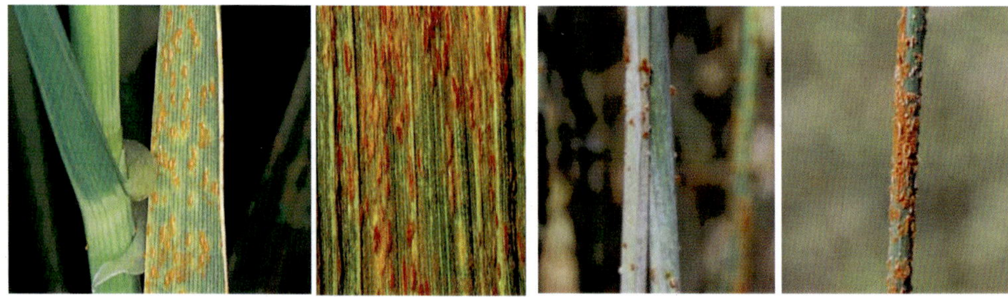

Figure 3.7: Brown Rust *(Source: cropscience. bayer.co.uk).* (*pg.* 58)

Figure 3.9: Stem Rust (*pg.* 60)

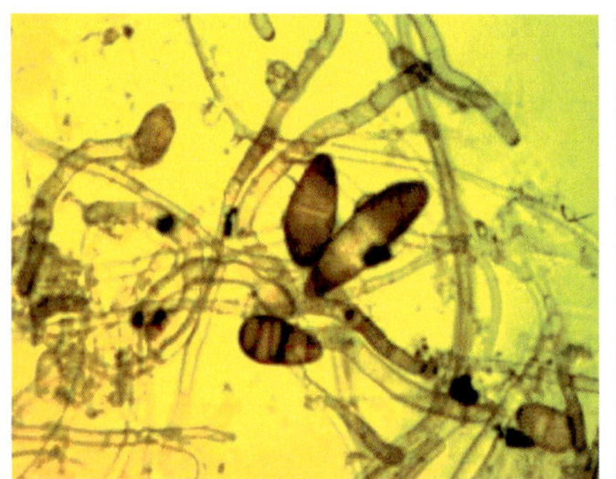

Figure 3.11: Spot Blotch (*pg.* 62)

Figure 3.12: Spot Blotch (*pg.* 63)

Figure 3.15: Covered Smut (*pg.* 66)

Figure 3.17: Molya Disease (*pg.* 69)

Figure 4.1: Crown Rust of Oat (*pg.* 76)

Figure 4.2: Stem Rust of Oat (*pg.* 76)

Figure 4.3: A. Septoria Blight; B. Victoria Blight (*pg.* 78)

Figure 4.4: Bacterial Stripe of Oat Blight (*pg.* 78)

Figure 4.5. A. Culm Rot; B. Covered Smut (*pg.* 80)

Figure 4.6: Powdery Mildew of Wheat. (*pg.* 81)

Figure 4.7: A. Barley Yellow Dwarf; B. Bacterial Leaf Spot. (*pg.* 83)

Figure 5.1. Nitrogen Deficiency (*pg.* 89)

Figure 5.2: Phosphorous Deficiency (*pg*. 90)

Figure 5.3: Zinc Deficiency (*pg*. 90)

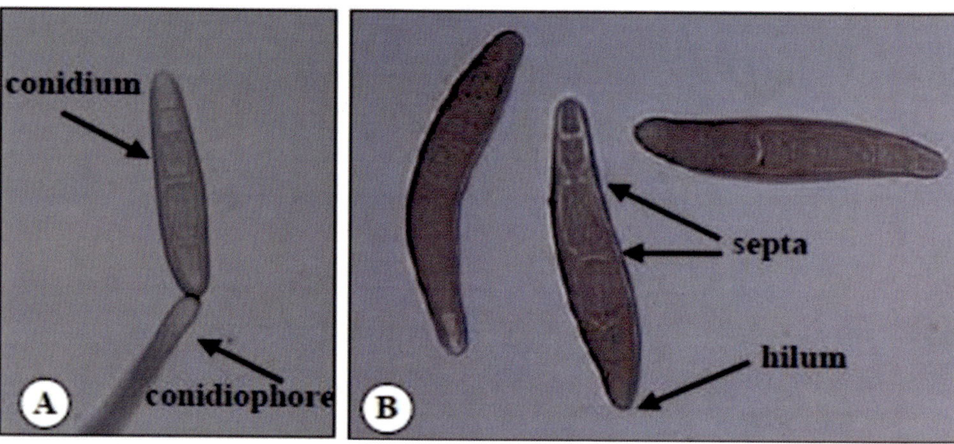

Figure 5.4: A. Conidium on conidiophore; B. Characteristics of Conidium. (*pg*. 91)

Figure 5.5: Turcicum Leaf Blight (TLB)/Northern Corn Leaf Blight (NCLB). (*pg*. 92)

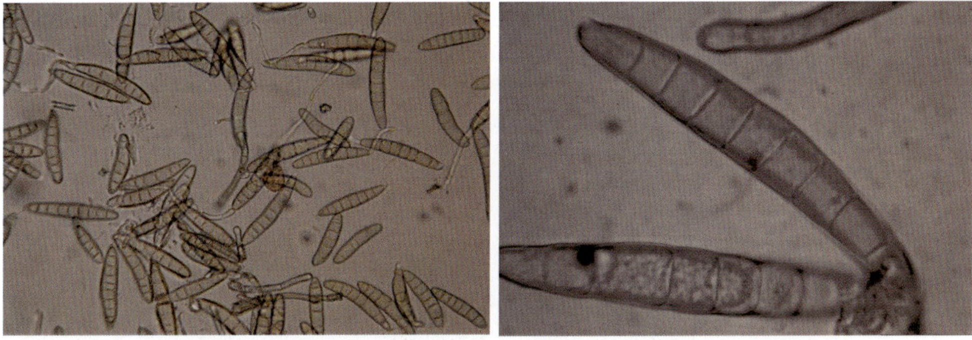

Figure 5.7: Conidia of Bipolaris maydis. (NCLB). (*pg.* 94)

Figure 5.8: Symptoms of Maydis Leaf Blight (MLB)/ (*pg.* 95)

5.10: Conidia of *Curvularia lunata.* (*pg.* 97)

Figure 5.11: A. Curvularia Leaf Spot Symptoms on a Corn Leaf; B. Curvularia Leaf Spot Lesions. **(*pg.* 98)**

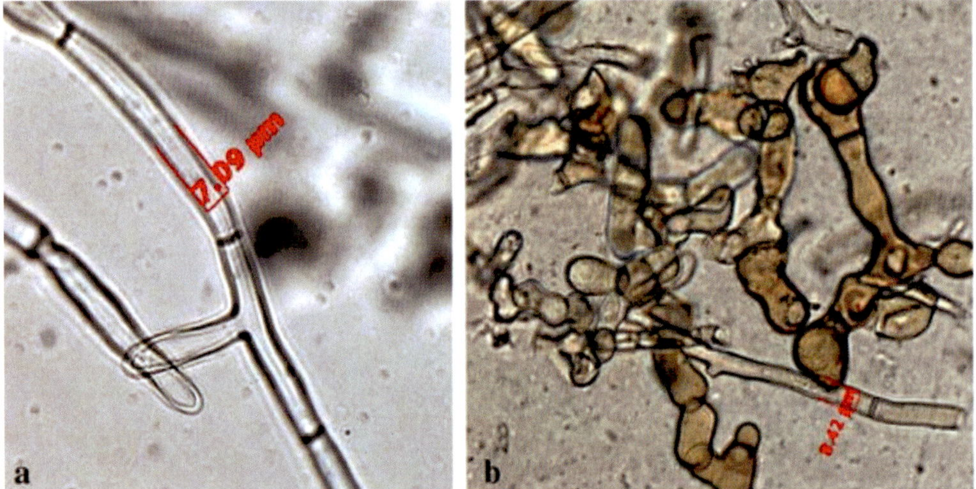

Figure 5.13: Morphological Characteristics of *R. solani*. **(*pg.* 100)**

Figure 5.14: Banded Leaf and Sheath Blight on Lower Maize Leaves and Large Brown Sclerotia Produced by *Rhizoctonia solani* on Infected Leaves and Stem **(*pg.* 101)**

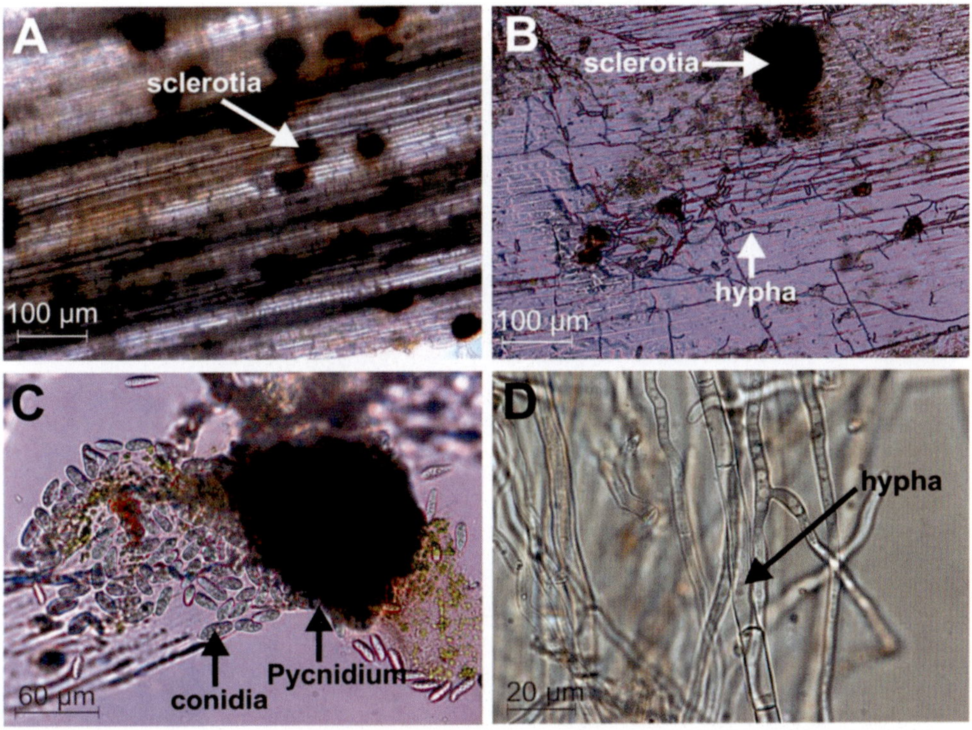

Figure 5.16: Characteristics of M. phaseolina: (A) Sclerotia; (B) Hyphal network; (C) conidia from the pycnidium; (D) Aerial hyphae. (*pg.* 104)

Figure 5.17: Shredding of Stalk. (*pg.* 105)

Figure 5.19: Pith Shredding Indicative of Fusarium Stalk Rot. (*pg.* 107)

Figure 6.1: A. White downy growth on lower surface of leaf; B. Leaf shredding along the white streaks. (*pg.* 114)

Figure 6.3: Leaf Anthracnose. (*pg.* 117)

Figure 6.5: Ergot of Sorghum (Honeydew stage). (*pg.* 120)

Figure 6.7: Kernel Smut of Sorghum. (*pg.* 122)

Figure 6.9: Long Smut of Sorghum. (*pg.* 124)

Figure 6.10: Head Smut of Sorghum. (*pg.* 125)

Figure 6.11: Grain Mould of Sorghum. (*pg.* 126)

Figure 6.12: Charcoal Rot (*pg.* 127)

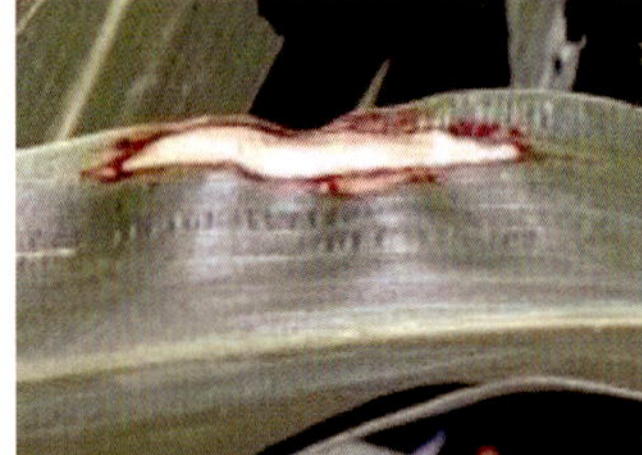

Figure 6.13: Leaf Blight. (*pg.* 129)

Figure 6.14: Grey Leaf Spot (*pg.* 131)

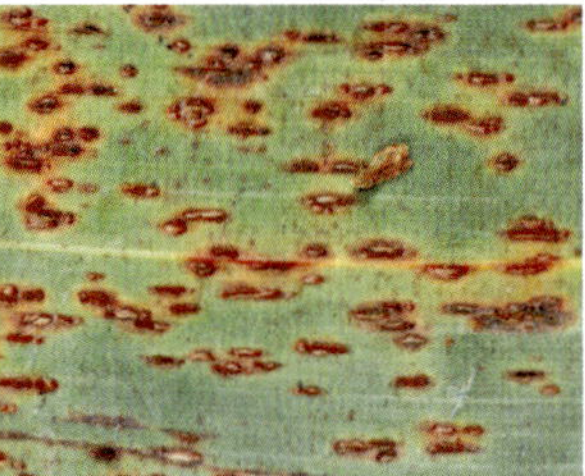

Figure 6.15: Sorghum Rust. (*pg.* 132)

Figure 6.16: Stalk Rot. (*pg.* 133)

Figure 6.17: Striga asiatica. (*pg.* 134)

Figure 7.1: Ergot Symptoms in Pearl Millet (*pg.* 146)

Figure 7.3: Downy Mildew Symptoms in Pearl Millet **(*pg.* 148)**

Figure 7.5: Smut Symptoms in Pearl Millet **(*pg.* 151)**

Figure 7.7: Blast Symptoms in Pearl Millet **(*pg.* 153)**

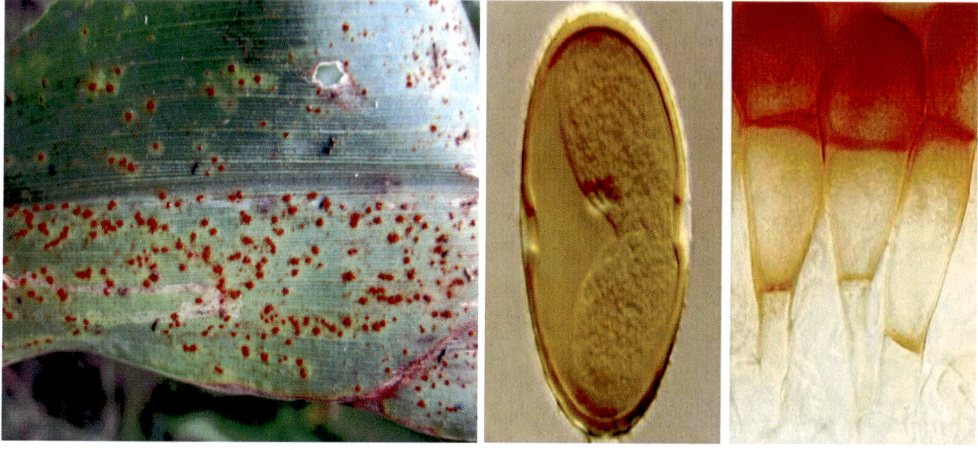

Figure 7.9: A. Rust symptoms in pearl millet (CABI, 2023) with B. Uredospore, and C. Teliospore (de Carvalho et al., 2006). **(*pg.* 155)**

Figure 8.1: Grain and Head Smut of Small Millets viz., (a) Finger millets; (b) Barnyard millets; (c) Little millets; (d) Foxtail millets. (*pg*. 165)

Figure 8.2: Udbatta Disease of Millets. (*pg*. 166)

Figure 8.3: Downy Mildew of Small Millet. (*pg*. 168)

Figure 8.4: Finger Millet Blast. (*pg*. 170)

14. Maize Mosaic Virus

Transmission: By *Peregrinus maidis* (Plant hopper)

Symptoms

Maize mosaic virus (MMV) infection is characterized by the appearance of thin, discontinuous, chlorotic streaks between the leaf veins and plant stunting (Figure 8.13a). As the condition grows more severe, the lesions get necrotic. The early-infected plant ultimately dies without an earhead emerging. Later-stage infected plants might have grainless or grain-filled earheads. When more than one virus is present, the symptoms can differ, making identification in the field more difficult.

Figure 8.13: a. Maize Mosaic (MMV-S); b. Red Stripe (SRSD).

15. Red Stripe

Transmission: Aphid and sap transmitted (mechanically).

Symptoms

The disease is recognized by mosaic-like systemic symptoms, followed by necrotic red stripe, and red leaves affected by temperature (Figure 8.13b). Sugarcane mosaic virus-Jg (SCMV-Jg) is belongs to the group potyvirus that infects sorghum. Subsequently, the virus was identified as the Indian isolate of the sorghum red stripe virus (SRSV-Ind). Potyvirus SCMV was related to SRSV.

References

Extension Folder-1 compiled by: Sangappa, Amasiddha B., Shiwesh Kumar and Vilas A. Tonapi, ICAR-IIMR, Rajendranagar, Hyderabad (Website: www. Millets.res.in)

http: //www.agritech.tnau.ac.in/

https: //plantix.net/en/library/plant-diseases

https: //www.millets.res.in/books/DISEASES_OF_MILLETS.pdf

http: //cattheni.edu.in/wp-content/uploads/2018/09/Diseases-of-minor-millets.pdf

Chapter 9

Diseases of Chickpea Crop

Puskar Shukla[1]*, S.K. Biswas[1], Mukul Kumar[2] and Anoop Kumar[3]

[1]Department of Plant Pathology, Chandra Shekhar Azad University of Agriculture and Technology, Kanpur – 208 002, Uttar Pradesh
[2]Subjest Matter Specialist, Krishi Vigyan Kendra, Jamui – 811 305, Bihar
[3]Department of Plant Pathology, Sam Higginbottom University of Agriculture Technology and Science, Prayagraj – 211 007, Uttar Pradesh
**e-mail: puskarshukla8461@gmail.com*

Introduction

Chickpea is the most important winter season pulses crop, and it is botanically known as *Cicer arietinum* L. mainly two type chickpea occurs on the basis of size one is kabuli chana (big size) and another is desi chana (small size). It have diploid (2n) Chromosome number is 16. Chickpeas grown in semi-arid parts of the world by resource-poor farmers on marginal soils. In addition to boosting soil fertility, it is essential for meeting the population's protein needs. A highly nutritious pulse crop, chickpeas have the following nutritional values: 18–22 per cent protein, 61-62 per cent carbohydrate, 4.5 per cent fat, 280 mg calcium, 12.3 mg iron, 301 mg phosphorus, and energy 396 kcal/100g. Around the world, chickpeas, or chana as they are known in Hindi, are utilized as both an edible seed and to make flour. It is also eaten raw, carbonized, roasted as a snack, or in broth, among other ways. The leaf exudates' acid can be utilized as vinegar or applied medicinally, and its starch can be used as textile sizing to give cotton, silk, and wool fabrics a subtle sheen. Evidence from archaeology and botany suggests that "hunter-gatherer societies" utilized it for sustenance and food even before 10,000 B.C. (Ladizinsky 1975). The chickpea originated in southeast Turkey and spread throughout the Mediterranean, India, and Ethiopia after being domesticated in the Middle East. (Ladizinsky 1975; Van der Maesan 1987). Despite being grown on over 13.98 million hectares of land in 59 countries, fewer than one ton of chickpeas are produced per hectare (FAO 2016). Chickpea production potential has been hindered by a number of biotic and abiotic stressors. Furthermore, the frequency of disease, pests (pod borer), and droughts has risen due to recent climate change. (Fusarium wilt, Ascochyta blight, dry root rot) *etc.* Due to a lack of knowledge about the genetics of the qualities, breeding efforts were not as successful in the past because of the intricacy of these pressures. 2016 has been proclaimed the International Year of Pulses by the Food and Agriculture Organization (FAO) in an effort to raise awareness of the economic significance of pulses in food and nutritional security. As the most significant pulse crop, chickpeas face a difficult problem in the scientific community as they attempt to produce climate-resilient types with sustainable production to fulfill the rising demand to alleviate hunger. Before 2005, chickpeas were regarded as an orphan crop because they lacked the genetic resources necessary to adopt genomics-assisted breeding, which augments traditional breeding strategies for agricultural development (Varshney *et al.,* 2005). The chickpea is currently regarded as a crop with abundant genetic resources due to several national and international efforts (Thudi *et al.,* 2014; Varshney *et al.,* 2016). This book offers current knowledge on a variety of topics related to chickpeas, such as their economic significance, biology, the creation of genetic resources, and their draft genomes and their potential for improving chickpeas.

Sl.No.	*Name of Disease*	*Causal Organism*
A. Fungal Diseases		
1.	Chickpea wilt	*Fusarium oxysporum* f.sp. *ciceri*
2.	Acrophialophora wilt	*Acrophialophora fusispora*
3.	Alternaria blight	*Alternaria alternate*
4.	Aphanomyces root rot	*Aphanomyces euteiches*
5	Ascochyta blight	*Ascochyta rabiei*
6.	Black root rot	*Fusarium solani*
7.	Botrytis gray mould	*Botrytis cinerea*
8.	Collar rot	*Sclerotium rolfsii*
9.	Downey mildew	*Peronospora spp.*
10.	Dry root rot	*Macrophomina phaseolina*
11.	Rust	*Uromyces ciceris-arietini*
B. Bacterial Diseases		
11.	Bacterial blight	*Xanthomonas campestris* pv.*cassaie*
12.	Bacterial leaf spot	*Robbsia andropogonis*

Sl.No.	Name of Disease	Causal Organism
13.	Bacterial Wilt	*Xanthomonas campestris* (Pam.) Dowson
C. Viral Diseases		
14.	Bushy stunt	Chickpea bushy stunt virus
15.	Distortion mosaic	Chickpea distortion mosaic virus
16.	Yellowing	Pea enation mosaic virus
17.	Proliferation	Cucumber mosaic virus

Major Diseases of Chickpea

1. Wilt of Chickpea

Economic Importance and Occurrence

Wilt is the most important disease of chickpea that is caused by fungus, it is soil and seed borne nature that can cause 10 – 15 per cent yield losses depend upon severity of disease.

Causal Organism

Disease is caused by *Fusarium oxysporum* f. sp. *ciceri* which belongs the kingdom fungi, Phylum- Ascomycotina, Class- Sordariomycetes, Order- Hypocreales, and Family- Nectriacea. Macro-conidia are borne on the branched conidiophore while micro-conidia are borne simple short conidiophore. Chlamydospores occur singly, in pairs, or rarely in chains. They have rough or smooth walls and can be intercalary or terminal.

Symptoms

Figure 9.1: Disease Symptoms of Chickpea Wilt in Infected Field.

Three to five weeks after planting, a serious situation sets in when the entire seedling collapses and lies flat on the ground. The stems above and below the collar region shrink unevenly when a seedling is uprooted. Black discoloration of interior tissue is clearly visible from the collar area and down word when affected plant is spilt vertically.

Disease Cycle and Epidemiology

If there is no host, the pathogen can live for up to six years in soil and plant detritus. In light sandy soil as opposed to heavy clay soil, disease intensity is often higher. Chickpea wilt is caused by soil temperatures between 24 and 26 °C and soil moisture levels exceeding 25 per cent within the soil's water-holding capacity.The Pathogen is soil as well as seed borne in nature and survive through chlamydospore in seed and in dead plant debris in the soil. Primary infection take place through chamydospore or mycelium and fungus may survive more than 6 years in the absence of chickpea. Optimum temperature for the fungus infection around 25 degree Celsius.

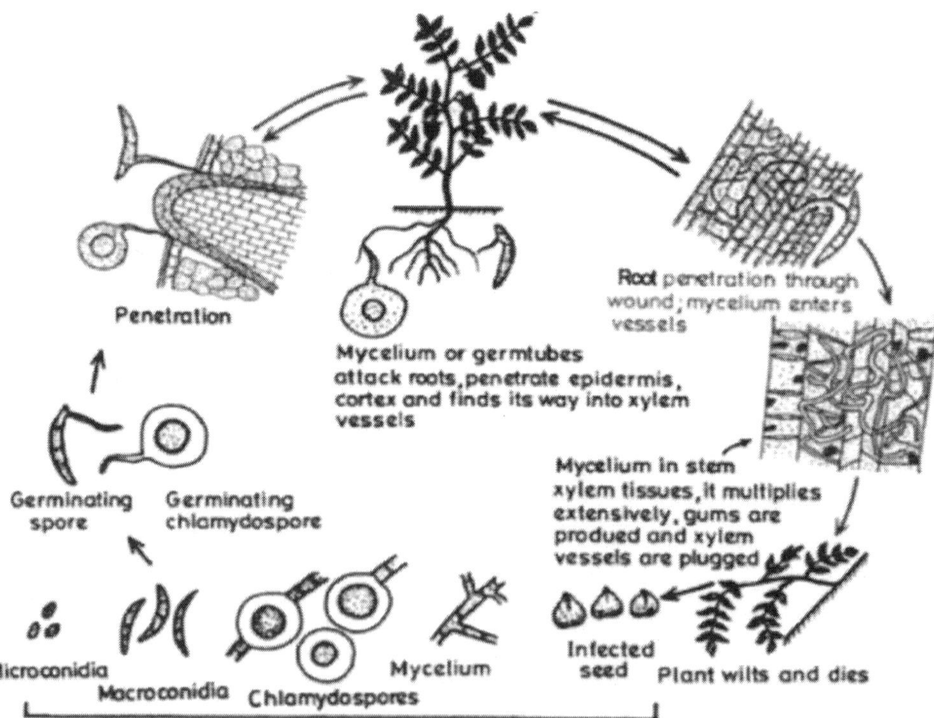

Figure 9.2: Disease Cycle of Chickpea Wilt (*Source*: Jalali and Chand, 1992).

Management

1. Deep summer ploughing.
2. Deep sowing approximately 8-10 cm suppress disease severity.
3. Follow crop rotation measures continuously.
4. Always use disease free and healthy seeds.
5. Steer clear of sowing during extremely hot weather.
6. Follow 6-year crop rotations with sorghum.
7. Grow resistant variety like Awrodhi, Gaurav, Udai, BG-261 *etc.*
8. Before sowing the Seed treated with Bio agent like *T. viride* @ 4gram per kg seed or *Pseudomonas fluorescens* @ 10gram per kg of seed.
9. Spot drenching with Carbendazim fungicide @1g/lit or soil application bio agent *Pseudomonas fluorescens* or *T. viride* 2.5 kg/ha with 50 kg FYM.
10. Seed treatment with Thiram + Carbandizm @ 1g+2g per kg of seeds.

2. Collar Rot of Chickpea

One significant soil-borne disease, collar rot, is pervasive throughout many chickpea-growing regions. The seedling stage is when losses are the greatest, particularly in paddy chickpea based cropping systems. Between 60 and 90 per cent of chickpea plants in India die from the collar rot disease of chickpea.

Symptoms

Figure 9.3: Collar Rot in Chickpea (*Source*: ICAR-IIPR).

It appears early on, up to six weeks after seeding. The presence of dried plants in the field that have a minor yellowing of the leaf before they die is a sign of the disease and Seedlings start to become green. The junction between the stem and the root softens somewhat, compresses, and decomposes. Affected areas become brown and white Sclerotia, or black specks that resemble mustard seeds, can be seen developing on the white-infected plant portions.

Causal organism: *Sclerotium rolfsii* and *Athelia rolfsii.*

Disease Cycle and Epidemiology

A facultative parasite is the organism causing the disease Without chickpea sclerotia, masses of hyphe can live for up to six years in soil as a secondary host or on plant waste. High soil moisture content, organic matter that hasn't decomposed at the soil's surface, low soil pH, and temperatures between 28 and 31 °C are all conducive to the development of disease. The disease mostly affects seedlings when paddy chickpea-based farming methods are utilized.

Management

1. Deep plowing throughout the summer.
2. Steer clear of excessive wetness during sowing.
3. Seedlings need to be shielded from too much dampness.
4. Eliminate weeds and leftover crop residue both before and after harvest.
5. The field should be cleared of any undecomposed material before the ground is prepared.
6. Apply a combination of carbendazim (1 g per kilogram of seed) to the seeds.

3. Ascochyta Blight of Chickpea

The most common and dangerous disease in most regions of the world where chickpeas are grown is Ascochyta blight. Crop growth season weather is characterized by cool, overcast, and humid conditions.

Symptoms

A. rabiei infects the leaves, petioles, stems, pods, and seeds of plants that are above ground. The first indications of the illness are water-soaked lesions that turn brown or dark brown on the leaves and stems. Early signs include oval-shaped brown lesions on the stem and small, circular necrotic lesions that are apparent on the leaves. Later in the sickness, when the lesions have girdled the stem, the portion above the girdle may fall off. During the reproductive stage, as the disease worsens, asexual entities called pycnidia cause round lesions with concentric rings to form on leaves, stems, and pods. The tainted pod becomes blighted, which often results in the production of shriveled and sick seeds.

Figure 9.4: Symptoms of Ascochyta Blight (*Source*: graincentral.com).

Causal Organism

The Ascochyta blight causative agent is *A. rabiei,* can infect all areas of the plant, causing necrosis, tissue collapse, and ultimately the loss of organs or perhaps the entire plant. This might result in overall or significant yield reductions. *Anthrax rabiei* (Pass.) The anamorph of the fungus during its imperfect or asexual stage is known as labr. The anamorph is also known as *Phoma rabiei* (Pass.) or *Phyllosticta rabiei* (Pass.). It also possesses a teleomorph, commonly known as a sexual reproductive stage, called *Didymella rabiei* (Kovachevski) v. Arx (also known as *Mycosphaerella rabiei* Kovachevski). Class Dothideomycetes, order Pleosporales, family Didymellaceae, and phylum Ascomycota are the major taxonomic classifications for this fungus.

Disease Cycle and Epidemiology

The disease is both inwardly and externally transmitted via seeds in nature. The disease may persist in contaminated seed, infected plant waste, and on other host plants where secondary infections occur via conidia and ascospores that can be spread by rain splash, wind, water movement, and contaminated machinery. 20 to 25°C temperature, high relative humidity, a 10- to 17 hr period of leaf wetness, and genotypes with tall and fern-shaped leaves are all favorable conditions for the development of ascochyta blight.

Management

Careful IDM measures can prevent crop failures because of:

1. Using cultivars that are tolerant or resistant.
2. Planting chickpea in regions where the establishment of the Ascochyta blight disease is least likely.

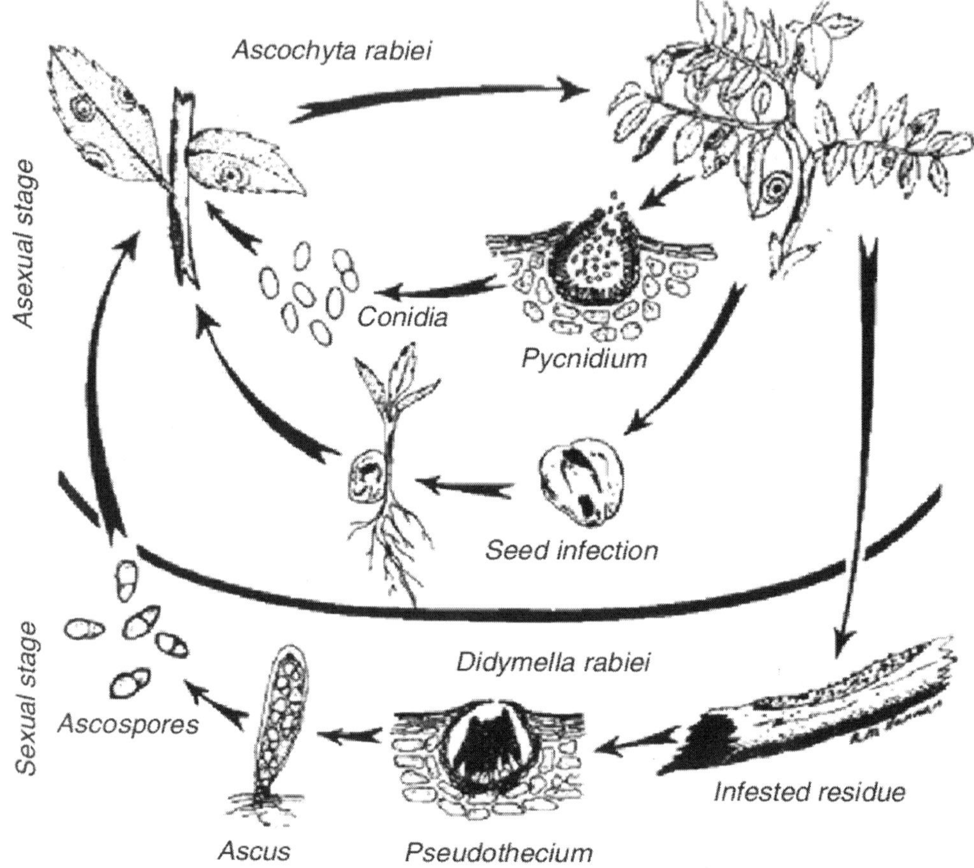

Figure 9.5: Disease Cycle of Ascochyta Blight.

3. Using healthy seeds devoid of *A. rabiei.*
4. Rotating crops that are not hosts is essential to limiting the disease.
5. Appropriately eliminating diseased stubble can also lessen the amount of fungal inoculum.
6. T. harzianum at a rate of 4 grams per kilogram of seed.
7. Carbendazim + Mancozeb 2 g/Kg seed treatment.

4. Botrytis Grey Mold

Economic Importance and Occurrence

The fungus Botrytis cinerea is what causes botrytis grey mould (BGM) in chickpeas. *B cinerea* is a major disease that affects fruit, including grapes, strawberries, and apples, as well as decorative plants produced in greenhouses and pulse crops, especially lentils. Flowers are particularly susceptible to BGM infection; however, grasses and grains are unaffected by *B. cinerea.*

Symptoms

If groups of plants are infected, the first sign of Botrytis Grey Mold infection in a crop is frequently drooping of the terminal branches, which may manifest as yellow spots in the crop. The telltale sign is a grey fuzz that appears on flowers, pods, stems, dead leaves, and petioles when there is excessive humidity. Although lesions can appear anywhere on the stem, they are typically initially noticed on the lower portion of the stem, frequently beginning in the leaf axils (Figure 6). Typically, infected seeds are smaller than normal and frequently have white to grey fungal growth on them. Spore clouds are seen when a heavily BGM-infected canopy is opened (avoid breathing them). The fuzz is hardly noticeable in dry weather, but it reappears when wet weather returns. Small *B. cinerea* resting bodies (sclerotes), which can produce spores on their surface, may form on infected dead tissue. They are dark brown or black in color. The stem lesions brought on by Sclerotinia sclerotiorum (at and above ground level) and Sclerotinia minor (at ground level) can be mistaken for those brought on by botrytis grey mold, although neither of these diseases produces the gray fuzz that is distinctive of botrytis grey mold. Additionally, lesions caused by sclerotinia typically remain white and are covered in a thick, cottony fungal growth in which irregularly shaped black sclerotes form.

Figure 9.6: Damaging Symptoms of Botrytis Grey Mold.

Epidemiology

Diffuse white fungal growth is produced by B. cinerea, however it eventually turns grey as a result of the massive amount of spores that are produced and carried in clusters at the tips of black stalks. On a chickpea stem, a single 2 cm long lesion can release more than 10 million spores. As a result, under favorable meteorological circumstances, B. cinerea can grow quickly. The spores may travel over long distances by air, and if they land on chickpea plants, they will latent until the right circumstances arise for spore germination. Both infection and germination require free moisture. Under ideal circumstances, lesions and the gray "fuzz" become visible 5-7 days after infection. Frequent rainfall events and moderate temperatures (20–25°C) are ideal for B. cinerea. Until the average daily temperature (ADT) reaches 15°C or more, it is not considered dangerous. B. cinerea thrives in environments with high relative humidity and quick leaf loss in the canopy, which are produced by early canopy closure, extended plant wetness, and cloudy weather. B. cinerea can live as sclerotes, on and in infected seeds, infected stubble, on other hosts, and in dead plant tissue. Although the proportional significance of these in Australia is uncertain, recent studies conducted in Victoria have shown that B. cinerea may live in the field for up to 18 months on infected stubble. According to another studies from Western Australia, B. cinerea sclerotes become less viable in hot conditions, which makes it impossible for them to survive the summer.

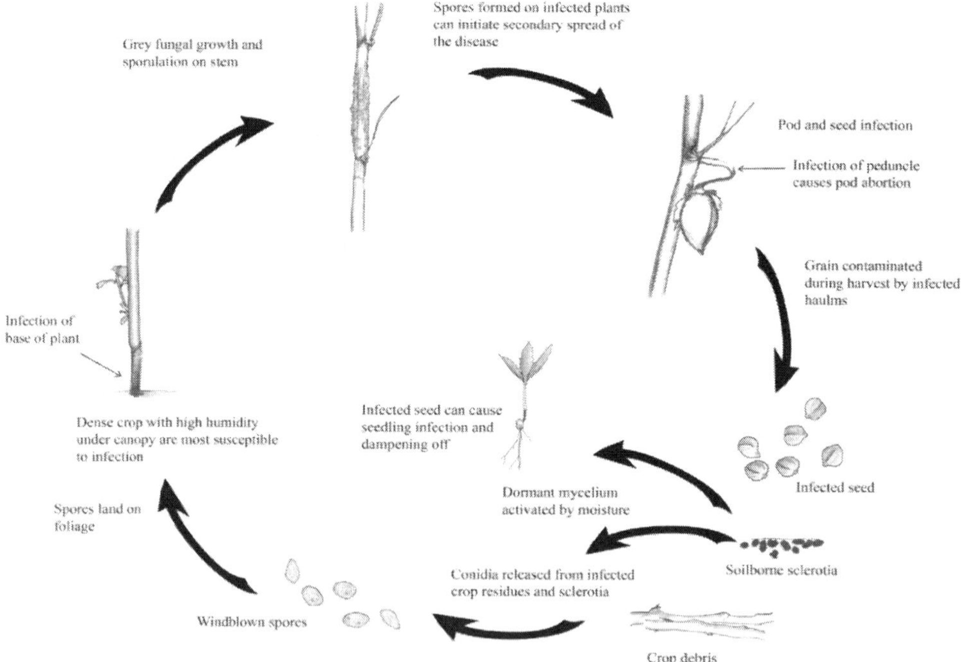

Figure 9.7: Disease Cycle of Botrytis Grey Mold.

Management

Stubble management: As long as contaminated stubble is present on the soil's surface, the pathogen is probably going to be able to survive and spread. The capacity of B. cinerea to generate airborne spores is eliminated by stubble burial, slowing the rate at which the illness spreads.

Planting: Chickpea plants growing spontaneously in or close to the growing area where BGM was a major issue and needed to be controlled by cultivation or pesticide application

Source and handling of seed: Purchase seed from a commercial provider or one whose BGM content is known to be minimal. All seed has to be adequately treated with a licensed fungicide seed dressing, regardless of the source. Seed dressings using thiam-based fungicides are efficient in lowering BGM from contaminated seed, but they do not completely eradicate it.

Row distance: Wider rows of plants promote better air circulation throughout the crop and lower canopy humidity.

5. Chickpea Stunt Disease

Multiple viruses are the cause of chickpea stunt disease. In India, chickpea stunt disease is mostly linked to luteovirus and chickpea clototic dwarf virus giminivirus.

Symptoms

Sorting of internode small and brittle leaves with marginal browning and thickening of leafletes and stems are characteristics symptoms of this disease. Yellowing may note in infected plants and some cultivar show leaf, petioles, and stem reddening. Most of the infected plant are stunted, which have no flower and pod.

Figure 9.8: Damaging Symptoms of Chickpea Stunt Disease.

Transmission

Luteovirus and giminivirus are not transmitted by mechanically and they are not seed borne, they transmitted by aphids and graft transmissible.

Management

1. Crop should be sown early.
2. Use resistant varieties like ICCC- 10, GG- 669, SAK- I9561 have tolerant to this disease.
3. Spray metasystox @ 1ml/lit water 2 to 3 time at 15 days interval.

References

Amule, R., Gupta, O. and Mishra, M. (2014). Techniques for screening of chickpea genotypes against collar rot, its management through host plant resistance and fungicides. *Legume Research-An International Journal,* **37(1)**: 110-114.

Bekele, B., Damate, T., Bedassa, T., Fikre, A., Tadesse, N. and Mihirate, T. (2021). Complex Viral Diseases Threatening Lentil and Chickpea Production in Ethiopia: *Pest Managment Journal of Ethiopia,* **24**(01): 21-40.

Chongo, G., Buchwaldt, L., Gossen, B. D., Lafond, G. P., May, W. E., Johnson, E. N. and Hogg, T. (2003). Foliar fungicides to manage ascochyta blight [*Ascochyta rabiei*] of chickpea in Canada. *Canadian Journal of Plant Pathology,* **25**(2): 135-142.

Dhaliwal, G. S. and Arora, R. (2001). Role of phytochemicals in integrated pest management. *Phytochemical biopesticides,* 97-117.

Dubey, S. C. (2016). Race profiling, genetic diversity, diagnostics and management of *Fusarium oxysporum* f. sp. *ciceris* causing wilt in chickpea. *Indian Phytopathology,* **69**(3): 210-217.

Duzdemir, O., Selvi, B., Yanar, Y. and Yildirimi, A. (2014). Sources of resistance in chickpea (*Cicer arietinum* L.) land races against *Ascochyta rabiei* causal agent of ascochyta blight disease. *Pak. J. Bot,* **46**(4): 1479-1483.

Gan, Y. T., Siddique, K. H. M., MacLeod, W. J. and Jayakumar, P. (2006). Management options for minimizing the damage by ascochyta blight (*Ascochyta rabiei*) in chickpea (*Cicer arietinum* L.). *Field Crops Research,* **97**(2): 121-134.

Gorfu, D. and Hiskias, Y. (2001). Yield losses of crops due to plant diseases in Ethiopia. *Pest Managment Journal of Ethiopia,* **5:** 55-67

Jiménez-Díaz, R. M., Castillo, P., del Mar Jiménez-Gasco, M., Landa, B. B. and Navas-Cortés, J. A. (2015). Fusarium wilt of chickpeas: Biology, ecology and management. *Crop Protection,* **73**: 16-27.

Motagi, B. N., Rao, M. L. and Mathad, A. (2020). Integrated and sustainable management of fungal diseases of chickpea: Current status and challenges. *Management of Fungal Pathogens in Pulses: Current Status and Future Challenges,* **73-91**.

Sahni, S. and Prasad, B. D. (2020). Management of collar rot disease using vermicompost and a PGPR strain Pseudomonas sp. and their effect on defense-related enzymes in chickpea. *Indian Phytopathology,* **73**(2): 301-311.

Sankar, P. M., Vanitha, S., Kamalakannan, A., Raju, P. A. and Jeyakumar, P. (2019). Biological Management of Fusarium Wilt in Chickpea (*Cicer arietinum* L.) Caused by *Fusarium oxysporum* f. sp. *ciceris. Legume Research-An International Journal,* **42**(6): 838-843.

Shahid, A. A., Husnain, T. and Riazuddin, S. (2008). Ascochyta blight of chickpea: Production of phytotoxins and disease management. *Biotechnology advances,* **26**(6): 511-515.

Tewari, A. K. and Mukhopadhyay, A. N. (2003). Management of chickpea root rot and collar rot by integration of biological and chemical seed treatment. *Indian Phytopathology,* **56**(1): 39-42.

Chapter 10

Diseases of Pigeonpea Crop

Smrati Katiyar

Department of Plant Pathology, CSJM University, Kanpur, Uttar Pradesh
Email: smratikatiyar4932@gmail.com

Introduction

A crucial pulse crop grown for commercial purposes in India, the pigeon pea (*Cajanus cajan*) belongs to the *Fabaceae* family. The term "katjang" in Malay, which refers to a legume, is whence the species "cajan" gets its name. In other language, it is also known as arhar (Hindi), tuver (Gujrati), tori (Konkani), tur (Marathi), dinger (Punjabi), kandi/togari/tuvaramu (Telugu) andrahar (Nepali). It is considered as "Poor man's meat" and significant sources of proteins, vitamins, minerals and essential amino acids that significantly contribute to global nutritional security (Singh *et al.*, 2015. The seeds of the pigeon pea are native to India and contain a high content of protein (22.3 per cent), fat (1.7 per cent), carbohydrates (576.6 per cent), fiber (1.5 per cent), calcium (73 mg/100 g), phosphorus (304 mg/100 g), iron (5.8 mg/100 g), moisture (13.4 per cent), and kalorific value (335 Kcal/100 g). Pigeon peas are rich in nutrients, but they also have medicinal qualities because they contain a lot of polyphenols and flavonoids, which have been shown to prevent and treat a variety of illnesses in people, including diabetes, menstrual problems, abdominal pain, bronchitis, pneumonia, cough, and diarrhea. Just behind chickpea, pigeonpea is the second-most important pulse crop in India. India leads the globe in pigeon pea production, representing 90 per cent of global production (Mespogu *et al.*, 2012). Due to its ability to adapt to mixed or intercropping systems and ability to grow in arid climate, this common Kharif pulse crop is highly valued by Indian farmers. In addition, it increases soil fertility by fixing nitrogen and recycling nutrients from fallen leaves. Depending on the kind, it is 2 meters tall, highly branching shrub with trifoliate leaves, a woody base, and slender stems. Plant leaves are alternating and spirally organized, with oblong to elliptical shaped leaflets. Flowers on the plant are typically yellow, but they can occasionally include purple or red striations. The seed pods are flat, 5-9 cm long, and either straight or sickle-shaped. Two to nine seeds, each of which might be any color from the list below, could fit inside each pod.i.e., brown, yellow, purple, black, white, cream, or mottled with any combination of these hues. In many sustainable farming methods, it is also grown as intercrops, cover crops, *etc.*The large number of diseases that attack the pigeon pea crop is perhaps the main constraints to imposing significant yield decline and overall quality deterioration. The output of pigeon peas is severely reduced by a number of biotic and abiotic diseases. Pigeon pea crops have been found to be affected by over 50 diseases that are caused by bacteria, viruses, nematodes, fungus, and other organisms. Wilt, Sterility mosaic disease, Alternaria blight, Phytophthora blight, Powdery mildew, Bacterial lef spot, and Pearly root (pigeon pea cyst nematode) are among the major pigeon pea diseases. Young plants, typically five to six weeks old, are most commonly affected by seedling wilting, whereas adult plants exhibit the disease's usual early symptoms, such as premature yellowing of the leaves. This chapter covers the occurrence, economic significance, causes, symptomatology, etiology, spread, and management practices of major diseases that affect pigeon peas. These include sterility mosaic virus, yellow vein mosaic, dry root rot, powdery mildew, stem blight, bacterial leaf spot and stem canker, seedling blight, anthracnose, leaf spot, halo blight, and stem rot *etc.*

Diseases classified as:

A. Fungal Diseases

1. *Fusarium oxysporum* f.sp. *udum* causes pigeon pea wilt.
2. Pigeon pea stem blight (*Phytophthora drechsleri* f.sp. *cajani*).
3. Root rot: *Rhizoctonia bataticola.*
4. Powdery mildew: *Levellula taurica.*
5. Alternaria blight: *Alternaria alternate.*

B. Bacterial Diseases

1. *Xanthomonas compestris* pv. *cajani* is the causative organism of stem canker and leaf spot.

C. Viral Diseases

1. Sterility mosaic virus
2. Yellow vein mosaic virus

D. Nematode Diseases

1. *Heterodera cajani,* the pearly root (pigeon pea cyst nematode).

Fungal Diseases

1. Pigeon Pea Wilt

Occurrence and Economic Importance

Wilt, also known as Fusarium wilt, is a significant and dangerous pigeon pea crop disease that is caused by the soil- and seed-borne fungus Fusarium udum (Perfect stage – Gibberella indica), which severely reduces production in vulnerable cultivars. The disease is more common while the crop is in its blossom and pod development stages.

Causal Organism

The wilt disease of pigeon pea caused by *Fusarium oxysporum* f.sp. *udum*. It is belong to Kingdom- *Fungi,* Phylum- *Ascomycota,* Class- *Sordariomycetes,* Order- *Hypocreales,* Family- *Nectriaceae,* Genus- *Fusarium,* and Species- *oxysporum***(Kirk *et al.,* 2008).**

Symptoms

Figure 10.1: Wilt Disease of Pigeon Pea (*Source*: O.P. Sharma).

Primary symptoms of wilt disease appears patches of infected during flowering and pod filling stage. The fungus enter through wound in the vascular system of plant which formed at the root tip and starts invading xylem vessel.

The symptoms starts with loss of leaf turgidity and interveinal chlorosis of leaves and branches. Black dark purple bands and streaks are formed on the stem surface due to the collapse of the vascular and root system which extended from the base to upward portion of crop. The fungus also attacks the crop at seedling stage while the symptom becomes prominent during the mature stage due to a shortage of water or low soil moisture. In humid condition, pinkish or whitish mycelia growth is observed at the basal part of the stem, and intense black mycelium growth is found is in the xylem when the stem split longitudinally.

Epidemiology

The pathogen is soil and seed borne and can survive in infected plant debris in the soil for about 3 years. The soil temperature of 17-25°C facilitates the germination of the spores. Wilt is also favoured by the continuous growth of pigeon pea on the same land (monocropping) over the years.

Disease Cycle

Fusarium udum, saprophytically overwinters in soil or infected pant parts and survives in various forms of spores *viz.,* chlamydospores, macroconidia, and microconidia, while in the infected seed the pathogen occurs dormant in the form of chlamydospores. Under the favorable conditions, the spores starts germinating and gradually penetrate the host root tissue or on to the epidermal layer and later move to the xylem tissues and invade the xylem vessels. The infection spread in the whole plant due to the rapid and extensive multiplication of fungal mycelium which response to gums production and blockage of the xylem vessels resulting in complete wilting of the plant. Infected plant parts and seeds get incorporated as residue into the soil and the resting spores continue the cycle further for subsequent infection. The chlamydospores remain viable for 10-20 years in the field.

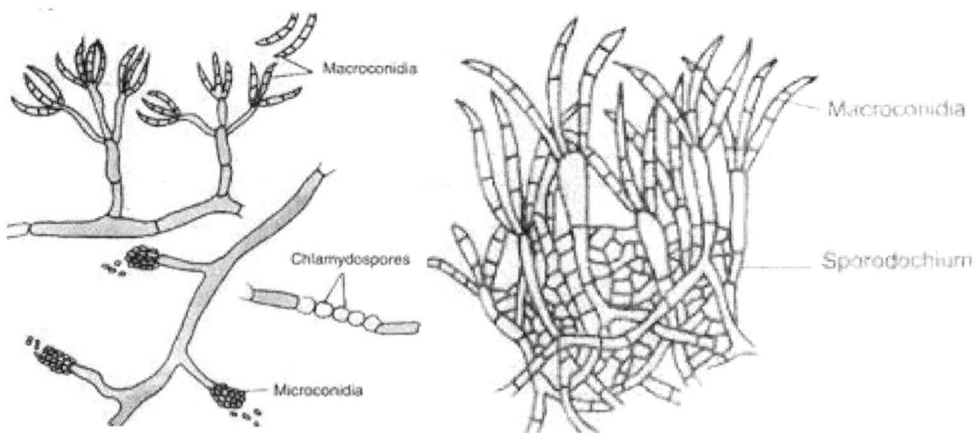

Figure 10.2: Microconidia, Macroconidia, Chlamydospres, Sporodochium of Fusarium (*Source*: www.biologydiscussion.com).

Management

1. Proper soil sanitation practices, destruction of infected and residual plant parts along with seeds to reduce inoculums. Intercropping with sorghum reduced the incidence of wilt disease.
2. Crop rotation with tobacco, sorghum and castor for three years completely eradicates the pathogen from the field.
3. Soil solarization in summer reduces the inoculums.
4. Basal soil application of neem cake @150 kg/ha.
5. Seed treatment with *T.viridae* @ 4g/kg or*P.fluorescens*@10g/kgand Carbendazim or Thiram @ 2g/kg.
6. Treat the seed with Thiram 0.3 per cent or Carbendazim 0.2 per cent.
7. Soil application of *Pseudomonas fluorescens*or *T. viridae* @ 2.5kg/ha +50 kg of well decomposed FYM or sand at 30 days after sowing.
8. The implication of varieties like Maruthi, Durga, Pusa-9, and Sharad or genotype resistance to the pathogen.

2. Stem Blight of Pigeon Pea

Occurrence and Economic Importance

Another name for phytopthora blight is pigeon pea stem blight. It is the third most serious fungal disease of pigeon pea, after sterility mosaic and Fusarium wilt. It is caused by the fungus Phytophthora drechsleri and results in large financial losses. Williams, Grewal, and Amin originally described the condition at IARI in 1966.

Causal Organism

Stem blight of pigeon pea caused by *Phytophthora drechsleri* f.sp. *cajani*which belongs to Kingdom – *Chromista*, Phylum- *Oomycota*, Order- *Perenosporales*, Family- *Perenosporaceae*, Genus- *Phytophthora*, Species- *drechsleri*, and f.sp. *cajani* **(Kirk *et al.*, 2008).**

Symptoms

The symptoms initially appear as dark brown to the purple necrotic lesion that subsequently becomes enlarges in size and forms depression and girdles at the basal part of the stem. Later the visible symptom extended to the aerial portion of the crop like leaves and branches which caused softening of the tissue and loss in turgidity resulting in dry of young plant. Due to softening of the stem or bark of mature plants, the whole plant falls down from the point of infection which ultimately leads to the death of the plant.

Figure 10.3: Characteristics Symptom of Stem Blight of Pigeon Pea (Source: Pande *et al.,* 2011).

Epidemiology

The development and spread of the disease are enhanced and favored by poor soil drainage system and low-lying areas for sowing. A short shower of heavy rains and the prevalence of hot weather (up to 30°C) during July - September causing leaf wetness for almost 7-8 hours favors the rapid and intense progress and spread of the disease. Alfisol soil is found more suitable for disease progress than vertisol soil. The wild pigeon pea relative *Cajanus scarabaeoides* var. *scarabaeoides,* is considered a host of the blight pathogen.

Disease Cycle

The pathogen is soil born and may survive in the resting stage as dormant mycelium and in infected plant parts or debris as oospores. Under the congenial environmental condition, the primary infection spread from oospores when it comes in contact with the host plant through rain splash and irrigation channels. The secondary spread occurred by zoospores from sporangia under proper moisture and humidity.

Management

1. Proper field sanitation and drainage should be done and low-lying areas for sowing should be avoided.
2. Crop rotation should be followed.
3. Adaptation of wider intercropping space.

4. Always prepared raised seed bed with proper and good drainage facilities.
5. Sowing time should be managed such that it doesn't coincide with heavy rainfalls.
6. Seed treatment with Metalaxyl 35 per cent WS @ 3gram/kg seeds and also spray of Metalaxyl @ 500g per hectare is recommended.
7. Resistant varieties like ICPL 12055, ICPL 12114, JKM - 189, and JA - 4 should be grown.

3. Root Rot

Occurrence and Economic Importance

The root rot or dry root rot is considered an important disease of pigeon pea causing huge economical losses ranging up to 10-100 per cent under suitable environmental conditions. The pathogen appears more severely when the crop is grown in the off-season or on black soil. The disease is widely distributed in India, Myanmar, Nepal, Sri Lanka, and Tobago.

Causal Organism

Root rot of pigeon pea caused by *Rhizoctonia bataticola* (Sclerotial stage) or *Macrophomina phaseolina* (Pycnidial stage), which belongs to Kingdom- *Fungi,* Phylum – *Basidiomycota,* Class- *Agaricomycetes,* Order- *Cantharellales,* Family - *Ceratobasidiaceae,* Ginus- *Rhizoctonia* and Species- *bataticola* **(Kirk *et al.,* 2008).**

Symptoms

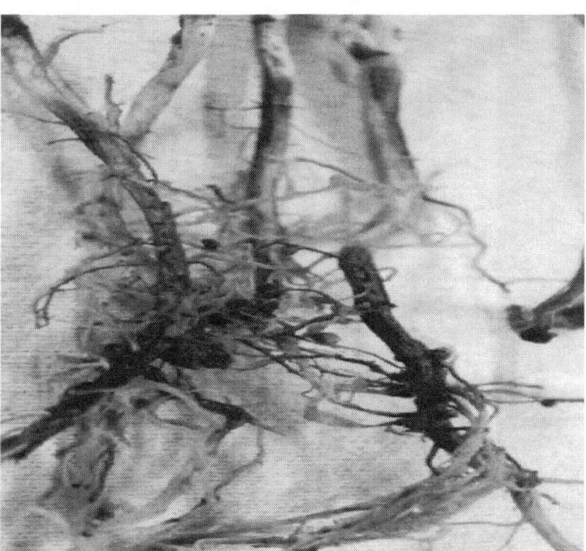

Figure 10.4: Root Rot of Pigeon Pea.

The early seedling stage of crops are when root rot symptoms first appear. The disease begins with leaf yellowing and drooping, which causes the leaves to prematurely lose their pigment. The diseased plants may be easily removed from the ground, and the uprooted roots show signs of rot and shreds. Inspection reveals a decaying root structure with severely damaged lateral roots and weak, brittle bark. In the roots area, tiny, black-colored microsclerotia can also be discovered. Congenial conditions cause the rotting to spread to the stem's base, resulting in the formation of spindle-shaped lesions with brown edges and light grey centers that contain pycnidial bodies. Later, the lesions merged, causing total dryness and the demise of the whole plant.

Favourable Conditions

1. Extended periods of draught interspersed with irrigation.
2. High temperatures between 28 and 35 °C.

Epidemiology

The pathogen is both soil as well as seed borne. The disease symptom is maximum in dry regions followed by irrigation. Prolonged dry spell after rainfall during the crop maturity stage, favours higher production of sclerotia. High temperature is also associated with higher disease intensity. High day temperature (more than 30 °C) during flowering and pod formation stage favour high disease severity.

Disease Cycle

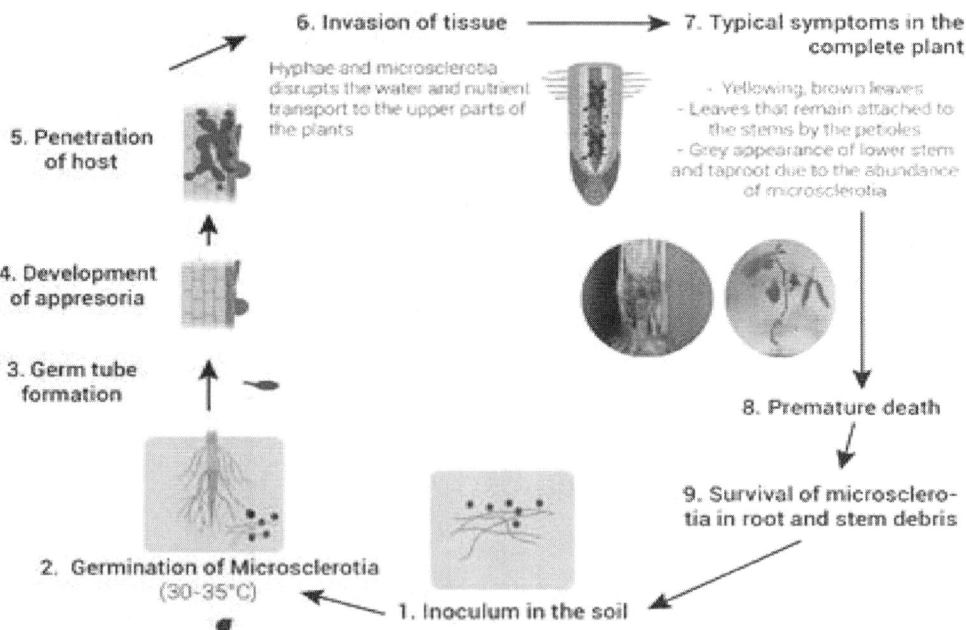

Figure 10.5: Disease Cycle of Root Rot of Pigeon Pea.

Management

1. Selection of disease-free field with no previous record.
2. Irrigation timing should be adjusted and soil moisture should be maintained at an optimum level throughout the crop season.
3. Avoid late sowing to escape drought and high temperature.
4. Seed treatment with *P. fluoresecens* @ 1kg/ha or thiram @ 2kg/ha.
5. Soil application of *T. viridae* or *P. fluorescence* @2.5 kg/ha + 50 kg of FYM at 30 days after sowing of the crop.
6. Spot application of *Carbendazim* @1-1.5gm/lit is also effective for disease control.

4. Powdery Mildew

Occurrence and Economic Importance

Powdery mildew of pigeon pea disease is distributed in a wide range causing severe economic damage in various countries like India, Ethiopia, Sri Lanka, Tanzania, Uganda, and Zambia.

Causal Organism

Powdery mildew of Pigeon pea caused by *Odiopsistaurica*(Teleomorphic stage- *Leveillulataurica*), which belongs to Domain- *Eukaryta*, Kingdom- *Fungi*, Phylum- *Ascomycota*, Class- *Leotiomycetes*, Order- *Erysiphales*, Family- *Erisiphaceae*, Genus- *Levellula* and Species- *Odiopsistaurica* **(Kirk *et al.*, 2008).**

Symptoms

Figure 10.6: Symptoms of Powdery Mildew (*Source*: agritech.tnau.ac.in).

White, powdery growths on all aerial plant parts, including leaves, flowers, and pods, are the symptoms. Tiny chlorotic spots appear on the upper surface of individual leaves as the first symptoms, and then white, powdery patches appear on the corresponding bottom surfaces. The fungal sporulation that causes these patches begins as a tiny discolored speck from which a powdery mass radiates outward on all sides. The heavily contaminated leaves prematurely drop and turn yellow. The upper leaf surface of the older leaves has white spots with powdery development.

Epidemiology

A cool and humid climate favours infection and colonization, though a warm humid climate favours sporulation and dispersion of spores. Sporulation of the pathogen is more prevalent on the young leaves than on the older leaves. The conidia of the pathogen germinate at a very wide temperature range between 13.5- 35°C. The conidia contain high moisture (70 per cent) so they are capable of germinating at low moisture levels. A prolonged temperature of 25°C accelerates the severity of the disease. Varieties with thin and succulent leaves are more vulnerable to attack by the pathogen. Ratoon of harvested stubbles or volunteer plants acts as a shelter for the survival of the pathogen. In India, early sowing of the plants and irrigation favour the establishment of the disease.

Disease Cycle

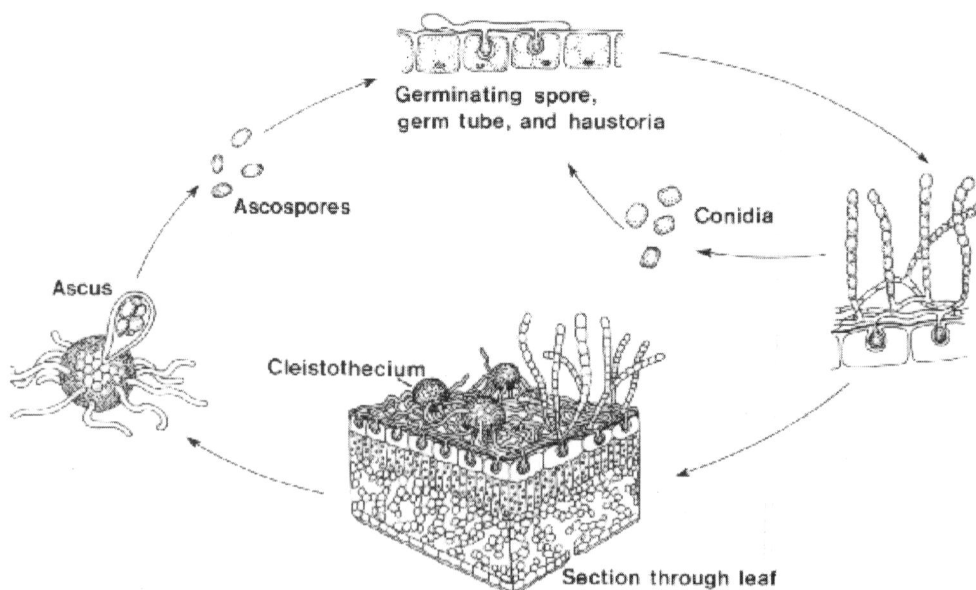

Figure 10.11: Disease Cycle of Pigeon Pea Powdery Mildew (*Source*: G.L. Schumann).

Odiopsis taurica is an obligate parasite that requires a living host for growth and reproduction. At the stage of crop maturity, the pathogen produces sexual spores called ascospores in a sac – like fruiting body known as chasmothecium. Under the congenial condition, chasmothesium releases ascospores. These ascospores penetrate the new host through stomata snd colonize the host tissues or germinate on the leaf surface and the appressorium develops a penetration peg which penetrating upon the epidermis produces haustoria. The haustoria help in nutrient uptake from the epidermis layer. The branch come out and the majority of the fungal body remains outside and new conidiophores arise from superficial hyphae present on the leaf surface. The conidia that are produced from conidiophores are then dispersed with the help of wind.

Management

1. Foliar application of Carbendazim@1g/l is effective against the disease.
2. Spray of wettable sulphur @ 2.5g/l or Neem seed kernel extract @ 5 per cent or Neem oil @ 3 per cent.
3. Spray Triadimefon (Bayletan 25 per cent EC) @ 0.03 per cent is found effective against the disease.
4. Combined application of antagonists *viz.; Trichoderma viride* or *Trichoderma harzianum* with *Pseudomonas fluorescence* or *B. subtilis* or the fungicide carbendazim reduced the incidence of dry root rot.
5. The use of synthetic fungicides such as Thiophanate-methyl at 1 g a.i. per kg seeds was found effective against dry root rot.

5. Alternaria Blight

Causal Organism

Leaf spot of Pigeon pea caused by *Alternaria alternate, which* belongs to Domain- *Eukaryota*, Kingdom- *Fungi*, Phylum- *Ascomycota*, Class – *Dothideomycetes*, Order- *Pleosporales*, Family- *Pleosporaceae*, Genus- *Alternaria*, and Species- *A.alternata* **(Kirk *et al.*, 2008).**

Symptoms

The leaves first develop tiny necrotic spots, which enlarge over time to produce distinctive lesions with a dark and light brown concentric ring and a wavy, purple border. The lesions converge and blight leaves as the infection increases. The disease primarily affects older leaves of adult plants, though it can sometimes infect young plants' new leaves, especially after the rainy weather.

Management

1. Spray Mancozeb 1kg/ha.

Figure 10.12: Alternaria Leaf Spot Symptoms (*Source*: agritech.tnau.ac.in).

6. Botrytis Grey Mold

Occurrence and Economic Importance

The disease is caused by *Botrytis cinerea* and was first time reported by Shaw and Ajrekar from India in 1915 and later by Butler and Bisby in 1931. In 1979, the outbreak of the disease occurred over large areas in Tal area of Bihar and also appear in severe form in Haryana and some parts of Punjab.

Causal organism: *Botrytis cinerea,* (Perfect stage- *Botryotinia fuckeliana*)

Symptoms

Under high relative humidity and low temperature, the disease initially appears in patches during the crop vegetative growth stage. The pathogen attacked all aerial plants parts *viz.;* leaves, petioles, tender, branches, flowers, and fruits. On the surface of the infected plant part, the symptoms initiated as grayish lesions which later turn to dirty grey or black color and increase rapidly in size under the favourabls condition. The lesion area is covered by pathogen conidia and conidiophores. The affected branches hang down, and the thicker stem even breaks apart from the damaged area. Under the congenial weather condition, the lesions, on the leaves become enlarged rapidly and the entire plant becomes dry. In case of severe infection, water- soaked, irregular grayish white lesions appear on the pods which later turn to dirty grey or black color when black sporodichia start to develop on lesion. The infected pods small and shriveled.

Epidemiology

The pathogen overwinters in the soil as mycelium and sclerotia in decaying plant debris. The temperature of 18- 23°C and damp weather is congenial for growth, sporulation, spore release, germination, and the establishment of the infection. At low temperatures, the pathogen becomes highly active and caused heavy economic crop losses.

Disease Cycle

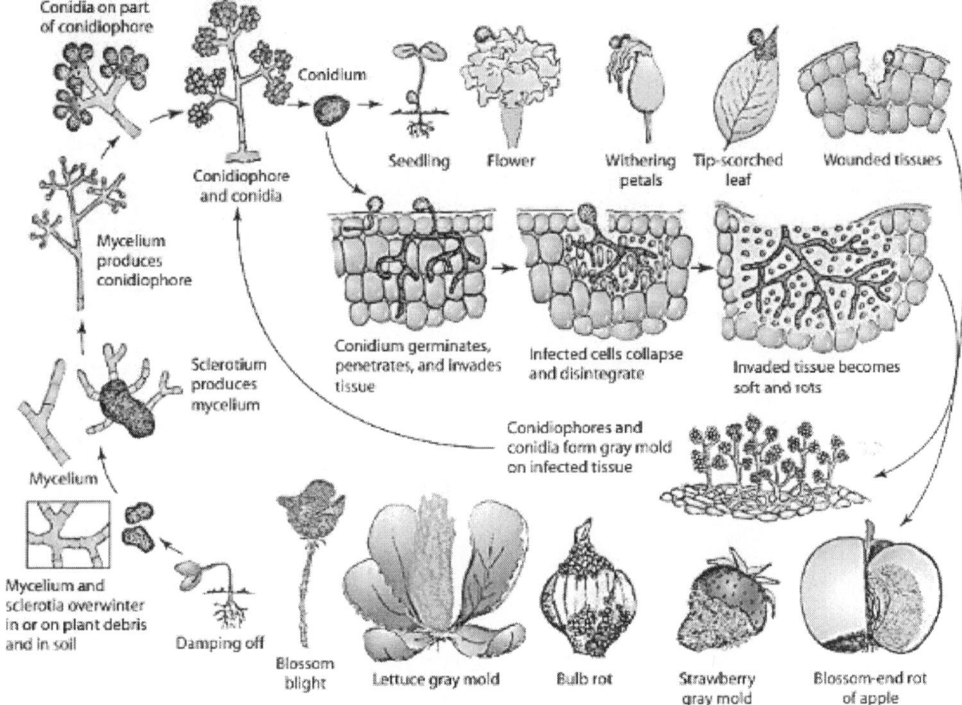

Figure 10.13: Disease Cycle of Botrytis Graymold (*Source*: Agrios, 2006).

Management

1. Field sanitation reduces the pathogen inoculum.
2. Avoid the use of excessive irrigation.
3. Sowing of *Botrytis* resistance varieties of pigeon pea reduced the incidence of the disease.
4. Treat the seed with Carbendazim (Bavistin) + Thiram (1:1) @ 3g/kg seed.

5. The spray of Chlorothalonil @ 3gram/L of water is found effective against the disease.
6. Foliar application of Captan @ 5 - 6 kg/ha at 15 days intervals or Carbendazim @ 1.5g/L.
7. Mancozeb @3 g/lit of water is found effective.

Bacterial Disease

7. Bacterial Leaf Spot and Stem Canker

Occurrence and Economic Importance

In 1950 the disease was first reported from India by Kulkarni *et al.*, There is no specific period and the disease can occur anytime in the crop.

Causal Organism

Bacterial leaf spot and stem canke r caused by *Xanthomonas campestris* pv. *Cajani*, which belong to Domain- *Bacteria*, Phylum- *Pseudomonas*, Class- *Gammaproteo bacteria*, Order- *Xanthomonales*, Family – *Xanthomonadaceae*, Genus- *Xanthomonas* and Species- *X compestris* pv. *cajani*. It is rod-shaped, strictly aerobic, gram -ve, a non-spore-forming bacterium with a single polar flagellum at one end surrounded by a capsule. On the culture growth medium, the bacteria produce yellow colour colonies.

Symptoms

Figure 10.14: Bacterial Leaf Spot and Stem Canker

Plants are infected by the pathogen *Xanthomonas campestris* pv. *cajani* at every stage of growth, however stem infection is primarily observed in younger plants. The disease is more common in hot and humid climates, hence in India, it mostly spreads during the rainy season, which lasts from July to September. The symptoms start off on the lower surface of the leaves as a tiny, water-soaked lesion with a black center and a black center and bright yellow halos surrounding it. The lesions combined to create blotches that were about 2 mm across. Rough, elevated, cankerous sores develop on the stem after a subsequent infection. While canker caused stem breaking at the place where primary leaves connect, and the cracked part normally connects to the plant, leaf spot causes the leaf to stay attached to the plant since defoliation does not occur. The infected plants frequently do not break, and the stem cankers grow larger until they are 15 to 25 centimeters long. The branches that are impacted by the severe infection dry off.

Mode of Spread

The seed-borne bacterium is mostly responsible for the first infection. The wind, rain splashed by the wind, irrigation water, insects, and other tools can all contribute to the secondary spread of the bacterium. The bacteria enters through wounds created by insects or by natural holes.

Epidemiology

A warm 25–28 °C temperature and 85 per cent humidity are conducive to the disease's development. The disease incidence is more common in the field's low-lying wet portions than in its well-irrigated sections.

Management

1. Plant debris that is infested needs to be removed and destroyed.
2. Prevent irrigation water from the contaminated field from entering the healthy one.
3. The affected stubbles must be burned or turned under by plowing.
4. Keep away of using nitrogenous fertilizers rarely.
5. The self-exhibited volunteer pigeon pea plant and weed hosts are rogue.
6. Immerse seeds in hot water for 10 minutes at 50–60°C.
7. Use antibiotics, such as streptocycline at 100 ppm, two to three times every ten days.

Viral Diseases

8. Sterility Mosaic (Sterility mosaic virus)

Occurrence and Economic Importance

The disease is mostly prevalent within the Indian subcontinent and is epidemic in India, Myanmar, Bangladesh Sri Lanka, Thailand, and Nepal. In India, the disease was first reported in Bihar in 1931. It occurs frequently in the state of Maharashtra, Bihar, Uttar Pradesh, Tamil Nadu, Gujarat, and Punjab. Significant yield loss can be recorded up to 90 per cent when the infection starts within 45-50 days after sowing as the plants completely fail to produce flowers and pods.

Causal Organism

Pigeon pea Sterility Mosaic Virus (PPSMV) of pigeon pea had been known in India since 1931. The actual causal agent has been identified recently as a virus with highly flexuous filamentous particles of 3-10 nm width and undefined length. It is an RNA virus in which the RNA genome is divided into 5-7 RNA species.

Symptoms

Figure 10.15: Symptoms of Sterility Mosaic of Pigeon Pea (*Source*: Patil *et al.*, 2015).

The disease's most noticeable symptoms are a bushy look and pale green leaves. The sickness is easily seen in the field, even from a distance, because of the recognized obvious symptoms. At every stage of crop growth, the disease infection is visible. The diseased plants have a pale green, bushy appearance and lack blooms and pods. The diseased leaves exhibit characteristic mosaic patterns of light and dark green and shrink in size. Usually, the afflicted plants grow stunted and lose their ability to develop pods. They feature some secondary and tertiary branches that emerge from the leaf axils, as well as short internodes. Young leaves first exhibit vein cleaning signs, whereas older leaves have mosaic symptoms. Afterwards, the leaves gathered into a group on the upper section of the tip and the floral portions have grown abnormally, turning into a twisted structure of leaves. Plants who acquire the virus in their early stages–up to 45 days–become totally sterile, with a 95 per cent reduction in yield. However, the plants become less susceptible to the illness and may even become partially sterile after 45 days of maturity.

Epidemiology

The disease's occurrence is influenced by pigeon pea intercropping or mixed cropping with millets or sorghum. The crop's tendency to rot also makes diseases more common. During hot summer months, shade and humidity encourage a high mite population multiplication. Relationship between pathogen and novelty: PPSMV is similar to viruses in the genus Tenuivirus in terms of the size and appearance of its VLPs, as well as the quantity and size of its protein and RNA components. All of the viruses in the genus Tenuiviruses, however, primarily affect a broad variety of plant species within the *Poaceae* family and are phloem-limited. They are spread by plant hoppers, or Delphacids. The ultrastructural analysis of pigeon pea infected with PPSMV revealed quasispherical membrane-bound bodies (MBBs) and fibrous inclusions (FIs) bodies measuring 100 to 150 nm. These membrane-bound entities were tagged with PPSMV antiserum in situ, indicating that the 32-kDa antigen unique to PPSMV is the primary antigen present. The PPSMV virus's non-structural inclusion protein is called fibrous inclusions (FIs). The eriophyid mite is the primary vector of virus transmission, acting in a semi-persistent way.

Mode of Spread

The Eriophyid mite (*Aceria cajani)* act as the vector for transmission of PPSMV from one plant to another. The virus transmits through the semi-persistent method. There is very high host specificity seen between the virus and the mite. The vectors *Aceria cajani* are very small in size ranging between 200- 250 um and are pale crimson to orange in colour. They have a short life cycle of about 2 weeks. The mites are minute and occupy the lower portion of the leaves. Feeding does not cause any direct damage to the plant but once they are established on the lower portion of the plant, the mites can multiply to very high densities within a very small period. Both young and adults can transmit the virus. The minimum acquisition feeding time is 5 -10 minutes and the time required for the transmission of the virus to a new host is 30 minutes of feeding time. The mites can transmit the virus for about 48 hours after the acquisition of the virus. The disease can be transmitted by grafting but is not transmitted either through sap or seed. Some species of nematode have also been identified to transmit the virus. The crop ratooning also somehow favours the disease incidence.

Disease Cycle

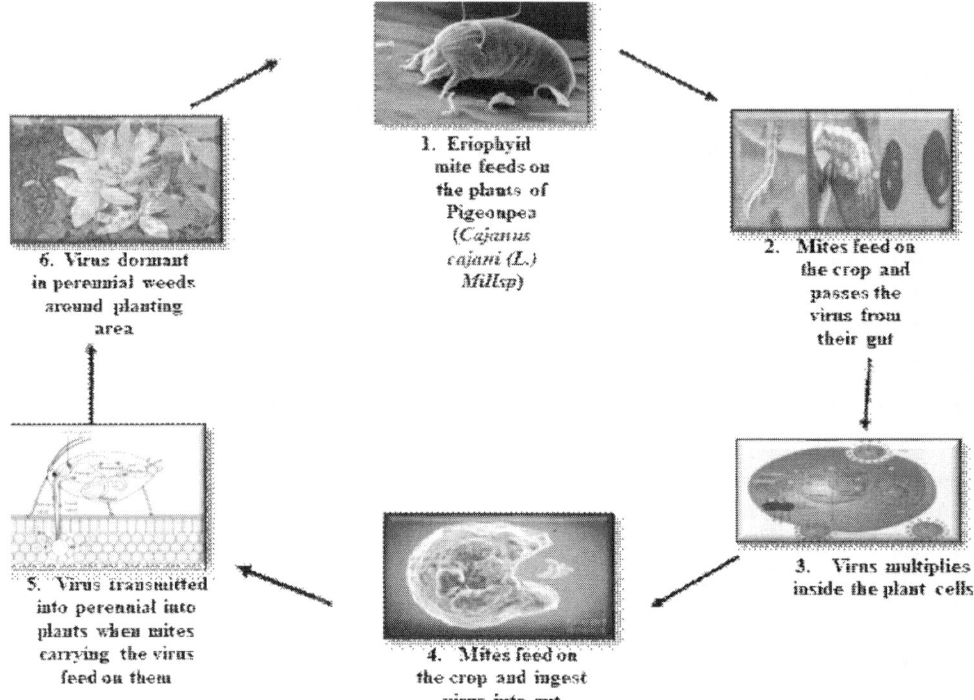

Figure 10.16: Disease Cycle of Pigeon Pea Sterility Mosaic Virus (*Source*: Singh, N. *et al.*, 2021).

Management

1. Wild relatives like *Cajanus scarabaeoides* and *Cajanus cajanifolius* as well as perennial pigeon pea plants which harbour the mites during the off-season should be removed and destroyed.
2. Ratooning of pigeon pea should be minimized.
3. Pigeon pea should not be interred or mixed crop with sorghum and millets.
4. Species of plants like Attilocia should be destroyed because it harbours mites during the off-season.
5. Roughing of plants that show symptoms during the early stages of growth should be done.
6. Spraying the crop with Dicofol (Kelthane) – 450 ml, Phosalone -600 ml, or Monocrotophos – 450 ml in 300 liters of water per acre, twice, 15 and 30 days after sowing for the vector control.

7. Soil application of chemicals like Carbofuron 3G -12 kg or Phorate 10 G – 4kg should be mixed properly with 10 kg sand around the stem portion of the plant.

9. Yellow Vein Mosaic

Occurrence and Economic Importance

Yellow mosaic disease of pigeon pae occurs in different countries like India, Jamaica, Nepal, Sri Lanka. It has been reported to be spreading throughout different agroclimatic zones of India. Williams *et al.* (2008) first described the occurrence of yellow mosaic disease of pigeon pea (PYMD). Based on whitefly (*Bemisia tabaci*) transmission and symptomatology, Nene *et al.* (1971) reported that the yellow mosaic of pigeon pea was caused by mung bean yellow mosaic virus (MYMV). It has been reported in northern and southern Delhi, Uttar Pradesh, Andhra Pradesh, and Karnataka. The virus was found in naturally infected pigeon pea plants, with geminate particles sized 15- 18 X 30 nm.

Causal Organism

Bipartite begomoviruses (genus Begomovirus, family Geminiviridae) have been identified as the four virus types causing yellow mosaic disease in legumes in South Asia. The species are Dolichos yellow mosaic virus, Horsegram yellow mosaic virus, Mungbean yellow mosaic virus, and Mungbean yellow mosaic virus India. The disease is brought on in India by the Mungbean Yellow Mosaic India Virus (MYMIV). In addition to pigeon pea, its host soybean, lima bean, French bean, mungbean, urdbean, common bean, Brachiaria ramose, Xanthium strumamum, Eclyptaalba, and Cosmos bipinatus also. There are differences between the MYMV strain of soybeans prevalent in north India and the strains found in southern and western India. It is spread by *Bemisia tabaci* and has a narrow host range.

Symptoms

The apparent symptoms start off as little yellow spots on the young leaves, which subsequently develop into a yellow mosaic pattern and finally cause the leaves to completely yellow, dry out, and wither. Two basic forms of symptoms are seen, depending on the host response: yellow mosaic, which manifests as yellow A more prevalent and severe symptom on the sensitive pigeon pea variety is mottling. The symptoms of necrotic mottle show a host's resistant response, however in necrotic mottle cases 30, the center of the yellow spot develops a necrosis that is distinguished by finer veins. Seeds per pod, number and size of pods per plant, and leaf size are all reduced by yellow mottling, either by itself or in conjunction with necrotic. Typically, the produce pods are small, withered, and distorted.

Figure 10.17: Yellow Vein Mosaic of Pigeon Pea (*Source*: G.V. Ranga and Rao, 1999).

Management

1. Avoid late sowing of crop.
2. Uproot and burn infected plant parts from the field.
3. Removal of weed from the vicinity of the field.
4. Priming of seed for 8 hours before sowing resulting in the reduction of disease severity.
5. Oil spray is also recommended as it kills the insect within 15 min.
6. Spray of Metasystox @ 0.1 per cent from when the crop is about 1 month old.

Nematodes Disease

10. Pearly Root (Cyst Nematode)

Occurrence and Economic Importance

Reports of *Heterodera cajani* have come from Pakistan, India, and Egypt. It is extensively found in black cotton soil in western and southern India, as well as sandy loamy soil in northern India (Reddy *et al.*, 1993; Varaprasad *et al.*, 1997). Infections have been shown to cause yield losses of up to 30 per cent. Three juveniles per 3 cm of soil at the beginning of the population density can result in a 25 per cent decrease in plant biomass and yield.

Causal Organism

Pigeon pea cyst disease caused by *Heterodera cajani,* which belons to Kingdom – *Animalia,* Phylum- *Nematoda,* Class – *Secernentea,* Order – *Tylenchida,* Family – *Heteroderidae,* Genus- *Heterodera* and *Species- H. cajani*

Symptoms

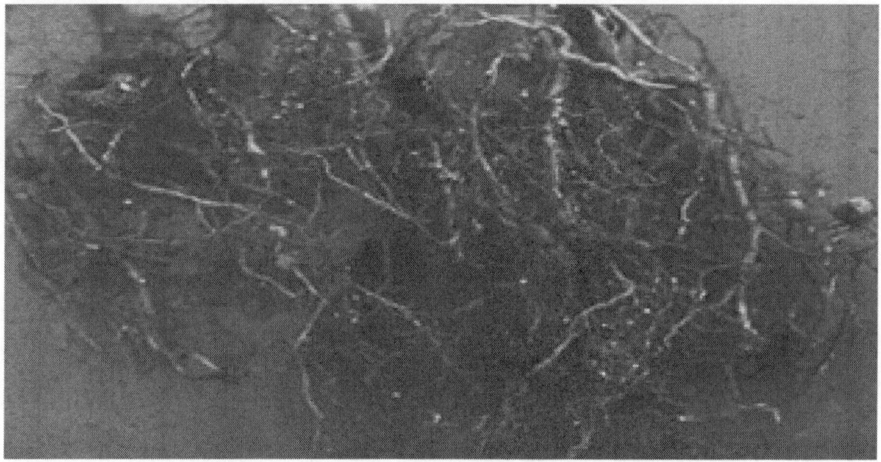

Figure 10.18: Pearly Root Symptoms of Pigeon Pea (*Source*: ICRISAT).

Heterodera cajani is mostly a parasite of roots. The appearance of cysts on the root surface is one of the most significant defining symptoms. When H. cajani is present in pigeon pea during the vegetative stage, identifying the "pearly root" that results from the presence of white females is a helpful symptom (Sharma 1993). Stunting, decreased leaf lamina size, and yellowing of the cotyledon leaves are signs of nematode damage (Gaur and Singh 1977). There are fewer and smaller flowers and pods, and the root system might not be as well-developed. Even in severely infected soils, foliage symptoms are rarely visible; nonetheless, a careful comparison of the diseased plant's height and vigor with that of healthy plants will reveal the infection.

Management

1. Summer ploughing of fields during hot months.
2. Crop rotation with nonhost crops for 2-3 years.
3. Soil application *Pseudomonas fluorescens* or *Trichoderma viride* @ 2.5 kg/ ha at the time of sowing.
4. Seed treatment with *Pseudomonas fluorescens* and *Trichodermaviride* @ 5+5 g/kg seed

References

Mesapogu, S., Bakshi, A., Kishore, B., Babu, S., Reddy, S., Saxena, S. and Arora, D. K. (2012). Genetic diversity and pathogenicity variability among India isolates of *Fusarium udum*infecting pigeonpea [*Cajanus cajan*(L.) Millsp]. *International Research Journal of Agricultural Science and Soil Science,* **2**(1): 051-057.

Prabhavathi, N. M. and Ramappa, H. K. (2018). Field evaluation of Pigeonpea Genotypes againstPigeonpea sterility mosaic virus (PPSMV). *International Journal of Pure and AppliedBioscience.* **6**(5): 57-61.

Singh, B. P., Singh, G., Kumar, K., Nayak, S. C. and Srinivasa, N. (2020). Management of fungal pathogens in pulses: current status and future challenges. *Springer International Publishing.*

Singh, I. P. (2016). Nutritional benefits of pigeon pea in *Biofortification of Food Crops,* pp. 73-81.

Singh, N., Narula, B., Ujinwal, M. and Langyan, S. (2021). Pigeonpea sterility mosaic virus a greenplague-Current status of available drug and new potential targets.

Sinha, P., Rizvi, G., and Parashar, R. (2018). Management of wilt disease of pulses: areview. *International Journal of Pure and Applied Bioscience.,* **6**: 696-708.

Verma, D. K., Srivastava, S., Kumar, V. and Srivastav, P. P. (2016). Biological Agents in FusariumWilt (FW) diagnostic for sustainable pigeon pea production, opportunities and challenges.*Current Trends in Plant Disease Diagnostics and Management Practices,* pp. 91-109.

Williams, F. J., Grewal, J. S. and Amin, K. S. (1968). Serious and new diseases of pulse crops in India in 1966. *Plant Disease Reporter.* **52**(4): 300-304.

Chapter 11

Diseases of Lentil Crop

Anju Shukla[1], Dipake Shital[1] and Sandeep Kumar[1]*

[1]Department of Plant Pathology, Chandra Shekhar Azad University of Agriculture and Technology, Kanpur – 208 002, Utttar Pradesh
**e-mail: shukla32111@gmail.com*

Introduction

Lentil, scientifically known as *Lens culinaris* or Lens esculenta, is a leguminous plant that thrives in cool seasons and is cultivated for its edible seeds. It is a member of the Fabaceae family. It is believed to have originated in the Near East or Central Asia. It is a rich source of dietary protein and other essential micronutrients in human health. It is most popularly consumed as dal in the Indian subcontinent by removing the outer peel and separating the cotyledons. It plays a very crucial role in many nation's agricultural systems. Lentil ranks sixth after dry bean, pea, chickpea, faba bean and cowpea on the basis of total global production. Lentils had became widely cultivated throughout history due to their hardy nature, adaptability to different climates and nutritional value. Seeds of linseed are lens shaped. It is around 40 cm tall, and the seeds develop in pods, usually with two seeds in each pod. Lentil plants develop beautiful flowers ranging from white to purple. It thrives well in properly drained sandy loam or loamy soil with good fertility and an ideal pH of 6.0 to 7.5. Lentil crops require full sunshine hours for optimal growth and yield. It is a valuable human food that is primarily eaten as dry seeds (whole, split, and decorticated seeds). It is simple to prepare and has a great nutritional value. Dry leaves, stems, left over broken pods are used as supplement to dairy animals. It comprises carbohydrates in the range of 57-60 per cent, protein at 24-26 per cent, and 1.3 per cent fiber. Additionally, it is abundant in essential minerals, with 100 grams of seeds containing approximately 69 mg of calcium, 300 mg of phosphorus, and 7 mg of iron(Erskine and Sarker, 2004). India and Canada are the largest producers of lentils in the world. India is currently ranked second in the world for both area and output of lentils, accounting for 28.24 per cent and 16.71 per cent of the total area and production, respectively (FAOSTAT, 2018). Lentil crops can be affected by various diseases stemming from fungi, viruses, nematodes, insect pests, parasitic plants, and environmental stresses. The main factors which is responsible for low yield of lentil are the bio aggressors such as fungal diseases. Major constraints affecting lentil production includes wilt, rust, botrytis grey mold and ascochyta blight. Disease management can be achieved by integrating two or more disease management options. To address this, it's necessary to employ strategies such as planting cultivars that are resistant or partially resistant, modifying the timing of sowing, utilizing bio-control methods, and applying chemical treatments to seeds.

Lentil Diseases and their Causal Agent

Fungal Diseases

1. Wilt: *Fusarium oxysporum* f.sp. *lini*
2. Botrytis grey mold (BGM): *Botrytis cinerea*
3. Rust: *Uromyces viciae-fabae*
4. Ascochyta blight: *Ascochyta lentis*
5. Stem mold and rot: *Sclerotinia sclerotiorum*
6. Stemphylium blight: *Stemphylium botryosum*
7. Anthracnose: *Colletotrichum lini*
8. Damping-off: Disease complex (*Alternaria* spp., *Colletotrichum lini, Fusarium* spp., *Pythium* spp: *Rhizoctonia solani*
9. Septoria leaf spot: *Septoria linicola* (*Mycosphaerella linicola*)

Viral Diseases

1. Lentil yellows disease: Bean leaf roll virus (BLRV)
2. Leaf Crinkle disease: Oat blue dwarf virus

Phytoplasma Diseases

1. Aster yellows: Phytoplasma

Fungal Diseases

1. Wilt

Occurrence and Economic Importance

Fusarium wilt stands out as the foremost biological limitation to lentil productivity on a global scale. In 1934, undivided Bengal in India became the first region to report lentil wilt. The pathogen is prevalent in India and results in significant harm, thereby constraining lentil production in various Indian states including Uttar Pradesh, Madhya Pradesh, Himachal Pradesh, Bihar, West Bengal, Assam, Rajasthan, Haryana, and Punjab (Chaudhary *et al.*, 2009). The extent of yield losses varies based on the timing of plant wilting, with potential losses reaching 100 percent when wilting happens before the pod stage and approximately 67 percent if it occurs before harvest.

Causal Organism

Wilt disease of lentil is caused by *Fusarium oxysporum* f. sp. *lentis*, which belonging to Kingdom- *Fungi*, Division- *Ascomycotina*, Class- *Ascomycetes*, Subclass- *Sordariomycetidae*, Order- *Hypocreales*, Family- *Nectriaceae*, Genus- *Fusarium*, Species- *oxysporum*, and forma specialis - *lentis*. The pathogen is known to produce three distinct types of asexual spores: microconidia, macroconidia, and chlamydospores. Microconidia are typically transparent, single-celled, and have an oval or kidney-shaped appearance. Macroconidia are generally elongated, consisting of two to seven cells, with a pointed tip and a notched base. Chlamydospores are single-celled, either oval or spherical, and possess a thick wall. They are formed individually in macroconidia or at the tip or within the hyphal strands. When cultured, the pathogen responsible for wilt has a hyaline appearance, is segmented, and exhibits extensively branched mycelium. The growth pattern on media can vary, ranging from fluffy to closely packed, and the color may range from transparent to pink (Khare, 1980).

Symptoms

In the field, the disease manifests itself in patches at both the seedling and adult stages. Abrupt drooping is the hallmark of seedling wilt, which is followed by leaf drying and seedling mortality. There is no internal vascular bundle discolouration visible upon breaking, and the roots seem healthy with fewer side branches. On mature plants, wilt symptoms include rapid drooping of the top leaflets, leaflet closure with dull green foliage, followed by wilting of the whole plant or specialized branches. Diseased plants often yield shriveled seeds which are of inferior quality both for consumption and seed purpose.

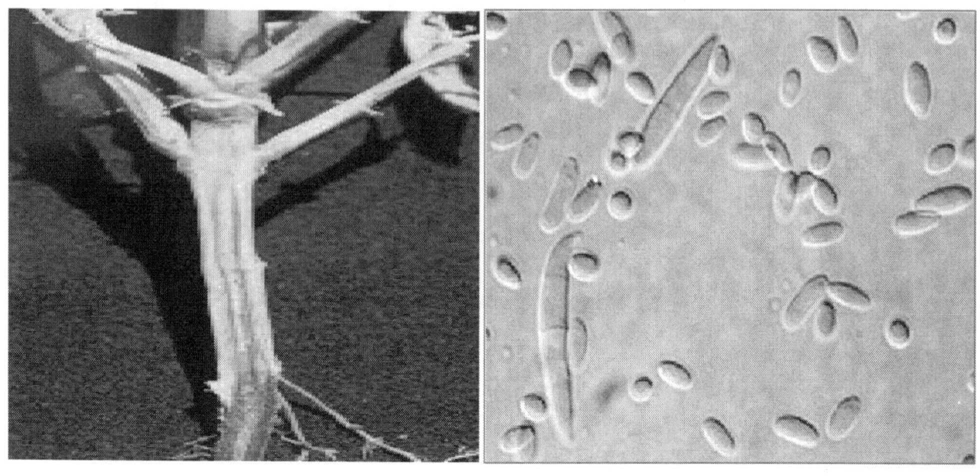

Figure 11.1: Wilt Symptoms **Macroconidia and Microconidia**

Epidemiology and Disease Cycle

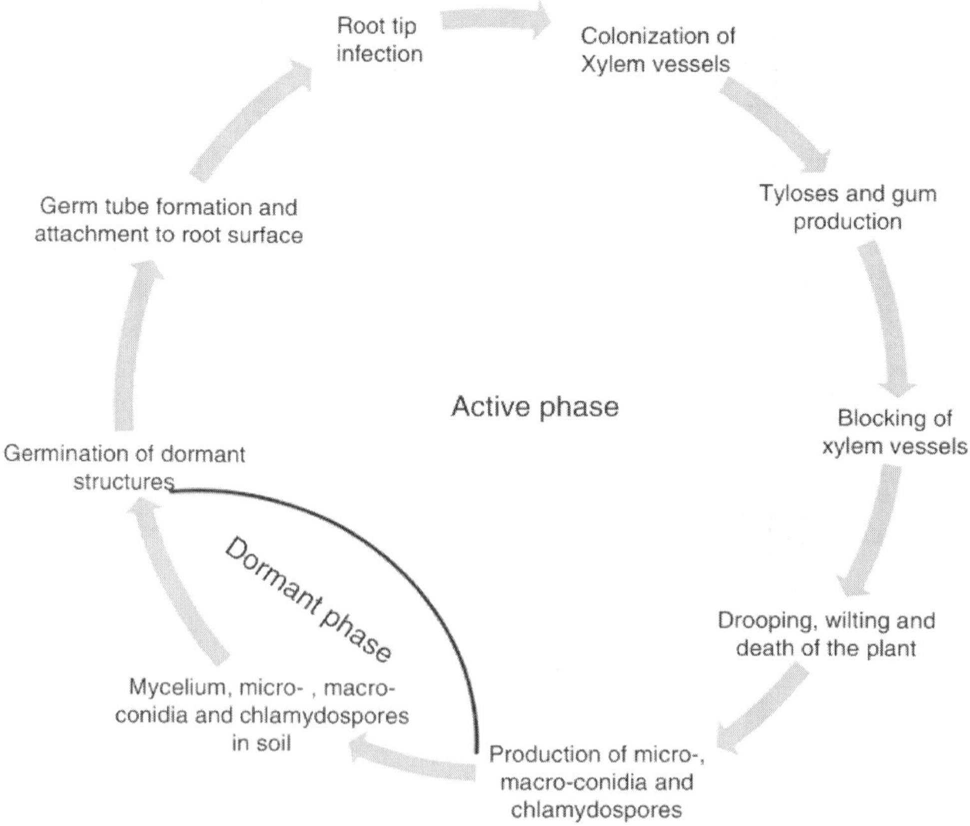

Root tip infection

Colonization of Xylem vessels

Tyloses and gum production

Blocking of xylem vessels

Drooping, wilting and death of the plant

Production of micro-, macro-conidia and chlamydospores

Mycelium, micro- , macro-conidia and chlamydospores in soil

Germination of dormant structures

Germ tube formation and attachment to root surface

Active phase

Dormant phase

Figure 11.2: Disease Cycle of Fusarium Wilt. (*Fusarium oxysporum* f. sp. *lentis*)

In the absence of its host, the pathogen *Fusarium oxysporum* f. sp. *lentis* can be found in soil and plant detritus for several years in a latent state. It causes wilting by completely blocking the xylem vessels. The disease is favoured by low soil temperature and disease incidence chances increases as plant matures. The Fusarium wilt pathogen primarily initiates the disease by invading the roots of the host plant. It enters the plant's xylem tissues after crossing the cortex. Subsequently, it rapidly spreads throughout the plant's vascular system, becoming systemic within the host's tissues, and may even directly infect the seeds. The dissemination of infected seeds plays a crucial role in the long-distance dispersal and transmission of Fusarium wilt diseases to new regions. In contaminated soil, healthy plant root tips are penetrated by the germ tube of spores or fungal mycelium. The pathogen is mainly restricted to the xylem vessels, where it branches and produces microconidia. This severely obstructs the water supply to various parts of the plant, leading to stomatal closure, wilting, and leaf death, often followed by the demise of the entire plant. The fungus subsequently invades all plant tissues, eventually reaching the surface where it produces an abundance of spores. These spores can be dispersed by wind, water, or the movement of soil or plant debris. *Fusarium oxysporum* can endure as mycelium and chlamydospores in seeds and soil, as well as on infected crop residues, roots, and stem tissue buried in the soil for periods exceeding six years (Singh *et al.*, 2007). An optimal temperature range of 22 to 25 °C, together with dry and warm soil, are conducive to the development of disease.

Management

1. Deep ploughing of the field during summer destroys the dormant structures surviving in soil.
2. Crop rotation with crops of other families such as cereals and legumes.
3. Adding organic matter to the soil improves its ability to interact antagonistically with other microorganisms present in the soil.
4. Using antagonistic microflora like *Bacillus subtilis*, *Trichoderma harzianum*, *T. viride* @ 4 g/kg seed *etc.*
5. Grow resistant varieties such as Pant L-4, Pant L-6, Pant L-8 and Noori.
6. Treating seeds with benornyl at a concentration of 0.3 per cent, or a combination of thiram and benomyl in equal parts at 0.3 per cent, or applying COC at a rate of 0.25 per cent, leads to a decrease in the occurrence of wilt and results in higher grain yields.

2. Botrytis Grey Mold

Occurrence and Economic Importance

Botrytis grey mold of lentil is a serious disease wherever the crop is grown. It has been observed to occur sporadically. This disease is predominantly found in regions where the annual rainfall exceeds 400mm. The extent of crop damage resulting from the disease can vary significantly, ranging from minor to severe. The severity of the impact depends on factors such as the crop variety, crop location, timing of the infection, and the volume of rainfall received. This pathogen is particularly destructive in North Indian climates and is also a significant concern in various areas of Australia, Argentina, Nepal, Myanmar, Bangladesh, and Pakistan. Under favorable conditions, it can lead to yield losses of 70-80 per cent. (Haware and McDonald, 1992). In addition to direct losses in terms of yield, the pathogen causes discoloration of seed which reduces its market value.

Causal Organism

Botrytis grey mold of lentil is caused by *Botrytis cinerea.* The pathogen is known to produce numerous colourless conidia (asexual spores) borne on greyish, branched tree-like conidiophores. The conidia produced by asexual means is dispersed by wind and by rain-water which further causes fresh infections. During the summer season, the fungus undergoes an asexual reproduction cycle. It survives the winter by forming robust structures known as sclerotia or by keeping its mycelium intact. In the spring, these structures sprout and give rise to conidiophores.

Symptoms

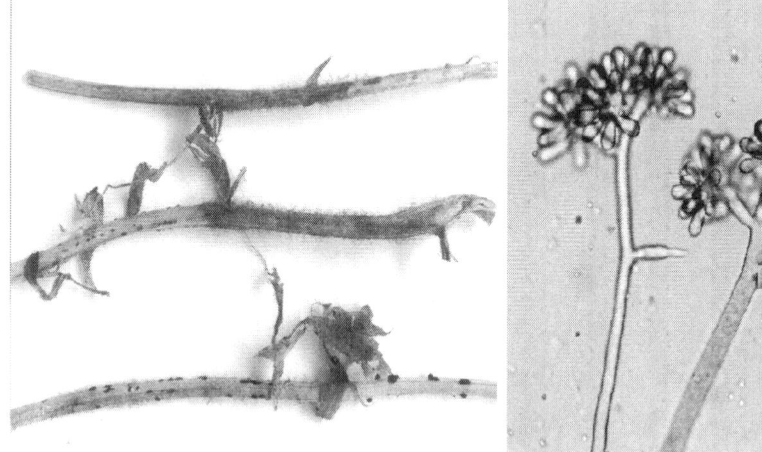

Figure 11.3: Lesions on Diseased Stem **Conidiophore of Pathogen**

The pathogen affects all the aboveground plant parts of lentil. When the disease reaches its maturity, the initially dark green lesions on the lower leaves turn grayish-brown. Afterwards, a number of lesions combine to infect entire leaflets. Leaves that have been severely diseased wither and fall to the ground. Lesions when appears on stem girdles the stem and cover it with a layer of masses of grey mold. The most damaging symptoms are observed when full vegetative growth is obtained and a humid microclimate is produced under the crop canopy. Pods which become infected with grey mould will rot and dried out. Diseased pods are often chaffy with seeds infected seeds further give rise to infected crops.

Epidemiology and Disease Cycle

Botrytis grey mold inoculum propagates through infected seeds, sclerotial bodies, mycelium surviving in infected plant debris, and alternate hosts plants. High moisture content and RH favours the disease. RH > 95 per cent and temperatures between 15 and 25°C are ideal for the onset and progress in of disease, particularly during flowering and following canopy closure. Botrytis grey mold pathogen survives on infected plant debris, diseased seeds, and as sclerotial bodies in the soil or on alternate hosts. Botrytis cinerea infects a wide variety of hosts, which encompass lupin, field pea, faba bean, chickpea, as well as pasture legumes like lucerne and clover. A lot of ornamental and horticultural crops are also affected by the pathogen. Fungal spores produced on infected plant debris and alternate hosts are spread through wind. The disease can spread quickly under high humidity and ideal temperatures, generating spores on freshly affected tissue and subsequently dispersing the disease within crops.

Management

1. Disease free seeds must be selected for sowing. Sowing must be done on recommended date. Optimum seed rate must be used for proper crop establishment.
2. Crop rotation with crops of other families such as cereals and legumes.
3. Grow resistant varieties such as Pant L-639 and Pant L-406.
4. Seed treatments with fungicides like thiabendazole, carboxin, benomyl, or chlorothalonil (0.1 per cent) can lower the amount of inoculum carried by the seeds.
5. Fungicides may harm rhizobia, so it's advisable to apply fungicide to seeds and then separately inoculate them with rhizobia. When dealing with acidic soils, it's especially important to ensure that rhizobia are applied to the seeds right before planting.

3. Rust

Rust is a severe disease caused by the fungus *Uromyces viciae-fabae*. It affects lentil crops cultivated worldwide and reduces productivity in a number of countries by causing partial defoliation and, in cases of severe infection, early plant death. Abortion of flowers and pods results in a loss of harvest.

Occurrence and Economic Importance

The disease can cause total crop failures and is a potential threat to the cultivation of lentils. The disease results in significant yield losses that vary between 60 per cent and 69 per cent. (Sepulveda, 1985). In the years 2008–2009 in the state of Uttarakhand, a major outbreak of lentil rust was documented in the Narmada Valley in Madhya Pradesh, resulting in yield losses of up to 100 per cent. Rust has been an important constraint that has negatively impacted yield in the tarai region of the state of Uttarakhand and its adjacent areas. In the past, disease in this region has largely shown as an epiphytotic form. (Khare and Agarwal, 1978). Lentil rust is an autoecious rust and produces all its five spore stages on a single host.

Causal Organism

Rust disease is caused by the fungus *Uromyces viciae-fabae*. Lentil rust is an autoecious rust and produces all its five spore stages on a single host. Spermogonia and aecia are produced mostly on abaxial surface of leaf, predominantly along veins. Uredinia and telia are amphigenous and present on both the leaf surface. Urediniospores exhibit a broadly ellipsoidal shape, with dimensions ranging from 22 to 32 µm in length and 17 to 25 µm in width. They are uniformly covered in small spines (echinulate), have 3 to 5 germ pores distributed either equatorially or scattered across the surface, and possess a light golden-brown wall that measures 1 to 2.5 µm in thickness.Teliospores, on the other hand, are ellipsoidal, obovoidal, or cylindrical in shape with a rounded or somewhat pointed tip. They have a size range of 24 to 40 µm in length and 17 to 26 µm in width. The spore wall is chestnut-brown, smooth, measuring 1 to 3 µm in thickness on the sides and widening to 5 to 12 µm at the apex. Additionally, the pedicels, which are extensions connecting the spores, are brownish in color and can reach lengths of up to 100 µm. (Negussie and Pretorius 2012).

Symptoms

Figure 11.4: Rust Symptoms on Foliage.

Initial symptoms appear as tiny, pale, slightly elevated dots on the upper surface of the leaves. These spots often have a lighter halo around them and become powdery and orange or brown in colour as they enlarge. Leaf drop, reduced plant growth, and early plant death are effects of a severe infection.

Epidemiology and Disease Cycle

Disease development is favoured by conditions of high humidity with temperatures between 20 and 22°C. The disease often shows symptoms in the initial podding/flowering stage. Primary inoculum that has survived in infected plant detritus in soil, weed hosts, or volunteer hosts as teliospores may contaminate lentil seeds and onsets the disease in new season. Uredospores from diseased crops are carried by the wind and propagate the disease (secondary). At temperatures between 17 and 22 °C, aeciospores germinate and infect neighbouring plants, producing either secondary aecia or, at a temperature of 25 °C, uredia. Telia develops quickly after uredosori later in the season. All spores except teliospores can withstand high temperatures.

Management

1. Select healthy seeds from certified sources for sowing.
2. Remove all the infected plant debris weed hosts and volunteer hosts which act as reservoir of inoculum of pathogen.
3. Crop rotation with crops of non-host families such as cereals for atleast three years.
4. Use of resistant varieties such as Pant L-639, Pant L-406, Pant L-6, pant L-7 and Pant L-8.
5. Use of foliar fungicides as Hexaferb and Dithane M-45 give best control. Fungicides as Mancozeb (0.2 per cent a.i.), Bayleton (0.05 per cent a.i) and Calixin (0.2 per cent a.i.) are also found effective against the pathogen.

4. Ascochyta Blight

The disease is caused by the fungus *Ascochyta lentis*. The pathogen attacks all aboveground plant components at any stage of growth in the presence of favourable conditions. Disease causes reduced production and poor seed quality.

Symptoms

Disease lesions occurs on the leaves, petioles, stems, and pods. Crops that are infected severely may become blighted, and the seed may shrivel, become smaller, and turn discoloured leading to yield loss.

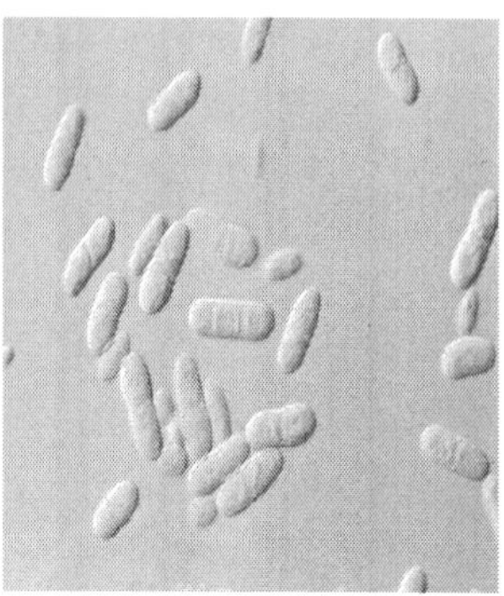

Figure 11.5: Lesions on Leaves Lesions on Pods Conidia of Fungus

Epidemiology and Disease Cycle

Ascochyta blight is a seed-borne disease. Cool, humid weather is conducive for commencement of disease cycle. Disease development requires 24 to 48 hours of dampness and leaf must be wet for a long time. Disease development happens with temperature ranging from 10°C to 20°C occurs while maximum disease development occurs at about 15°C. The fungal pathogen can overwinter in infected seed, and in infected plant debris of previously sown crops. Seed can remain infected for several years. Sowing infected seed can give rise to infected seedlings, and the appearance of symptoms at the seedling stage.

Management

1. Resistance breeding combined with conventional approaches is the best cost-effective and long-lasting way to control ascochyta blight.
2. At least four years of crop rotation should be maintained with non-host crops like cereals.
3. The use of certified, disease-free seed and timely sowing as recommended for that region can minimize disease.
4. Chemical control can be achieved with various fungicides such thiobendazole @ 0.1 per cent (Mertect), carbathiin, benomyl, and carbendazim.

5. Stem Mold and Rot

Causal organism: *Sclerotinia sclerotiorum*

Symptoms

In the initial stage of seedling emergence, the fungal pathogen causes blight or die back. Sunken lesions with brown discolouration occurs on the roots. The rotten plants further dries and die. On the diseased tissue, white fluffy mycelium is observed and slerotial bodies can also be observed in patches. Affected plant becomes stunted and shows chlorosis. Nodules formed on legume crop are less numerous, smaller, and pale in color.

Figure 11.6: White Mycelium on Stem and Roots.

Epidemiology

Sclerotinia sclerotiorum is a pathogenic fungus that thrives in wet, cool, and dark conditions. The fungus can totally invade a plant host in a wet area, colonising nearly all the plant's tissues with mycelium. The range of 15°C to 21°C is ideal for growth. The Pathogen produces a lot of mycelium and sclerotia when the environment is moist.

Disease Cycle

Life cycle of White mold is a monocyclic disease and produces no other secondary inoculum. Sclerotial bodies can survive in soil for long duration before germinating to produce cup shaped fruiting bodies with diameter of 3-6mm termed as apothecia. The ascospores are contained in the asci-lined cup of the apothecium. When the ascospores are released from the asci, they are transported to a suitable host by several methods. Introduction of the pathogen to the crop field occurs through wind borne ascospores released by infected plant debris of previous crop, by contaminated machinery, by movement of water from diseased field to healthy field, infected seeds, admixtures of sclerotial bodies with healthy seed lot. *S. sclerotiorum* can colonise all the organs of living plant. White fluffy mycelium can be observed on the surface of diseased tissues. The pathogen will start producing sclerotia after the growing season is over. The sclerotia will then overwinter until next season on the ground's surface or in the soil, on infected plant debris. The lifecycle will then proceed as before (Abawi and Grogan, 1979).

Management

1. Cultural practices includes planting crops at optimum spacing to promote air circulation.
2. Excessive irrigation should be avoided until flowering and proper drainage conditions should be ensured in field.
3. Follow crop-rotation with crops of other families such as cereals and pulses.
4. Use high-vigor seed, improve soil fertility, lime acidic soils, and rip up the ground to loosen up compacted soil.
5. Most economic way of disease control is by planting varieties resistant to stem mold and rot.
6. Up to 95 per cent of disease can be controlled by treating with protectant fungicide such as azoxystrobin (Quadris),and a mixture of the fungicides fludioxonil and cyprodinil applied on lentil roots.

6. Stemphylium Blight

Causal organism: *Stemphylium botryosum*

The disease is caused by the bacterium *Stemphylium botryosum*. It is primarily a foliar disease and lowers seed output by up to 100 per cent and further diminishes seed quality. The disease is seed borne and also spread through spores disseminated by infected plant debris.

Symptoms

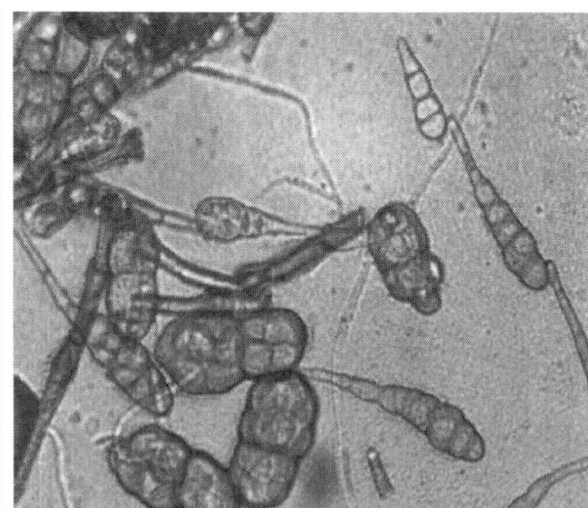

Figure 11.7: A. Blighted Leaves; B. Muriform Conidia of Pathogen.

At all phases of crop development, the disease can be seen on leaflets. Symptoms on seedlings are often mild. Tan-colored lesions that frequently have angular patterns of brighter and darker parts occurs on diseased leaves. They are larger and spread throughout the entire leaflet, which sets them apart from other lentil foliar lesions. Quality of seeds formed degrades and seed becomes smaller and stained if infections happen in the early phase of flowering.

Epidemiology

Wet weather conditions are favourable for sporulation of the pathogen as well as spore germination. When the temperature hits 25–30 °C and the humidity rises to above 85 per cent for 48 hours, the disease may spread rapidly (Mwakutuya and Banniza, 2010). Spores are airborne. After rain, plants may exhibit high infection levels that make maturing crops appear white. As is typical with other lentil diseases, diseased leaflets falls where they act as a source of spores for subsequent plant infections.

Management

1. An integrated approach with a combination of cultural, physical, biological and chemical method is best for disease management.
2. Non-host crop rotation, destruction of infected plant debris of old crops, cleaning the field, adjusting of the sowing time, and prior seed treatment can help in reducing disease inoculum load.
3. Fungicide spray of Rovral 50WP @ 0.2 per cent and after that with Dithane M-45 @ 0.2 per cent and Tilt 250EC @ 0.05 per cent is economic and beneficial for disease control.

7. Anthracnose

Causal organism: *Colletotrichum lini*

Symptoms

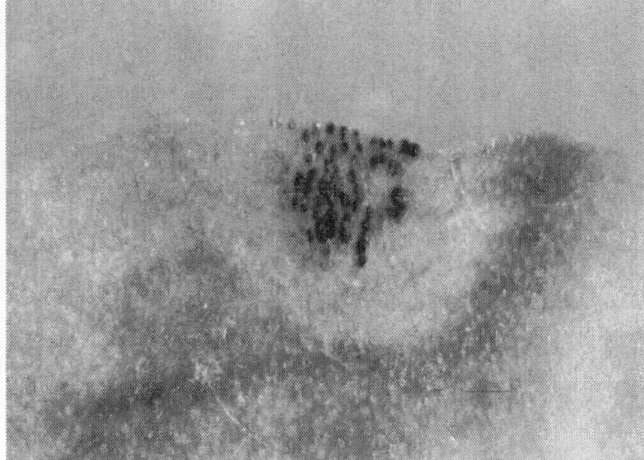

Figure 11.8: A. Stem Lesions; B. Microsclerotia.

Disease symptoms appears as spots ranging from white to grey on leaves and lesions on stem. A characteristic symptom of anthracnose is presence of pin-shaped black microsclerotia within the stem lesions. Starting on lower leaves and stems, they advance upwards. When a plant is mature, the stems frequently turn black and the entire plants may die back. All the leaves present on the plant droops down. The soils top is covered with leaf debris. Yellowing of crop foliage subsequently spreads to entire field.

Epidemiology and Disease Cycle

Anthracnose blight is a polycyclic disease and completes many cycles within one growing season with the help of asexual spores. A lot of wetness in the leaves, frequent rain, high humidity, a thick canopy, and warm temperatures of 20 to 24°C favours the disease. The pathogen overwinters in fields as microsclerotia on infected crops debris. When favourable environment conditions prevails and the lower stems and leaves come into contact with pathogen's inoculum, initial infection occurs. The spores of pathogen spreads from one plant to another with the help of rain splash and wind driven rain.

Management

1. Like other diseases of lentil, anthracnose blight can also be controlled with the help of integrated management strategy following cultural, physical, chemical and biological methods. Choosing resistant cultivar is the most economical method to prevent disease.
2. Follow proper crop rotation with crops of other families.
3. Seed treatment with *Trichoderma viridae* @4g/Kg seed.
4. Seed treatment with Carbendazim 50 WP, Carbendazim + Thiram, Thiram 75 WP, Mancozeb 75 WP at 2.5 g/Kg of seed.

8. Damping-off

Damping off is caused by various soil borne pathogens such as *Alternaria, Colletotrichum, Fusarium, Pythium, Rhizoctonia* spp., and various others.

Symptoms

Pre-emergence damping: Damping off can occur either before seed germination leading to rotting of seeds under soil and complete failure to germinate or when the germinating seeds are killed by pathogen due to formation of water-soaked lesions near the soil line. As the disease progresses, these lesions may turn darken to reddish-brown, or black. Expanding lesions quickly girdles tender stems.

Post-emergence damping off: This type of damping off occurs after emergence of seeds. Death of seedlings occurs in groups in roughly circular patches. The seedlings may also have ground-level stem injuries. Seedling vigour can be negatively impacted when stems start to deteriorate and become thin and tough (often referred to as '**wire stem**'). Leaf spotting and full root rot may accompany these symptoms as well.

Management

1. Seed treatment helps to obtain proper crop establishment and enhanced seedling vigor.
2. Deep summer ploughing followed by proper drainage after irrigation should be ensured.
3. The use of cultivars that are resistant to or tolerant of damping-off diseases
4. Chemical treatment with broad spectrum fungicide such as captan or ferbam or a prepared combination of fungicides (thiophanate methyl + etridiazole) should be used.

References

Abawi, G. S., and Grogan, R. G. (1979). Epidemiology of diseases caused by Sclerotinia species. *Phytopathology, 69*(8), 899-904.

Chaudhary, R. G., Dhar, V. and Singh, R. K. (2009). Association of fungi with wilt complex of lentil at different crop stages and moisture regimes. *Archives of Phytopathology and Plant Protection*; **42**: 340-343.

Erskine, W., Sarker, A., Lentil. I., Corke, H. and Walker, C. E. (2004). Encyclopedia of Grain Sciences. London. UK: *Elsevier;* pp. 142-150.

Garkoti, A., Kumar, S., Lal, M., and Singh, V. (2013). Major diseases of lentil: epidemiology and disease management-a review. *Agriways*, **1**(1): 62-64.

https: / /dpd.gov.in/Lentil.PDF

Khare, M. N. (1980). Wilt of Lentil. Jabalpur, India: Jawaharlal Nehru Krishi Vishwa Vidyalaya; pp. 155

Mwakutuya, E., and Banniza, S. (2010). Influence of temperature and wetness periods on the development of stemphylium blight on lentil. *Plant Disease*, **94**(10): 1219-1224.

Negussie, T. and Pretorius, Z. A. (2012). Lentil rust: Present status and future prospects. *Crop Protection*, **32:** 119-128.

Reddy, M. V., Haware, M. P., Nene, Y. L., and McDonald, D. (1992). Recent advances in fungal diseases of chickpea and pigeonpea.SEPULVEDA, P. (1985). Efecto de la roya de la lenteja, causada por Uromyces fabae Pers de Bary, en el rendimiento del cultivo. *Agricultura Tecnica (Santiago)*, **45**(4): 335-339.

Singh, G., Chen, W., Rubiales, D., Moore, K., Sharma Y. R. and Gan, Y. (2007). Diseases and their management. In Chickpea Breeding and Management (Eds Yadav, Redden, Chen and Sharma). *CAB International*, pp. 497–519.

Chapter 12

Diseases of Soybean Crop

Saurabh Saini[1]*, Girijesh Kumar Jaisval[1], Dipake Shital[1] and Arjun Rana[1]

[1]Department of Plant Pathology, Chandra Shekhar Azad University of Agriculture and Technology, Kanpur – 208 002, Uttar Pradesh
****e-mail: ankursaini723@gmail.com***

Introduction

The initial domestication of *Glycine max* (Lin.) Merrill, colloquially recognized as Soybean, transpired in ancient China between 1700 and 1100 B.C. Over epochs, East Asian societies have heavily depended on soybean for sustenance, fodder, and even medicinal purposes. Subsequently, in 1765, soybean found its way to North America, facilitated by its introduction via London, and later reached Africa through the endeavors of early 19th-century missionaries. In the modern era, global soybean production has surged, with estimations indicating a monumental 333.67 million metric tonnes harvested from an extensive agricultural expanse spanning 120.50 million hectares during the 2019-20 period. Notably, Brazil ascends as the preeminent soybean producer, boasting an impressive yield of 114.27 million metric tonnes, closely followed by the United States of America (96.79 million tonnes), Argentina (55.26 million tonnes), China (15.73 million tonnes), and India (13.27 million tonnes). Regarding India, it holds the fourth position in terms of cultivated area, encompassing 11.34 million hectares (28.02 million acres), constituting 9.41 per cent of the global soybean cultivation area, and secures fifth place in production, yielding 11.22 million tonnes during 2019-20. The primary soybean cultivating states within India are Madhya Pradesh, Maharashtra, Rajasthan, Karnataka, and Telangana. According to the initial advance estimates from the Ministry of Agriculture for the year 2021-22, soybean production is projected to reach 127.20 lakh tonnes, marking a slight decline from the 128.97 lakh tonnes recorded in the preceding year, 2020-21 (Soybean Outlook, January 2022). The swift expansion of soybean cultivation worldwide has engendered a corresponding rise in the prevalence and severity of soybean diseases. Notably, over a hundred pathogens are known to assail soybean plants, but only about a third of them present significant economic threats. The extent of yield diminishment varies contingent upon factors like the specific pathogen, prevailing climatic conditions, plant development stage, infection severity on individual plants, and the prevalence of infected plants within the population. It is crucial to note that certain pathogens are instigated to induce disease solely under precise environmental conditions. Furthermore, the severity of soybean maladies is influenced by the level of compatibility between the host plant and the pathogen, in addition to the environmental impacts on this interaction. Additionally, nematodes and insects exacerbate the issue by inflicting wounds on soybean plants during their feeding activities, which serve as entry points for various pathogens, rendering the plant susceptible to root infections, wilt-inducing agents, and seed-decaying organisms. In distinct regions of the country, over a hundred diseases have been documented to afflict soybean crops, with thirty-five of them being of particular significance in India. Annual yield losses attributable to these diseases in the country amount to nearly 12 percent of the total production. Globally, annual yield losses due to diseases in soybean alone range from approximately 10 to 30 percent of the total production. Therefore, a proactive approach to diagnosis is imperative for safeguarding and managing soybean crops. Consequently, this chapter encompasses the occurrence and economic significance, systematic classification of causal organisms, symptomatology, etiology, disease life cycle, and their management. This comprehensive understanding is vital in averting devastating diseases that lead to substantial yield losses in soybean crops.

Fungal Diseases

1. Leaf spot of soybean: *Cercospora kikuchii*
2. Wilt of soybean: *Fusarium oxysporum* f. sp. *tracheiphilum*
3. Anthracnose of soybean: *Colletotrichum truncatum*
4. Brown spot of soybean: *Septoria glycines*
5. Downey mildew of soybean: *Pernospora manshurica*
6. Powdery mildew of soybean: *Microsphaera diffusa*
7. Sudden death syndrome: *Fusarium virguliforme*
8. Frogeye leaf spot: *Cercospora sojina*
9. Target leaf spot: *Corynespora cassicola*

10. Phytophthora rot: *Phytophthora sojae*
11. Charcoal rot: *Macrophomina phaseolina*

Bacterial Diseases

1. Bacterial blight: *Pseudomonas savastanoi* pv. *glycinea*
2. Bacterial pustules: *Xanthomonas axonopodis* pv. *glycines*

Viral Diseases

1. Soybean mosaic: Soybean mosaic virus (SMV)

1. Leaf Spot of Soybean

Economic Importance and Occurrence

Cercospora leaf blight and purple seed stain are attributable to a shared fungal pathogen. The genesis of these maladies typically transpires towards the latter part of the growing season, spanning from the onset of seed development through pod maturation. While both afflictions have been documented across all soybean-cultivating regions within the United States, the southern states exhibit more pronounced yield detriments.

Causal Organism

The leaf spot disease in soybeans, induced by *Cercospora kikuchii*, pertains to the taxonomic hierarchy of Domain- Eukaryota, Kingdom- Fungi, Phylum- Ascomycota, Class- Dothideomycetes, Order- Capnodiales, Family- Mycosphaerellaceae, Genus- Cercospora, and Species- kikuchii. C. kikuchii serves as the primary etiological agent for both afflictions, displaying septate conidiophores of substantial dimensions, measuring 200-300 µm in length and 4-5 µm in width. The conidia produced by this pathogenic organism possess dimensions ranging from 170 to 190 µm in length and 3 µm in width. Notably, this pathogen is recognized for producing a photosensitive plant toxin termed Cercosporin, believed to contribute to the reddish-purple discoloration observed in infected tissues.

Symptoms

Cercospora leaf blight is primarily characterized by the emergence of purplish-brown lesions on the upper epidermal layer of the leaves. This chromatic shift from the verdant green hue progressively deepens into a more profound purple hue, ultimately enshrouding the entire upper leaf surface and bestowing a leathery texture upon the afflicted region. Repeated instances of infection prompt swift necrosis of the leaf tissues, culminating in defoliation, which initiates in the uppermost canopy. In contrast, purple seed stain manifests as irregular patches exhibiting a gamut of light to dark purple shades on the seed coat, frequently extending to envelop a substantial portion, if not the entirety, of the seed surface. Such infections can markedly compromise the quality, germination rate, and vitality of the seeds and ensuing seedlings. Importantly, protracted deferment of the harvest process can exacerbate the frequency of seed infection and discoloration.

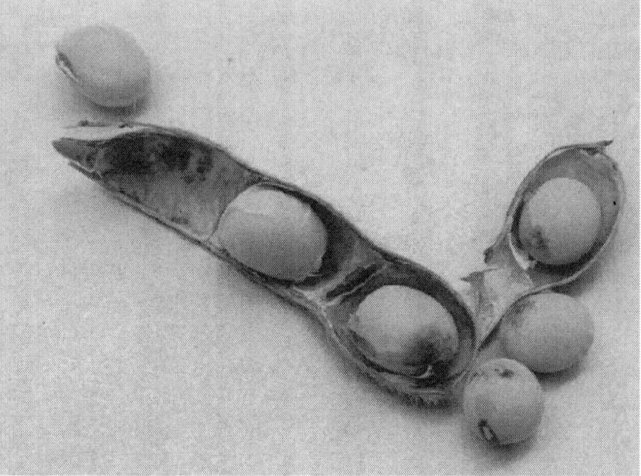

Figure 12.1: A. Cercospora Leaf Blight; B. Purple Seed Stain (Pat Miller).

Disease Cycle

Both afflictions stem from the fungus *Cercospora kikuchii*, which persists through the winter in remnants of the previous crop and contaminated seeds. As the weather transitions to a more humid and warmer state in the spring, the fungus initiates the production of spores, known as conidia. These minute propagules can be disseminated to newly sown soybeans, aided by the forces of wind or the impact of rain droplets. In some instances, symptoms may manifest on the seed coat, as the mycelium of *C. kikuchii* possesses the capacity to infiltrate the cotyledon and embryonic tissues.

Favorable Conditions

The most conducive circumstances for infection and disease progression materialize during prolonged periods of elevated humidity and temperatures ranging from 82° to 86°F, which create an optimal milieu for the pathogen.

Management Strategies

To proficiently combat the disease, several strategic measures can be implemented:

1. Opt for certified seeds devoid of disease pathogens and select later-maturing cultivars for optimum outcomes.
2. Implement a judicious crop rotation regimen, interchanging soybean cultivation with cereals to disrupt the pathogen's life cycle.

3. Execute a thorough eradication of plant remnants through meticulous plowing of the field immediately post-harvest.
4. Engage in chemical control methodologies, encompassing the application of foliar fungicides during the early pod development stage (R2-R4). Additionally, consider employing seed treatments if seed batches exhibit higher proportions of contamination.
5. For seed treatment, contemplate utilizing a blend of Thiram and Carbendazim (in a 1:1 ratio) at a rate of 2g per kilogram of seed.
6. A spray application of 1 per cent Bordeaux mixture or 2g of copper oxychloride or 2.5g of Zineb per litre of water has proven to be an effective control measure against leaf spot.

2. Wilt Disease of Soybean

Causal Organism

The Wilt disease of Soybean is brought about by *Fusarium oxysporum* f. sp. *tracheiphilum*, a member of the Kingdom Fungi, Phylum Ascomycota, Class Sordariomycetes, Order Hypocreales, Family Nectriaceae, Genus Fusarium, and Species *Fusarium oxysporum* f. sp. *tracheiphilum*. This fungus generates crescent-shaped macroconidia, characterized by their four to five septa, thin-walled structure, and translucent appearance. Additionally, it produces single-celled hyaline microconidia, which assume an oblong or oval morphology. Abundant chlamydospores are also produced.

Symptoms

Figure 12.2: Fusarium Wilt of Soybean.

Initial stages may not exhibit discernible symptoms, but they become apparent when the plant reaches approximately six weeks of age. Initially, certain plants display pallid green, flaccid leaves, which subsequently transition to a yellowish hue. The plants undergo stunted growth, chlorosis, drooping, premature leaf shedding, or wilting, ultimately leading to their demise within a short span. A brownish-purple discoloration in the cortical region is observed, often extending throughout the entire plant. In instances of severe root rot due to inadequate soil moisture, wilting may occur. Under such circumstances, disease-free plants continue to thrive, while diseased counterparts wilt and perish. This phase of the disease is most frequently observed in seedlings or young plants, and entire fields can be affected. When the stem base and taproot of an infected plant are longitudinally split, a brown or black discoloration of the vascular tissues becomes evident.

Favorable Conditions

The pathogen proliferates within temperatures ranging from 20-25 °C and necessitates a moist, humid environment for optimal spread.

Disease Cycle

The pathogen persists within infected crop residue in the field. Primary dissemination occurs through soil-borne chlamydospores and contaminated seeds. Secondary propagation takes place via conidia carried by irrigation water. Soybean cyst and sting nematodes render seedlings and young plants more susceptible to infection. *F. oxysporum* collaborates with *R. solani* to induce root rot in soybeans.

Management Strategies

1. Optimum for cultivars resistant to fusarium rot, soybean cyst nematode, and root-knot nematode.
2. Plant high-quality seeds in warm, well-drained soils.
3. Delay cultivation until soil moisture levels are low.
4. Treat seeds with Carbendazim or Thiram at a rate of 2 g/kg or treat with *Trichoderma viride* at 4 g/kg.
5. Administer spot drenching with Carbendazim at 0.5 g/litre.
6. Consider using Prothioconazole, an available fungicide with proven efficacy.

3. Anthracnose

Economic Importance and Occurrence

Anthracnose is a globally pervasive fungal ailment afflicting soybean crops wherever they are cultivated. In the Midwestern U.S., this malady is primarily induced by the fungal species *Colletotrichum truncatum*. However, it can also stem from several closely related species. *Colletotrichum* variants with an affinity for soybeans exhibit a broad host range, encompassing alfalfa, velvetleaf, and ragweed. It is important to note that the anthracnose that afflicts corn is attributed to a distinct pathogenic agent. Anthracnose has the capacity to infiltrate stems, leaves, and pods of soybean plants, and is typically present to varying degrees in soybean fields each growing season. While substantial reductions in yield are infrequent in the Midwestern U.S., they are more commonly observed in the Southern regions.

Causal Organism

Anthracnose of Soybean is instigated by *Colletotrichum truncatum*, classified within Domain-Eukaryota, Kingdom-Fungi, Phylum-Ascomycota, Class-Sordariomycetes, Order-Glomerellales, Family-Glomerellaceae, Genus-*Colletotrichum* and Species-*truncatum*. The causal fungus manifests numerous melanized fruiting bodies known as acervuli on the afflicted areas. These structures give rise to several abbreviated, dark spines (setae) emanating from each acervulus, imparting a dark, bristle-like aspect to the affected region. Seeds housed within affected pods may experience desiccation and become tainted with mold.

Symptoms

Figure 12.3: Anthracnose Infected Soybean Plants Displaying Rolled Leaves (*Source*: Muller).

Anthracnose can exert its influence on soybean plants at two distinct junctures: during the seedling phase and later, during the reproductive stage. Instances of pre- and post-emergence damping-off may transpire when planting contaminated seeds. However, the most prevalent infections occur during the bloom and pod development phases. Indications of the malady encompass irregularly contoured brown lesions manifesting on stems, petioles, and pods.

Severe manifestations entail leaf curling, untimely shedding, and hindered growth. Pods may display signs of desiccation and may house fewer, mold-infected, or even entirely non-viable seeds.

Disease Cycle

Colletotrichum species persist through the winter in the form of mycelia, taking residence on crop residues or within infected seeds. The presence of infected seeds can lead to instances of pre- and post-emergence damping off. Spores from infected plant residues are disseminated by wind and rain, resulting in infections occurring at various stages of soybean growth. Infections on leaves, stems, and pods typically arise later in the growing season, with the development of distinctive black fungal fruiting bodies on afflicted tissue, particularly as soybean plants near maturity. The disease endures on both plant and seed residues, serving as a potential source of infection for subsequent crop cycles.

Favorable Conditions

Anthracnose thrives under conditions of moisture and warmth, while disease severity tends to abate during dry spells. The likelihood of infection heightens when periods of leaf wetness, rainfall, or dew extend beyond 12 hours per day.

Disease Management

1. Employ crop rotation involving non-host crops to mitigate the presence of anthracnose inoculum within the field.
2. Implement tillage techniques that bury or shred crop residues, facilitating their decomposition in the soil and diminishing anthracnose inoculum, particularly in prior soybean fields or fields with other host crops.
3. Apply seed treatment using Carboxin + Thirum at a rate of 2g/kg of seed, followed by a foliar spray of trifloxystrobin + tebuconazole at 0.1 per cent.
4. Administer foliar fungicides between the initial pod development (R3) and early seed formation (R5) stages of soybean growth to curb anthracnose and mitigate seed infection.

4. Brown Spot of Soybean

Economic Importance and Occurrence

Brown spot, or Septoria leaf spot, has been documented throughout the southern U.S. soybean cultivation belt. This malady can induce premature defoliation, contributing to yield losses when susceptible cultivars are planted and conditions favor disease progression.

Causal Organism

Brown Spot of Soybean arises from *S. glycines*, a member of Domain-Eukaryota, Kingdom-Fungi, Phylum-Ascomycota, Class-Dothideomycetes, Order-Capnodiales, Family-Mycosphaerellaceae, Genus-*Septoria*, and Species-*glycines*.

Symptoms

Characteristic signs of brown spot encompass irregularly shaped, minute brown leaf spots, varying in size from minuscule specks to roughly 1/5 inch in diameter, evident on both the upper and lower leaf surfaces. Adjacent spots may coalesce, forming irregularly shaped patches. These lesions darken over time to a blackish-brown hue and often exhibit a yellowish halo around the spot. As the plant nears maturity, severely affected leaves take on a rusty brown appearance and may prematurely detach. When held up to light, infected leaves reveal opaque, dark lesions.

Figure 12.4: Septoria Leaf Spot of Soybean Leaf.

Disease Cycle

The brown spot disease is initiated by the fungus Septoria glycines, which overwinters on crop residues and within infected seeds. Initial infections occur on cotyledons and leaves through the release of conidia from pycnidia, flask-shaped fruiting structures found on crop debris. Conidia germinate on leaf surfaces and penetrate the plant through stomata. Secondary infections arise as conidia disperse upwards in the canopy, aided by wind or rain, impacting leaves, petioles, stems, and pods.

Favorable Conditions

Infection and disease development can occur within a temperature range of 60°F to 85°F, with an optimal range of 77°F. Longer durations of leaf wetness (up to 36 hours) correlate with increased disease incidence and severity.

Management

To effectively manage brown spot disease, consider implementing the following measures:

1. Utilize soybean varieties exhibiting resistance to the disease.
2. Practice crop rotation with non-host crops like corn, cotton, rice, or grain sorghum for a minimum of two years.
3. Employ tillage practices that reduce the accumulation of crop residue on the field's surface.
4. Avoid planting soybeans ultra-early (mid to late-April) and refrain from consecutive years of soybean cultivation in the same field.
5. Apply a foliar spray of fungicides belonging to the Strobilurin or triazole class at the R3 to R5 growth stages for optimal efficacy.

6. Downy Mildew of Soybean

Economic Significance and Occurrence

Downy mildew, while distributed globally, is predominantly observed in Arkansas and seldom leads to significant yield losses. However, under exceptionally favorable environmental conditions, the disease may escalate, causing premature defoliation, thereby contributing to diminished seed quality and smaller seed sizes.

Causal Organism

Downy mildew of soybean is instigated by *Pernospora manshurica*, categorized within Kingdom-Chromista, Phylum-Oomycota, Order-Peronosporales, Family-Peronosporaceae, Genus-*Peronospora*, and Species-*manshurica*.

Symptoms

Early indications of downy mildew may emerge on young plants, but the disease does not proliferate across the field until the late vegetative or early reproductive growth stages. Initially, upper leaf surfaces exhibit faint green to pale yellow spots that gradually expand into vibrant yellow patches. On the underside of the leaf, a gray, velvety fuzziness becomes discernible. The centers of these spots assume a brownish hue, bordered by a yellow rim. Moreover, infected seeds display a lackluster white appearance, often partially or entirely covered in fungal spores.

Figure 12.5: Upper and Lower Leaf Surface Leaves with Downey Mildew (*Source*: T. R. Faske).

Favorable Conditions

Optimal conditions for disease development involve cooler temperatures ranging from 68 to 72 °F, coupled with heightened humidity levels. Wind and rain act as vehicles for spore dissemination on the plants, facilitating infection and enabling swift spread throughout the field, particularly during periods characterized by cool, wet, or humid weather.

Disease Cycle

The disease cycle of downy mildew is initiated by the pathogen *Pernospora manshurica*, which overwinters within leaf debris, and infrequently, on seeds. Within the leaf debris, the fungus exists as oospores, characterized by their thick-walled, dormant state. Come spring, these oospores germinate and proceed to infect susceptible seedlings. Once infection takes hold, the fungus generates sporangia, which are infectious spores, on the newly infected leaves. These sporangia emerge from sporangiophores, acting as spore-bearing stalks. The visible appearance of these fungal structures is the downy-like growth evident on the lower leaf surface, thus lending the name downy mildew. Subsequently, the sporangia are disseminated via wind, finding their way to infect other plants.

Management Strategies

1. Opt for resistant soybean varieties against the disease.
2. Employ crop rotation with non-host crops, reducing the inoculum level in the field.

3. Implement meticulous crop residue management techniques, such as burying residues, effectively diminishing the quantity of inoculum present.
4. Administer foliar spray of Chlorothalonil 75 per cent WP at a rate of 1-2 ml per litre of water

7. Powdery Mildew of Soybean

Occurrence and Economic Significance

Powdery mildew epidemics occur sporadically in soybean growing regions approximately every 10-15 years, with the initial observation dating back to 1975. This disease manifests sporadically on soybeans, typically appearing late in the season during periods of cooler temperatures. While it rarely leads to significant yield loss, it can potentially induce defoliation and reduce yields in susceptible varieties.

Causal Organism

Powdery mildew of soybean is attributed to the fungus *Microsphaera diffusa*, classified within the kingdom-Fungi, Phylum-Ascomycota, Class-Leotiomycetes, Subclass-Leotiomycetidae, Order-Erysiphales, Family-Erysiphaceae, Genus-*Microsphaera*, and Species-*diffusa*.

Symptoms

Figure 12.6: White Powder like Patches Observed on Infected Leaf (*Source*: Dean Muller).

Powdery mildew of soybean is contingent upon cool air temperatures and low relative humidity. Characteristic symptoms include the emergence of white, powdery patches comprised of mycelium and conidia. These primarily develop on cotyledons, stems, pods, and notably, the upper leaf surface. Initially, small colonies form, which progressively expand and merge until the entire surface of the infected plant parts is enveloped with mycelium and conidia. It's worth noting that symptom manifestation is less frequent than the presence of observable signs of the pathogen. Leaf symptoms include the presence of green and yellow islands, interveinal necrosis, necrotic specks, and leaf blade crinkling, culminating in defoliation. Interestingly, these symptoms may be almost imperceptible when mycelial growth is particularly vigorous.

Disease Cycle

The propagation of powdery mildew fungi does not involve seeds or stubble. Instead, initial infection by the powdery mildew pathogen is likely facilitated by airborne spores originating from volunteer soybean plants or other alternative hosts. These spores do not require free water for germination. In periods of heightened humidity, the spores germinate, and the resulting fungal strands traverse the surface of the plant part. Subsequently, conidiophores emerge from the fungal strands, perpendicular to the leaf surface. The swift development of powdery mildew in a crop can be attributed to the brief time span between infection and the production of conidia, which can be as short as 7 days, generating millions of airborne conidia from a single leaf.

Management Strategies

1. Avoid planting susceptible varieties in shaded low-lying areas, as these locations often exhibit high humidity levels.
2. Refrain from excessive application of nitrogen fertilizer to prevent stimulating succulent growth, which is particularly susceptible to infection.
3. Implement proper spacing and thinning practices to encourage air movement, light penetration, and leaf drying.
4. Consider the application of sulphur-based fungicides like Calaxin at a rate of 0.5 ml per litre of water (foliar spray), as they have demonstrated efficacy in effectively managing powdery mildew.

7. Sudden Death Syndrome (SDS)

Occurrence and Economic Importance

Occurrence and Economic Significance: Sudden Death Syndrome (SDS), initially reported in Arkansas in 1971, has since been identified in most major soybean production regions of the U.S. This disease is often observed in well-managed, high-yield potential, irrigated fields grown under optimal conditions. Yield losses can range from minimal to as high as 100 per cent, depending on the timing of infection, cultivar susceptibility, and disease severity.

Causal Organism: *Fusarium virguliforme*

Symptoms

The symptoms of Sudden Death Syndrome (SDS) manifest in both the roots and foliage of soybean plants. On the foliage, initial signs include small, pale green, circular spots on the leaves during the early reproductive growth stages. These spots subsequently expand into striking yellow irregular blotches positioned between veins, while the veins themselves retain their green color. The yellow blotches eventually transition to a brown hue. In severe cases, the leaves prematurely drop, leaving the petioles still attached to the stem. As for the roots, an SDS-infected soybean plant experiences rotting and discoloration. Diseased plants can be effortlessly pulled from the soil due to the decay of their lateral roots.

Figure 12.7: Chlorotic and Necrotic Blotches between Central Leaf Veins on Plants Infected with Sudden Death Syndrome (*Source*: T. R. Faske).

Disease Cycle

Sudden Death Syndrome (SDS) is attributed to the soil-borne fungus *Fusarium virguliforme*, which overwinters in the form of thick-walled spores (chlamydospores) within the soil or on crop residue. Although infection may occur early in seedling development, symptoms only become discernible when plants have reached the mid-reproductive stages of growth. The most severe manifestation of symptoms typically occurs at temperatures ranging from 68 to 77°F. Interestingly, hot and dry weather appears to impede the progress of SDS, although instances of severe disease have been reported under these conditions. The disease can be particularly severe in fields also infested with soybean cyst nematodes, with its impact most pronounced in cultivars susceptible to both the fungus and the nematode. SDS is often most severe in saturated soils and is particularly prevalent in furrow-irrigated fields near the header pipe or in low-lying areas susceptible to waterlogging. Furthermore, high fertility and soil compaction exacerbate disease severity.

Management Strategies

1. Select soybean cultivars with lower susceptibility to SDS, as highly resistant cultivars are currently unavailable.
2. Implement cultural practices that enhance field drainage and consider crop rotation with a non-host crop for soybean cyst nematodes over a period of 2 years, which can help reduce SDS severity.
3. Note that foliar fungicides are ineffective in suppressing this disease, but seed treatments containing active ingredients like adepidyn or fluopyram have shown excellent efficacy in managing this disease.

8. Bacterial Blight of Soybean

Occurrence and Economic Importance

Bacterial blight is the most prevalent bacterial disease of soybeans, occurring in all soybean producing regions worldwide. It is one of the earliest leaf spot diseases to appear on young plants. Bacterial blight has been reported to cause significant yield reductions on susceptible cultivars under heavy disease pressure.

Causal Organism: *Pseudomonas savastanoi* pv. *glycinea*

Symptoms

Figure 12.8: Small Necrotic Lesion with Yellow Halos (*Source*: T.R. Faske).

Bacterial blight, an extensively prevalent bacterial ailment, impacts soybean crops on a global scale. While it is typically of limited concern in many soybean production regions in India, it stands out as one of the foremost leaf spot diseases to afflict nascent plants. It is worth noting that susceptible cultivars can endure substantial yield reductions under the weight of intense disease pressure. Although primarily a leaf-centric malady, bacterial blight's symptoms may also emerge on stems, petioles, and pods. Initially, leaf symptoms present as angular, translucent yellow to light brown spots, which exude a water-soaked appearance. As these spots progress, their cores assume a reddish-brown hue, sinking inward and surrounded by a water-soaked periphery, all encompassed within a yellowish-green halo. In instances of severe infection, premature defoliation may ensue.

Favorable Condition

The development of bacterial blight is facilitated by winds, coupled with cool and rainy weather, particularly at temperatures ranging from 75 to 79°F. Conversely, its progression is suppressed by hot and dry weather.

Disease Cycle

The pathogenic bacterium accountable for bacterial blight endures the winter within infected plant residue and seeds. Lesions originate on cotyledons due to infected seeds and may lead to secondary infections on leaves. The pathogen gains entry through stomata or wounds. Dissemination of the bacterium occurs via wind-blown rain and during cultivation or spraying episodes when the foliage remains damp. Seeds may also become infected through pods during the growing season.

Management Strategies

1. Employing disease-tolerant cultivars and utilizing pathogen-free seed.
2. Adopting crop rotation practices, alternating soybeans with non-legume crops.
3. Engaging in residue management and tillage practices that involve burying infested soybean residue, thus mitigating the disease risk. Exercise caution to avoid cultivation when foliage is moist.
4. Implementing seed treatment with streptocyclin @ 250 ppm (2.5 g/10 kg seed).
5. Considering the use of copper fungicides @ 2g/litre of water along with streptocyclin @ 250 ppm as an effective management approach for this pathogen.

9. Bacterial Pustule of Soybean

Occurrence and Economic Significance

Bacterial pustule has been identified in South Dakota. Due to the pustule formation, this disease is often mistaken for Asian soybean rust (which has not been observed in South Dakota). Unlike bacterial blight, it is important to note that bacterial pustule is not deterred by warm temperatures and will persist in spreading infection.

Causal Organism: *Xanthomonas axonopodis* pv. *glycines*

Symptoms

Early signs of this ailment involve the appearance of minute, light green spots with raised centers, visible on both leaf surfaces, although predominantly on the lower side. Disease progression leads to the formation of light-colored pustules, frequently centralized within the lesions. As these lesions merge, they give rise to extensive, irregular areas (patches) of necrotic tissue. Subsequent exposure to wind or rain dislodges the deceased tissue from the leaves, resulting in a ragged appearance.

Figure 12.9: Small Reddish Brown Lesions (*Source*: C. Coker).

Disease Cycle

The onset of bacterial pustule on soybeans is initiated by the pathogenic agent *Xanthomonas axonopodis* pv. *glycines*. This bacterial culprit endures the winter on crop residues and is disseminated through wind-blown rain, rain splashing from the soil, and often during cultivation when foliage remains moist. The bacteria secure entry into the plant through natural openings and wounds. Notably, the development of bacterial pustule thrives under warm, moist conditions, ideally within the temperature range of 85-90 °F.

Management Strategies

1. Utilizing disease-tolerant cultivars and employing pathogen-free seed.
2. Embracing crop rotation, alternating soybeans with non-host crops, such as corn.
3. Employing the practice of burning infested soybean residue to minimize the risk of disease. Exercise caution to avoid cultivation when foliage is damp.
4. Considering seed treatment with Streptocycline at 100 ppm, as this approach yields the most effective management of this pathogen.
5. Contemplating the use of copper fungicides @ 2g/litre of water along with streptocyclin @ 250 ppm as an effective management approach for this pathogen.

10. Soybean Mosaic Virus (SMV)

Occurrence and Economic Significance

Soybean mosaic virus (SMV) is ubiquitous across all soybean production regions worldwide. Yield decrement fluctuates between 8 per cent and 35 per cent, with some production systems experiencing a staggering 94 per cent reduction. While symptomatology may vary among different soybean cultivars, the most prevalent manifestation is a mosaic pattern characterized by shades of green and yellow. In advanced stages, this mosaic pattern transitions into a blend of yellow and brown, often culminating in premature defoliation.

Pathogen

The etiological agent responsible for this malady is the Soybean mosaic virus, classified within the Potyvirus family. This virus is characterized by its sinuous particles, measuring between 750 to 900 nanometers in length, and bearing a single-stranded RNA genome.

Symptoms

Figure 12.10: A. Mosaic Pattern on Infected Leaf (*Source*: J. Zhou) B. Leaf Yellowing at Late Infection Stage (*Source*: J. Zhou).

Plants afflicted by this virus display stunted growth, accompanied by leaves that exhibit puckering, crinkling, and a narrowed morphology. The leaf margins conspicuously curl downwards, and the leaf blades are profoundly undulated along the edges of the principal veins. Notably, the leaves appear more slender and possess a deeper shade of green compared to their healthy counterparts. As the disease advances, the pods yield fewer and smaller seeds, and the affected seeds display mottling and distortions. Consequently, infected seeds either fail to germinate or produce diseased seedlings.

Disease Cycle

Transmission of the Soybean mosaic virus occurs readily via sap, with or without the application of abrasive agents. Grafting is also a viable mode of transmission. Remarkably, a proportion of seeds from infected plants may carry the virus, which remains viable within seeds for at least two years. Dissemination of this virus is facilitated by at least 30 species of aphids in a non-persistent manner. For translocation and systemic infection to transpire, the viruses must undergo replication within infected cells. Once established, the Soybean mosaic virus can move both upwards and downwards within plants, being detectable in all components of systematically infected plants. The progression of the disease is facilitated by lower temperatures, approximately 18°C, and humid atmospheric conditions.

Management Strategies

1. Implementation of deep summer ploughing to mitigate disease incidence.
2. Utilizing soybean varieties that exhibit resistance or tolerance to the virus to mitigate its impact.
3. Employing healthy and certified seeds to diminish the risk of introducing the virus.
4. Maintaining weed-free fields to minimize potential virus vectors.
5. Removing infected plants through rouging and subsequent incineration to prevent further spread.
6. Administering pre-sowing soil application of Phorate at a rate of 10 kg/ha to impede disease progression.
7. Employing two foliar applications of Thiamethoxam 25 WG at 100 g/ha or Methyl demeton at 800 ml/ha, executed at 30 and 45 days post-sowing, to further manage the disease.

References

Akem, C. N. (1996). Management of soybean diseases. IITA Research Guide 40. Training Program, Intemational Institute of Tropical Agriculture (IITA), Ibadan, Nigeria. 36p.

Byamukama, E., Tande, C., Strunk, C. and Mathew, F. (2020). South Dakota state university of agronomy, horticulture and plant science department, 1-3

Dunleavy, Y. J. M., Chamberlain, d. W. and Ross, j. P. (1966). Soybean diseases 1-38

Faske, T. and Kirkpatrick, T. Cercospora Leaf Blight and Purple Seed Stain of Soybean, Division of agriculture and extension, university of Arkansas.

Faske, T., Kirkpatrick, T., Zhou, J. and tzanetakis, L. (2014) Soybean Diseases, Division of agriculture and extension, university of Arkansas. 1-18.

Giesler, L. J. (2011). Bacterial dsieases of soybean, university of Nebraska-Lincoln extension, institute of agriculture and natural resources.

Goodnight, M. and Tian, P. (2021). Cercospora Leaf Blight of Soybean, Plant Diagnostic Clinic, University of Missouri.

Grau, C. (2006). Powdery Mildew of Soybean, UW Madison Department of Plant Pathology, 1630 Linden Drive Madison, WI 53706

Groves, C. and Smith, D. (2021). Sudden Death Syndrome of Soybean. Universitry of Wisconsin-Madison Plant Pathology.

Hershman, E. D. (2012). Brown Spot of Soybean, Cooperation extension service university of Kentucky U.K.

Jeschke, M. (2020). Cercospora Leaf Blight and Purple Seed Stain of Soybeans, Pioneer Agronomy Sciences Vol. 12 No. 28

Jeschke, M. and Butzen, S. Anthracnose of Soybean, Crop focus, DuPont Pioneer Agronomy Sciences.

Kolhe, S. and Gupta, G. K., - Web based Soybean Disease Expert System, ICAR-Directorate of Soybean Research, Khandwa Road, Indore (M.P)-452001

Ryley M. J. (2015). Mildew diseases of soybean, Soybean fact sheet, Soy Australia Ltd. 1-4

Soyabean Outlook, January (2022). Agricultural Market Intelligence Centre, PJTSAU 1- 4.

Strunk, L. C. and Langham, m. (2019). Bacterial Diseases of Soybean. South Dakota Board of Regents, **58** 513-516.

Strunk, L. C. and Langham, m. (2019). Fungal and Fungal-like Diseases in Soybeans. South Dakota Board of Regents, **58** 517-542.

Sweets, E. L., Wrather, A., and Wright, S. (2008). Integrated pest management, soybean diseases, university of Missouri extension, 3-26.

Chapter 13

Diseases of Moong and Urd bean Crop

Prabha Siddharth*[1], **Vishwa Vijay Raghuvanshi*[2], ***Girijesh Kumar Jaisval*[1] *and* ***Arjun Rana*[1]**

[1]Department of Plant Pathology, Chandra Shekhar Azad University of Agriculture and Technology, Kanpur – 208 002, Uttar Pradesh
[2]Department of Plant Pathology, Acharya Narendra Deva University of Agriculture and Technology, Kumarganj – 224 229, Ayodhya, Uttar Pradesh
**e-mail: prabhasiddharth4864@gmail.com*

Introduction

In Asia, Mung bean (*Vigna radiata*) and urd bean (*Vigna mungo*) are significant pulse and catch crop. The introduction of short-duration cultivars has enabled the mung and urdbean to spread into new areas, including South America and South Africa. Abiotic and biotic factors both reduce mung and urdbean productivity. The main diseases that affect mung and urd bean include yellow mosaic, Cercospora leaf spot, anthracnose, bacterial leaf spot, powdery mildew, halo blight and tan spot. Drought, waterlogging, heat stress and salinity are important abiotic factors that affect mung and urd bean yield. In India, cultivation of mungbean is done in about area of 3 million hectare with the production of about 1 million tonnes. Mung bean seeds are a good source of protein (around 24 per cent of it is easily digested), fibre, phytonutrients, and antioxidants. Mung and urd bean is susceptible to many harmful diseases. According to Iqbal *et al.* (2014), the primary biotic constraint on yield is disease losses. In order to manage mung and urd bean fungal, bacterial, and viral diseases in the main South Asian growing regions, Pandey *et al*. (2018) and Naimuddin *et al*. (2016) give reviews. We have consulted a variety of sources. The International Mung bean Improvement Network (IMIN), which was founded in 2016, for instance, concentrates on gathering genetic resources. Mung bean and urd bean yields are significantly reduced by both abiotic and biotic stresses. The main diseases include Cercospora leaf spot, anthracnose, dry root rot, powdery mildew, web blight and mung bean yellow mosaic virus. Mung and urd bean diseases control by use a variety of techniques as a component of an integrated disease management (IDM) programme. The initial strategy for defense is cultural techniques. For a healthy crop, cultural practices related to time of sowing, soil preparation, trap crop, using disease free seed and crop rotation are crucial. Physical, biological and chemical strategies use for disease management have been studied and reported. The economic implications, sustainable management solutions and significant management issues for the foliar and soil-borne mung bean fungal infections are discussed here. This chapter contains detailed information regarding the significance of bacterial, fungal and viral diseases on mung and urdbea crop including its etiology, disease cycle and integrated disease management. Diseases of mung and urdbean categories as:

A. Fungal Diseases

1. Web blight: *Rhizoctonia solani*
2. Cercospora leaf spot: *Cercospora canescens*
3. Anthracnose: *Colletotrichum lindemuthianum*
4. Powdery mildew: *Podosphaera xanthii*
5. Root rot and leaf blight: *Rhizoctonia solani*
6. Rust: *Uromyces phaseolitypica*
7. Dry root rot: *Macrophomina phaseolina*

B. Viral Diseases

1. Mung bean yellow mosaic virus
2. Urd bean leaf crinkle virus
3. Leaf curl/Necrosis leaf bud necrosis virus

C. Nematode Disease

1. Root knot nematode: *Meloidogyne incognita*

Fungal Diseases

1. Web Blight

Economic Importance and Occurrence

Mungbean web blight levels in Eastern Uttar Pradesh ranged from 1.0 to 69.0 per cent, on average 12.7 per cent. At various levels of disease severity and in various varieties of mungbean, the disease was seen to diminish grain production by 33 to 40 per cent and 1000 grain weight by 28.6 per cent. One of the most critical issues with this crop in Northern India, where it has caused extensive damage to mungbean, is the web blight.

Causal Organism

The web blight of mung and urdbean is caused *Rhizoctonia solani* which belongs to Kingdom- Fungi, Phylum - *Basidiomycota*, Class- *Agaricomycetes*, Order- *Cantharellales* and Family- *Ceratobasidiaceae*. On the damaged plant portions, a lot of the coloured, spiderweb-like mycelial growth and white microsclerotia appear. Within a couple of days, the sclerotia change from brown to chestnut brown in color. *R. solani* colonies range in color from bright yellowish white. The color of the letter changes from yellowish brown to pale brown to dark brown.

Symptoms

Figure 13.1: Web Blight
(*Source*: Bal *et al.*, 2019).

The fungus affects all of the plant's above-ground tissues, including the leaves, petioles, stems, and pods, but it is most harmful to the foliage. Small, water soaked, irregular, greenish-pale patches that emerge damp on any part of the leaf or stem are the first signs of the condition. Leaflets shrink and dry out as the region becomes a dark brown color. The sickness spreads quickly when it is humid out there. It initially only affects the lower leaf's petiole and basal portions before spreading higher. This phase of the illness is particularly noticeable and harmful as the spot quickly spreads to cover larger areas of the leaf blade and stem.

Favourable Conditions

The development of the web blight disease brought on by *R. solani* is significantly influenced by environmental conditions. Higher soil temperatures (30–30°C), relative humidity close to 100 per cent, and higher aerial temperatures (26 – 32 °C) all favored the development of severe illness. Rainfall (91-97 mm) was a major factor in the web blight's rapid development.

Disease Cycle

The pathogen is mostly soil borne and produces sclerotia in soil and on plant tissue/seed coat, it can persist for many years. In addition to colonizing soil organic materials as saprophytes, *R. solani* also survives as mycelium, often as a result of plant pathogenic activity. *R. solani* causes web blight of mungbean, and has the ability to spread the disease through both soil- and air-borne modes of infection.

Management

1. Proper field sanitation should be done.
2. Removal of alternate host.
3. Rouging.
4. Use of resistant varieties such as Pusabaisakhi, SML 688 *etc.*
5. Crop rotation should be followed.
6. Seed treatment with Nativo @ 2.5 gm/kg of seed.
7. Spray of Propiconazol 3ml/l of water twice at 15 days interval.

2. Cercospora Leaf Spot

Economic Importance and Occurrence

It is a widespread and devastating disease caused by pathogen *Cercospora canescens* which causes loss of yield about 60 per cent. This disease was first reported in Delhi, India and suspected to occur in all parts of humid tropical area of India, Bangladesh, Phillipines *etc.* The disease causes quantitative and qualitative losses upto 95 per cent under natural epiphytotic conditions.

Causal Organism

Cercospora canescens is the causal agent of this disease. Clusters of dark brown septate conidiophores are produced by the fungus. The conidia are hyaline, linear, with thin wall and five to six septa. *C. canescens* is belongs to the Kingdom – *Fungi*, Phylum – *Ascomycota*, Class- *Dothideomycetes*, Order – *Capnodiales*, Family – *Mycosphaerellaceae*, Genus – *Crcospora*, and Species - *canescenes*.

Symptoms

On the leaves, tiny circular spots with a brown edge and a yellowish halo encircle a grey center. Brown irregular lesions are formed when several spots combine. In extreme situations, there is considerable defoliation. In severe cases, the stem and petioles may exhibit the brown lesions. The bacterium can also be found on uninfected plant waste and black gramme plants that are growing on their own. On the centers of the dots, the fungus may be seen as powdery growth.

Figure 13.2: Cercospora Leaf Spot (*Source*: Plantix).

Favourable Conditions

Splattering showers and humid conditions is favoured to the attack of pathogen. Excessive moisture or frequent dew leads to development of disease.

Disease Cycle

The fungus lives off of seeds and decaying plant matter. Conidia carried by the air are the secondary spread. Depending on the type of plant affected, a variety of Cercospora species can cause Cercospora leaf spots. Plant detritus, diseased plants, and contaminated seeds are the main sources of infection. Conidia carried by the air cause secondary spread. The severity of the illness and its spread are favored by high humidity. The fungus can survive the winter on plant waste and hangs around in the field for a while.

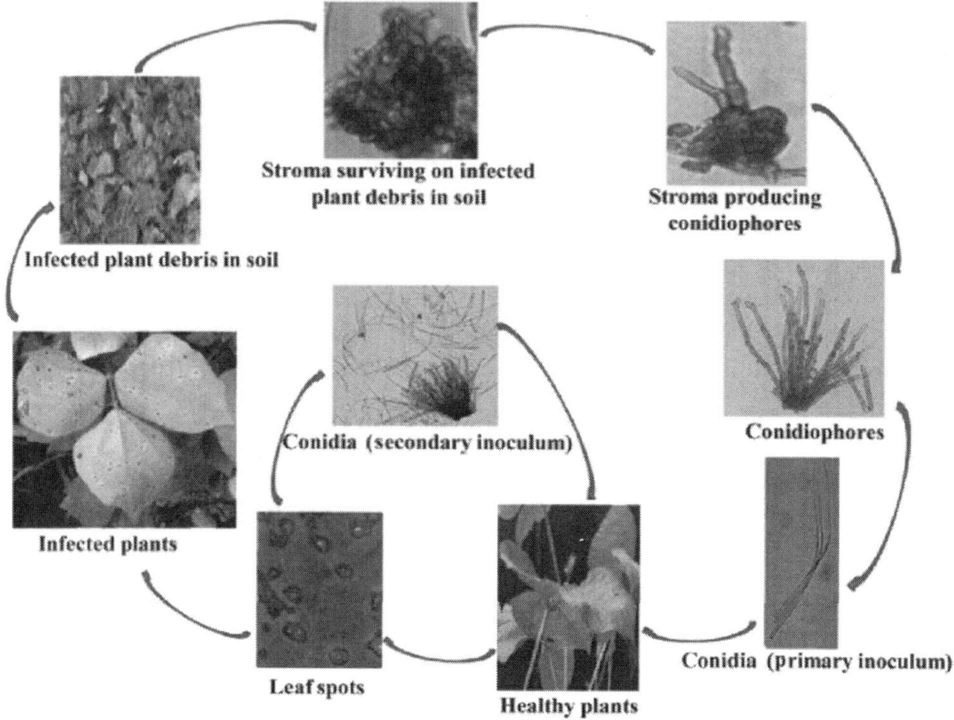

Figure 13.3: Disease Cycle of Cercospora Leaf Spot (*Source*: Chand *et al.,* 2015).

Survival and Spread

The fungus is soil-borne and spreads on damaged plant waste. It only survives by colonizing host tissues like a parasite. The fungus persists in plant debris that has been infected.

Management

1. Remove and burn infected plant debris.
2. Choose diseases resistant variety like T-9 and Pant Urd-19.
3. The crude extracts of cassava, garlic and zinger are applied for controlling the disease effectively.
4. Mulching reduces the disease incidence resulting in increased yield.
5. Spray of Mancozeb @ 2 kg/ha or Nativo at 500 g/ha.

3. Anthracnose Disease

Economic Importance and Occurrence

In 1875, plant specimens collected in Germany were used to describe anthracnose for the first time. Since then, the illness has grown to be one of the most significant and pervasive in the entire world. The occurrence of this serious disease, which is frequently seen in Gujarati areas where crops are grown. The anthracnose disease was originally discovered in India in Jorhat Assam, and it has since been discovered in all of India's major mungbean-growing states, ranging in severity from mild to severe in tropical and subtropical regions. The most dangerous illness that affects beans in cool climates in Latin America and Africa. Up to 90 per cent of field losses in these areas are attributable to seedling, leaf, stem, and pod infections when the climate is conducive to the disease.

Causal Organism

The causal organism of disease is *Colletotrichum lindemuthianum* (**Sexual stage**: *Glomerella lindemuthianum*) belongs to the Kingdom – *Fungi*, Phylum – *Ascomycota*, Class –*Sordariomycetes*, Order – *Glomerellales*, Family – *Glomerellaceae*, Genus – *Colletotrichum*, Species – *lindemuthianum*.

Symptoms

Figure 13.4: Anthracnose Disease
(*Source*: Rajaj, Annamalai university, Bugwood.org).

The symptoms are present in every plant aerial portion. The fungus damages seedlings by causing circular, black, sunken spots with a dark center and bright red or orange edges on the hypocotyl portion of leaves. The stem and petioles may both exhibit the lesions. On the pods, the noticeable symptom is seen. A small, water-soaked lesion that was first brown in colour then grew to the size of a circular, depressed area with a dark center and a vibrant red or yellow edge emerges on the pods. Acervuli-filled necrotic regions are created when many spots combine. Seeds from infected pods are stained.

Favourable Conditions

1. The disease is worse during the cool and rainy seasons.
2. High relative humidity (more than 90 per cent).
3. Low temperatures favour this disease (15-20 °C).

Disease Cycle

Conidia are generated in acervuli, emerge from the stroma under the epidermis, and eventually rupture to become erumpent. The fungus's mycelium is septate, hyaline, and branched. The acervulus has a few dark coloured, septate setae. Hyaline, short conidiophores with oblong or cylindrical, single-celled, hyaline, thin-walled conidia containing oil globules are present. The fungus's ideal stage results in perithecia with a small number of asci, usually eight of which are one- or two-celled ascospores with a central oil globule.

Survival and Spread

1. The pathogens are able to thrive on plant waste and seeds.
2. Conidia carried by the air cause disease to spread in the field.

Management

1. Remove and destroy infected plant debris in soil.
2. Choose disease resistant variety.
3. Seed treatment with *Trichodermaviride* @ 4gm/kg of seed or with Zineb at 2 g/kg.
4. Spray Nativo 500 g or Mancozeb@ 2kg/ha soon after the appearance of disease and repeat after 15 days.

4. Powdery Mildew

Powdery mildew is one of the common diseases of various legumes. On leaves and other green objects, white powdery spots first develop before they turn dull in colour. This round, progressively larger patches cover the lower surface as well. When the infection is severe, whitish powdery growth totally covers both surfaces of the leaves.

Economic Importance and Occurrence

From different countries the impact of powdery mildew disease of mungbean was reported which causes the yield loss upto 22 per cent in Philippines, upto 50 per cent in Australia and from 40-100 per cent in India. From Gujarat the yield loss was reported upto 40 per cent, 25 per cent in Maharashtra and about 50 per cent in Uttar Pradesh.

Causal Organism

Podosphaera xanthii is the causal agent of powdery mildew of mung and urdbean belongs to the Kingdom – *Fungi*, Phylum - *Ascomycota*, Class – *Leotiomycetes*, Order – *Erysiphales*, Family- *Erysiphaceae*, Genus- *Podosphaera*, and Species -*xanthii*. As it spreads across the leaf's surface and sends haustoria into the epidermal cells, the fungus, which is ectophytic, is active. Vertically emerging conidiophores with conidia in short chains appear on the leaf surface. Hyaline, single-celled conidia have thin walls and are elliptical, cylindrical, or barrel-shaped. Cleistothecia are tiny, globular, black entities with myceloid appendages that emerge later in the season. Each ascus has 3–8 elliptical, hyaline, single-celled ascospores, and each cleistothecium has 4–8 asci.

Symptoms

On the upper surface of the leaves, and occasionally on both surfaces, there are tiny, erratic powdery specks. During the stage of pod growth and flowering, the disease gets worse. The leaves, petioles, stem, and even the pods are entirely covered in the white powdery patches. The plant has a greyish white appearance, and its leaves eventually droop and fall off. Frequently, pods have few ill-filled seeds and are deformed and tiny.

Figure 13.5: Powdery Mildew (*Source*: Plantix).

Favourable Conditions

Cool temperature (10-20 °C) and moist weather (90 per cent relative humidity) favours to development of diseaset.

Disease Cycle

Infected plant waste contains cleistothecia, which are the pathogen's means of survival as an obligate parasite. Ascospores from persistent cleistothecia typically cause the primary infection. Conidia that are airborne are responsible for the secondary spread. The disease is spread by splashes of rain as well.

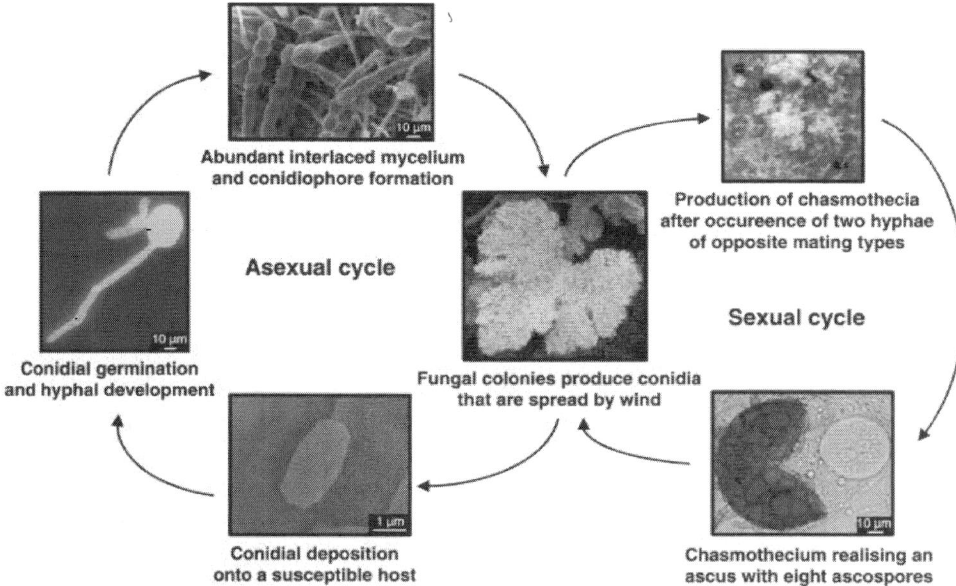

Figure 13.6: Disease Cycle of Powdery Mildew (*Source*: Garcia, 2008).

Survival and Spread

The disease may live on a variety of hosts and survives in conidial form off-season. Secondary dissemination occurs by airborne conidia formed during the season.

Management

1. Remove and destroy infected plant debris.
2. Use of hyperparasite *i.e., Ampelomyces quisqualis* for powdery mildew management.
3. Spray Wettablesulphur 2kg or Tridemorph 500 ml/ha at the initiation of disease and repeat 15 days later.

5. Root Rot and Leaf Blight

Economic Importance and Occurrence

It is one of the major seed-borne diseases of mung and urdbean. *Macrophomina phaseolina*, is responsible for a wide range of symptoms on different agricultural plants, such as seedling blight, root rot, charcoal rot, wilt or dry root rot, stalk rot, stem blight, fruit rot, seedling decay or damping-off, crown rot, and leaf blight. It is a devastating disease of beans and occurs in entire India.

Causal Organism

Rhizoctonia solani and *Macrophomina phaseolina* are the responsible fungus to cause root rot and leaf blight diseases of mung and urdbean. Systematic position of *Rhizoctonia* already mentioned under the disease web blight, The fungi *Macrophomina phaseolina* belongs to the Kingdom – *Fungi*, Phylum - *Ascomycota*, Class – *Dothideomycetes*, Order - *Botryosphaeriales*, Family – *Botryosphaeriaceae*, Genus – *Macrophomina*, and Species – *phaseolina*.

Symptoms

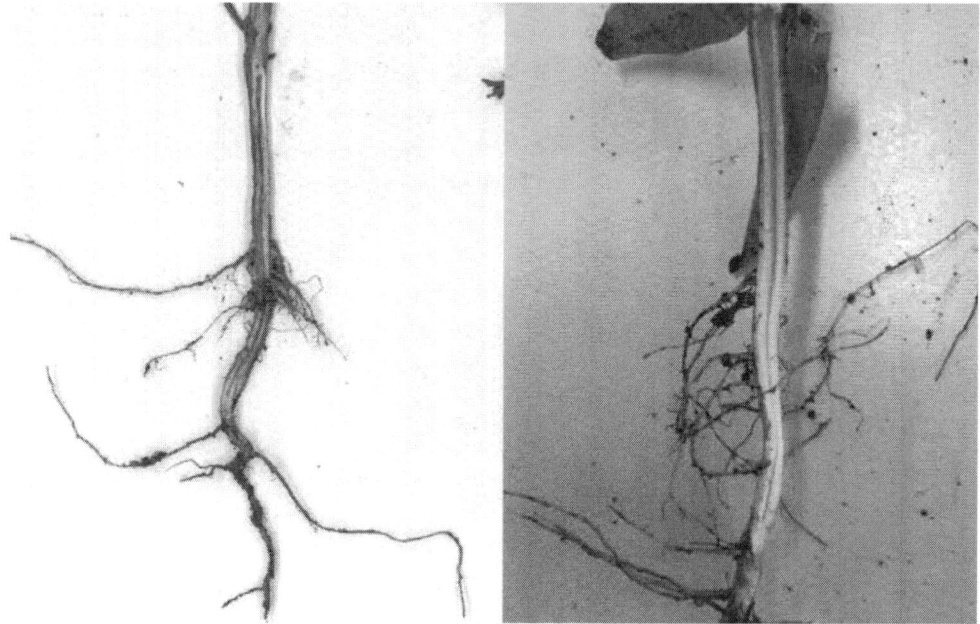

Figure 13.7: Root Rot and Leaf Blight (*Source*: The Beatsheet).

In green gram, the infections result in seed degradation, root rot, damping-off and seedling blight as well as stem canker and leaf blight. The condition frequently strikes during the pod development stage. The fungus initially manifests as signs of seed rot, seedling blight, and root rot. The afflicted leaves develop brown irregular blemishes and turn yellow in colour. Large blotches are created when these lesions consolidate, and the impacted leaves begin to dry up too soon. The bark easily peels off the stem's base and the roots, which turn black in colour. Internal tissue reddening is apparent when the tap root of the afflicted plant is cut open.

Survival and Spread

The original infection is caused by saprotrophic species that live in the soil, while secondary infection is brought on by asexual spores.

Favourable Conditions

Disease is more likely to occur in moist soil and humid environments. But *Macrophomina* even attack in less moist condition.

Disease Cycle

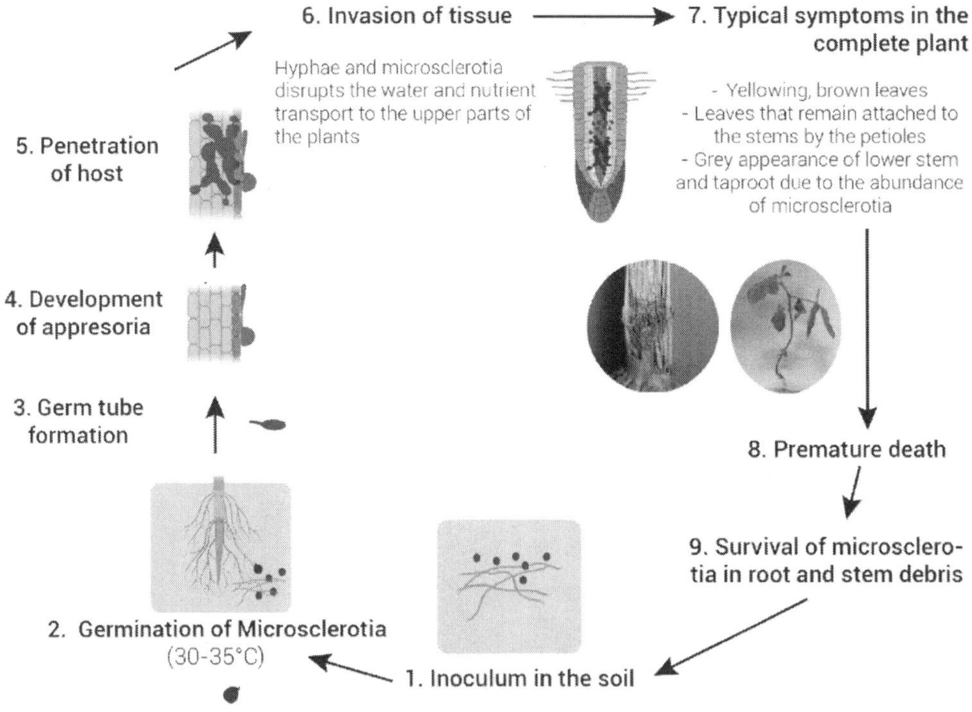

Figure 13.8: Disease Cycle of *Macrophomina phaseolina* (*Source*: Nathalie Marquez *et al.*, 2021).

Management

1. Pull out and burn the diseased plant.
2. Select resistant variety.

3. Destruction of weed.
4. Crop rotation should be adopted.
5. Seed treatment with Mancozeb @ 2.5 gm/kg of seed.
6. The root-rot disease complex was successfully reduced by the spray of Propiconazole @ 2 ml/water, which also increased the mungbean grain production in affected field.

6. Rust

Economic Importance and Occurrence

Rust of mung and urdbeans is mostly prevalent in southern and central part of India. It causes huge loss in beans crop like mung and urdbean.

Causal Organism

Uromyces phaseoli cause the disease in both mung and urdbean crops. The fungi belongs to the Kingdom – *Fungi,* Phylum – *Basidiomycota,* Subphylum – *Pucciniomycotina,* Class –*Pucciniomycetes,* Order – *Pucciniales,* Family – *Pucciniaceae,* Genus – *Uromyces,* and Specie –*phaseoli.* All the different spore stages in this autoecious, long-cycle rust happen on the same host. The single-celled, globoid or ellipsoid, brownish yellow uredospores contain echinulations. The teliospores are single-celled, pedicellate, globose or rectangular, chestnut brown in hue, and have warty the papilla on top. On the uppermost layer of leaves, yellow pycnia can be seen. On the underside of leaves, orange-colored cupulate aecia appear later.

Symptoms

Figure 13.9: Rust of Urd Bean.

Petioles, stems, and pods are rarely affected, whereas leaves are usually affected. Mostly on the lower surface, the fungus forms small, spherical, reddish-brown uredosori. They may show up in bunches, and a number of sori may combine to cover a sizable portion of the lamina. Late in the growing season, linear, dark brown teliosori develop on the leaves. The intense pustule production leads the leaves to dry out and fall off.

Total five spore stages in Puccinia life cycle:

Stage 'O' - Pycnial stage/spermatia

Stage 'I' - Aecial stage

Stage 'II' - Uredial stage

Stage 'III' - Telial stage

Stage 'IV' - Basidial stage

Favourable Conditions

1. Overcast and muggy with a temperature range of 18 to 24 °C.
2. Drizzled with mist at night is favorable condition for disease development.

Disease Cycle

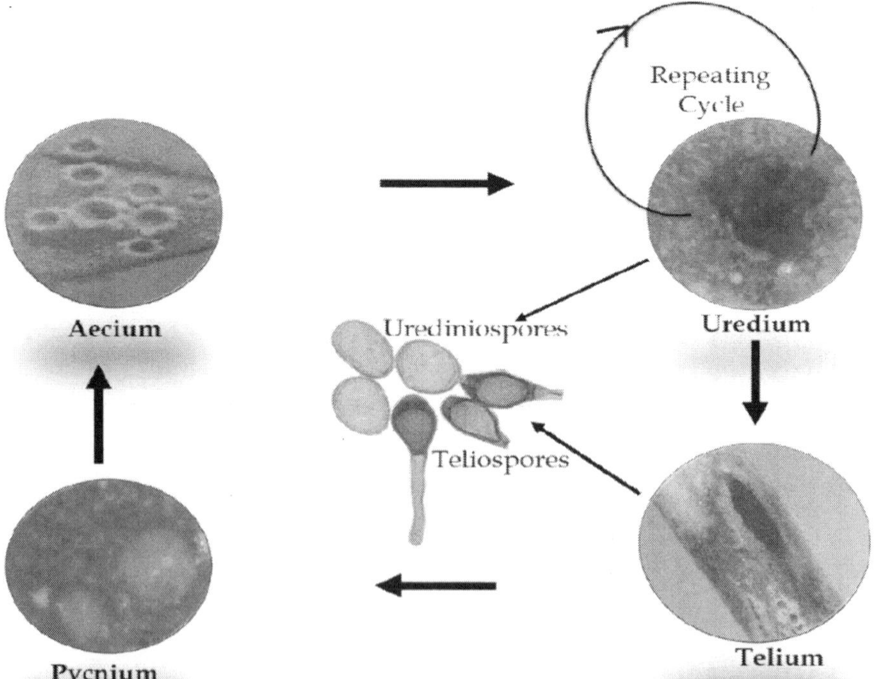

Figure 13.10: Disease Cyle of Rust (*Source*: Gautam *et al.,* 2022).

The pathogen lives on in crop waste as uredospores and teliospores in the soil. The sporidia produced from teliospores cause the initial infection. Uredospores carried by the wind cause secondary spread. On other hosts that are legumes, the fungus also persists.

Management

1. Remove and destroy the plant debris that is contaminated.
2. Diseased plant should uproot immediately and destroy.
3. Destruction of weed.
4. Crop rotation should be adopted.
5. Seed treatment with Vitavax powder @ 2.5 gm/kg of seed.
6. Immediately after the onset of the disease, spray Vitavax/carboxin at a rate of 2 gm per liter of water, or Hexaconazole 500 g or Propiconazole 1L/ha, and repeat after 15 days.

7. Dry Root Rot

Causal Organism

The causal agent of Dry root rot is *Macrophomina phaseolina*. Systematic position of fungi already mentioned under the disease root rot and leaf blight. In addition to constrictions at Hyphal branches, the fungus also generates dark brown, septate mycelium. There are numerous small, black, spherical sclerotia. On the tissues of the host, the fungus also creates dark brown, globular, ostiolated pycnidia. Thin-walled, hyaline, single-celled, and elliptical describe the pycnidiospores.

Symptoms

Yellowing and sagging of the leaves are the first signs of the disease. The plant dies a week after the leaves eventually fall off. At ground level, dark brown lesions can be observed on the stem, and the bark is showing signs of shredding. The afflicted plants can simply be taken out of the ground, leaving dried, rotting root fragments behind. Black minute sclerotia are abundant in the decaying tissues of the stem and root.

Favourable Conditions

Day time temperature should be 25°C. A protracted dry period followed by irrigation. Cycle of disease in addition to existing in infected debris, the fungus also exists in soil as a facultative parasite. Sclerotia in the soil and seeds are the main vectors of dissemination. Pycnidiospores, which are airborne, are the primary means of secondary transmission.

Figure 13.11: Dry Root Rot
(*Source*: Pandey *et al.*, 2019).

Management

1. Remove and destroy the plant debris that is contaminated.
2. Destruction of weed.
3. Crop rotation should be adopted.
4. Pelletize the seeds with *Trichoderma viride* at 4 g/kg or *Bacillus subtilis* at 10 gm/kg of seed.
5. Two spray of Tebuconazole @ 3 ml/liter of water at 15 days interval.

8. Mung Bean Yellow Mosaic Virus

A significant foreign disease of mungbeans that results in crop losses and leaf discoloration is the Mungbean Yellow Mosaic Virus (Begomovirus). The White Fly (*Bemisia tabaci*) is responsible for transmitting the virus throughout plants. The virus affects all legumes, including mung beans, urd bean and soybean *etc.*

Economic Importance and Occurrence

According to field observations, the prevalence of the Mungbean Yellow Mosaic Virus (MYMV) in mungbean varied from 4 to 40 per cent depending on the type of crop and the region. The incidence of MYMV in the instance of mungbean ranged from 5 to 100 per cent. It is quite concerning when viral illnesses affect commercial crops and research facilities on a large scale, especially when there are two or more viruses present and individual virus incidence is high. The only feasible means of preventing viral infections from spreading is to incorporate viral disease resistance into current cultivars.

Causal Organism

It is a viral disease caused by Mung bean yellow mosaic virus. The virus belongs to the Kingdom – Shotokuvirae, Phylum – Cressdnaviricota, Class - Repensiviricetes, Order – Geplafuvirales, Family – Geminiviridae, Genus – Begomovirus, and Species - Mungbean yellow mosaic virus. In the northern and central regions, Mungbean Yellow Mosaic India Virus (MYMIV) is the culprit, whereas the western and southern regions are affected by Mungbean Yellow Mosaic Virus. It is a member of the geminiviridae family of begomoviruses. viral particles with ssDNA, a bipartite genome, and the gem-like DNAs DNA-A and DNA-B.

Symptoms

Fresh leaves firstly develop in to little yellow spots or patches on their green blade. It quickly transforms into a distinctive golden yellow mosaic or brilliant yellow mosaic symptom. As the yellow tinge gradually gets worse, the leaves eventually turn entirely yellow. Plants that are infected mature later and produce fewer flowers and pods. The pods are deformed and tiny. Before the plant can set seed, an early ailment kills it.

Figure 13.12: Mung Bean Yellow Mosaic Virus.

Disease Cycle

Bemisia tabaci, a whitefly, is capable of spreading the disease in suitable circumstances. Virulence whiteflies that feed on plants carry disease. Crops sown in the summer are quite vulnerable. As a reservoir for inoculum, weed hosts like *Croton sparsiflorus*, *Eclipta alba* and other legume hosts are used.

Management

1. Up to 30 to 40 days after sowing, remove the unhealthy plants by rouging.
2. Occasionally get rid of the weed hosts.
3. Increase seeding by 20 to 25 kg per hectare.
4. Grow resistant mung bean varieties throughout the season, such as Pusa Baishakhi, Pusa Vishal, SML-668, ML-5.
5. When using the mixed cropping technique, plant rows of maize.
6. All the farm equipments should be sterilized with fire or hot water to stop the virus transmission.
7. Use of Botani Gard (biological insecticide) that contain *Beauveria bassiana*.
8. Apply the seed treatment of Imidacloprid at the rate of 2.5gm/kg of seed.
9. Spray Imidacloprid 17.8 per cent SL or Thiamethoxam 25WG at 75g/L in 500 liters of water.

9. Leaf Crinkle

Economic Importance and Occurrence

With an incidence of 5 to 28 per cent, the urdbean leaf crinkle virus (ULCV) was the second major viral disease, but it was more severe in mashbeans than mungbeans.

Causal Organism

Urd bean leaf crinkle virus (ULCV) is the causative agent of leaf crinkle.

Symptoms

Older leaves have significant crinkling and rugosity and thicken. Leaflet tips can be seen to crinkle and curl. Internodes and petioles are both compressed. Plant that has been infected appears bushy and stunted. Flowering is delayed, and any developed inflorescences have small, poorly-formed flower buds that do not open.

Figure 13.14: Leaf Crinkle (*Source*: Plantix).

Transmission and Favourable Conditions

The stigma of *Henosepilachna dodeca* acts as a virus transmitter. The disease spreads in fields primarily by seed or contact with infected leaves.

Disease Cycle

This virus is survived on host weeds including *Aristolochia bracteata* and *Digera arvensis*. The virus is seed-borne, and infected seeds are the principal source of infection. *Bemisia tabaci*, a silver leaf white fly, may aid in the secondary spread. The virus can spread through sap as well.

Management

1. Crop sanitation is very necessary to stop the transmission of virus.
2. Diseased plant should uproot immediately and destroy.
3. Eradication of collateral host.
4. Use a higher seeding rate of 25 kg/ha.
5. Seed borne infection was minimized by hot water seed treatment at 52 °C for 20–30 minutes.
6. Up to 40 days after seeding, pluck out the diseased plants at weekly intervals.
7. All the farm equipments should be sterilized with fire or hot water to stop the virus transmission.
8. Imidacloprid should be sprayed at 35 days after sowing.

10. Leaf Curl/Necrosis

Causal organism: Leaf bud necrosis virus

Symptoms

Caused by groundnut bud necrosis virus. Leaves cup and curve upward while veins clear. Brittle and occasionally extending to the petiole, vein necrosis can be seen on the underside of infected leaves. Early-stage plants that are impacted experience top necrosis and eventually die. The plant may only yield a few tiny, deformed pods.

Figure 13.15: Leaf Curl/Necrosis (*Source*: Plantix).

Disease Cycle

Frankliniella schultzii, Thrips tabaci, and *Scirtothrips dorsalis* are the three species of thrips that spread the virus. Tomato, petunia and chilli are weed hosts where the virus persists.

Management

1. Eliminate the weed hosts that are home to thrips and viruses.
2. Up to 40 days after seeding, pluck out the diseased plants at weekly intervals.
3. All the farm equipments should be sterilized with fire or hot water to stop the virus transmission.

4. 500 ml/ha of Imidacloprid should be sprayed 35 and 50 days after sowing.
5. Sowing of resistant varieties such as Pant Mung-6, SML-668, Pusa Vishal, PusaBaishakhi, Narendra mung-2.
6. Seed borne infection was minimized by hot water seed treatment at 52 °C for 20–30 minutes.
7. To eradicate nearby hosts for collateral weeds and pull out the infected plants.

11. Root-knot Nematode

Economic Importance and Occurrence

The production of the commercially significant leguminous crop mung bean (*Vigna radiata* L.) Wilczek) in Asia is severely hampered by root knot disease (RKN), which is brought on by the bacterium *Meloidogyne incognita*. Mung bean is grown in Thailand on an area of around 130,000 ha, with a seed production of roughly 90,000 tons. A good supply of minerals, high-quality carbohydrates, and protein is mung bean seeds. Because mung bean can fix atmospheric nitrogen in the soil through symbiotic association with *Rhizobium* species, mung bean cultivation can greatly improve soil fertility and productivity of succeeding crops. According to reports, *M. incognita* caused an 8.90 per cent crop loss in green gram. Black gram and green gram are being attacked by the nematodes *M. incognita, M. javanica, Heterodera cajani,* and *Rotylenchulus reniformis,* which results in an 8.90 per cent yield loss.

Causal Organism

Meloidogyne incognita nematode is cause knot on the root of mung and urdbean crop. *Meloidogyne* belongs to the Domain - Eukaryota, Kingdom - Animalia, Phylum - Nematoda, Class - Secernentea, Order – Tylenchida, Family - Heteroderidae, Genus - *Meloidogyne,* and Species - *incognita.*

Symptoms

The development of typical root galls on the roots of susceptible host plants is the primary symptom of root-knot nematode infection. Because of the damaged root structure, nutrient and water intake are significantly reduced, resulting in weak and low-yielding plants. Every crop reacts to root-knot nematode infection differently. Crop frequently experience significant forking, with lateral roots being the primary site of galling. Galls are often small, scarcely perceptible swellings like beads that can be found on onions and grasses. These symptoms frequently cause major financial losses for growers depending on the crop impacted and the degree of the illness.

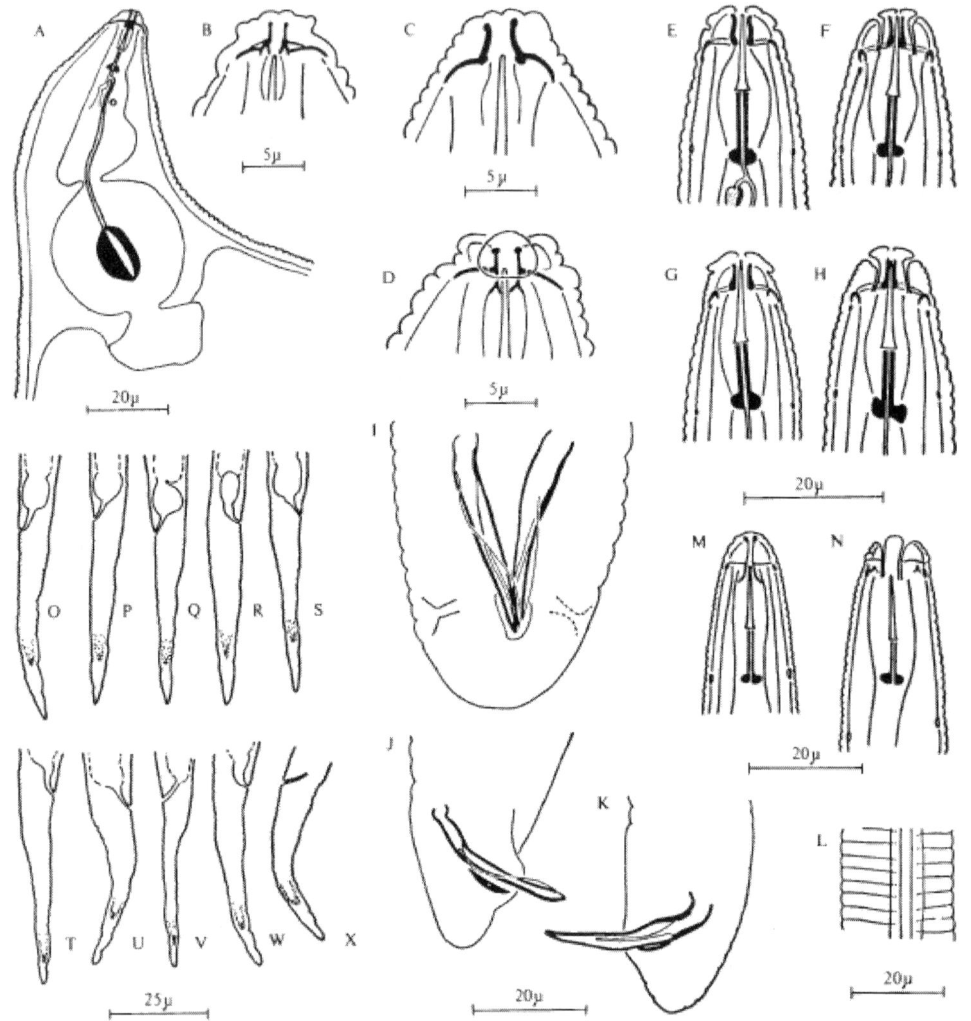

Figure 13.16: Female Nematode.

Disease Cycle

First appearing as eggs, root-knot nematodes quickly transform into J_1 (first-stage juvenile) worms. The J_1 stage molts into a J_2 nematode entirely inside the transparent egg shell. The only stage that can start infections is the motile J_2 stage. J_{2s} attack developing root tips and get inside the cells of the roots through the root cap. By injecting esophageal gland secretions into root cells, they start a feeding site as they travel to the region of cell elongation. The parasitized cells undergo significant physiological changes as a result of these nematode secretions, which turn them into gigantic cells. The giant-cells that the nematode feeds on will also perish if the nematode does. It takes 37 days at 21 °C (70 ° F) for *M. incognita* to complete its life cycle.

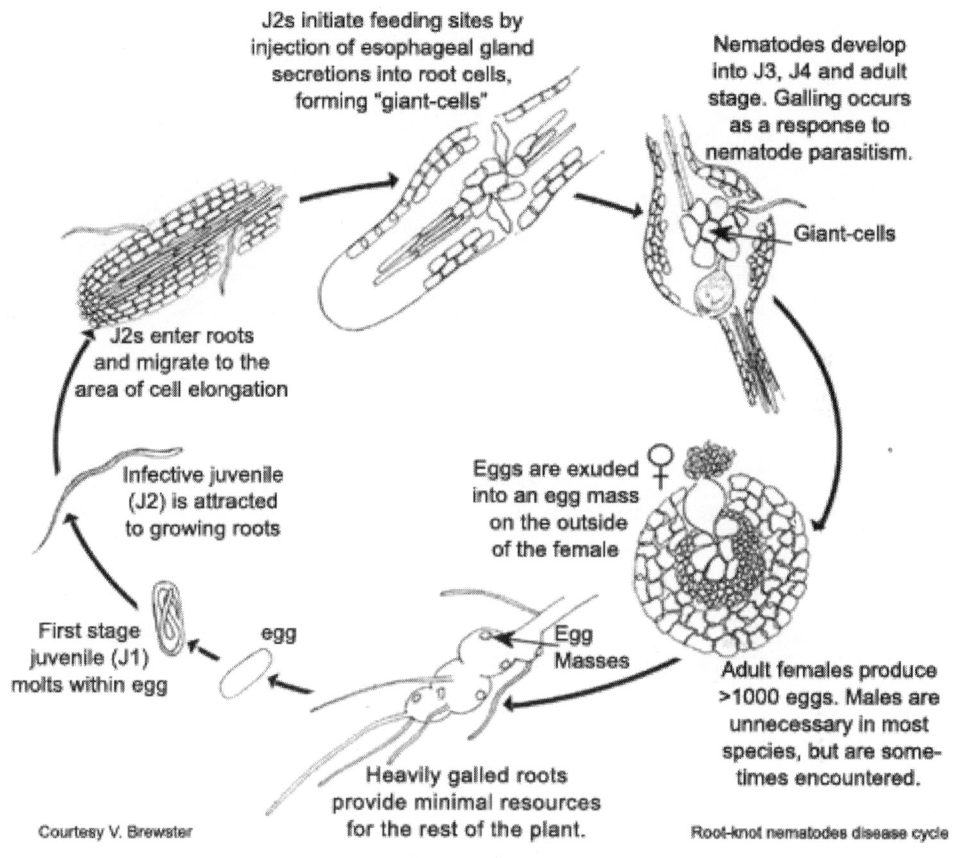

Figure 13.17: Disease Cycle (*Source*: Mitkowski *et al.*, 2003).

Management

The control of *M. incognita* involves a variety of tactics, including cultural, biological, and chemical control, and is mostly dependent on the crop that is damaged. Crop rotation using resistant or nonhost cultivars is one method of cultural control that can be utilized to keep the nematode population under control. Additionally, using organic amendments and antagonistic plants like *Crotalaria spectabilis* (Leguminosae) and a number of *Tagetes* species (Asteraceae) can successfully combat this worm. As a biological control, some parasitic fungus have been used, such as *Paecilomyces lilacinus*. Several nematicides are available on the market to manage *M. incognita*.

References

Bal, S. S., Kumar, S., Mishra, I. O. P., Panda, P.K., Panigrahi, R. K. and Prusti, A. M. (2019). Evaluation of mungbean and urdbean genotypes against web blight disease under natural field condition. *Int. Jr. of Chemical Studies*, **7**(5): 3374-3377.

Gupta, D. C. and Verma, K. K. (1990). Studies on avoidable losses in mung bean (*Vigna radiata*) due to root-knot nematode, *Meloidogyne javanica* and its control under field conditions. *Indian J. Nematol.*, **20**(2): 148-151.

Iqbal, U. and Mukhtar, T. (2014). Morphological and pathogenic variability among *Macrophomina phaseolina* isolates associated with mungbean (*Vigna radiata* L.) Wilczek from Pakistan. *Scientific world journal*, 95-175.

Itoh, T., Garcia, R. N., Adachi, M., Maruyama, Y., Tecson-Mendoza, E. M. and Mikami, B. (2006). Structure of 8Sα globulin, the major seed storage protein of mung bean. Acta Crystallogr. *D. Biol. Crystallogr*, **62**: 824–832.

Khaire, P. B. and Hake, L. G. (2018). Disease Management of Kharif Green gram and Black gram. *Popular Kheti*, **6**: 96-103

Naimuddin, A. M. and Singh, N. P. (2016). Yellow mosaic of mungbean and urdbean: current status and future strategies. *J. Food Legumes*, **29:** 77–93.

Pandey, A. K., Burlakoti, R. R., Kenyon, L. and Nair, R. M. (2018). Perspectives and challenges for sustainable management of fungal diseases of mungbean [*Vigna radiata* (L.) R. Wilczek var. radiata]: a review. *Front. Environ. Sci.*, **6**: 53.

Patel, M. and Patel, A. D. (2017). Management of *Meloidogyne incognita* in Mungbean by Seed Soaking in Different Chemicals under Pot Condition. *Int.J.Curr.Microbiol.App.*Sci., **6**(10): 1080-1084.

Siengchin, K., Ruanpanun, P. and Somta, P. (2020). Damage Potential of Root Knot Nematode (*Meloidogyne incognita*) population Density on Plant Growth Parameters related to Plant Age of Mung Bean (*Vigna radiate* (L.)Wilczek). *J. ISSAAS*, **26**: 111-122.

Singh, J., Mishra, K. K. and Singh, A. K. (2013). Current Status of Web Blight of Mung Bean. *An Asian J. of Soil Science*, **8**: 495-504.

Chapter 14

Diseases of Pea Crop

Arjun Rana, Girijesh Kumar Jaisval, Prabha Siddharth and Saurabh Saini*

Department of Plant Pathology, Chandra Shekhar Azad University of Agriculture and Technology, Kanpur – 208 002, Uttar Pradesh
****e-mail: arjunrana0719@gmail.com***

Introduction

Pea is highly nutritive, containing high percentage of digestible protein, along with carbohydrates and vitamins. It is also very rich in minerals. Garden peas are placed in group third on production efficiency. Pea (*Pisum sativum* L.) is a very common nutritious vegetable, grown in the cool season throughout the world (Robinson, 2019). The genus *Pisum* includes six to seven species mostly originated from Mediterranean area and west Asia, out of which only *Pisum sativum* is cultivated. Ethiopia is probably the main centre of origin of the garden peas. *Pisum sativum* having a chromosome number 2n=14. Cultivated form of pea was only one species *Pisum sativum* and subdivided it into two sub species (Ben-Ze'ev and Zohary 1973). The plant is an annual herbaceous plant with limited life that climbs via tendrils and leaflets. Cultivars might be tall, semi-tall, or dwarf. Except for taproot, the root system is not well developed. The stem is weak, round, and slender. With up to three pairs of leaflets and a terminal branching tendril, the leaves are pinnate. Pea is grown for the fresh market and also for export (Wardle, 1991). It is high value crop having more market demand but less in supply. The inflorescence is Raceme arising from the axils of leaf. The flowers are typically papillionaceous with green calyx comprising of five united petals (one standard, two wing and two keels). The stamens are in diadolphuos (9+1) condition. It is a self-pollinated crop due to presence of keel like structure on flowering part by which cross pollination is restricted (Lesins and Lesins, 2012). *Pisum sativum* is an annual plant, with a life cycle of one year. It is a cool season crop grown in many parts of the world. Depending on the area, planting can occur in the winter or early summer. About 22 °C is the ideal temperature for seed germination. Increased warmth speeds up germination, but plant stand deterioration affects plant growth. Optimum monthly mean temperature for growth of pea crop is 10-18.3 °C (Mukherjee *et al.*, 2016). Diseases of Pea can be classified as: fungal, bacterial and viral diseases of pea.

A) Fungal Disease

1. Ascochyta blight
2. Powdery mildew
3. Downey mildew
4. Pea rust
5. White rot
6. Root rot
7. Wilt

B) Bacterial Disease

1. Bacterial blight

C) Viral Disease

1. Pea early browning virus
2. Pea mosaic
3. Pea streak
4. Pea leaf roll
5. Cucumber mosaic virus

A) Fungal Disease

1. Ascochyta Blight of Pea

Economic Importance and Occurrence

Pea blight, or Ascochyta blight, is a major global disease. In 1830, Libert identified the disease as Ascochyta pisi and published the first description of it in Europe. Later, in 1861 and 1927, Berkeley and Bloxam as well as Jones described other fungus that produced sickness symptoms that were quite similar. Ascochyta blight is the second most significant disease in terms of afflicted acres. In Saskatchewan, yield reductions ranging from 5 to 15 per cent are typical when precipitation persists. Although losses of this size are uncommon in commercial fields, yield losses of up to 50 per cent have happened in research fields. A subsequent crop with high disease levels is contaminated with Ascochyta-infected seeds. Yields have increased 15 to 75 per cent when fungicides are applied (Skoglund *et al.,* 2011).

Causal Organism

Disease caused by *Ascochyta rabiei* f. sp. *pisi.*The fungi belongs to the Kingdom - *Mycota,* Phylum - *Ascomycota,* Class - *Dothideomycetes,* Sub-class - *Coelomycetes,* Order - *Pleosporales,* Family - *Didymellaceae,* Genus - *Ascochyta* and Species – *pisi.*

Symptoms

In areas with higher humidity levels, such as beneath the plant canopy on lower leaves, stems, and tendrils, early signs are most frequently seen. The initial signs are little, erratic, purplish-brown specks. When the humidity persists, the specks become larger and fuse together, causing the bottom leaves to turn entirely blighted. Foot rot, or girdling of the stem along the soil line, can result from severe diseases. Lesions caused by foot rot can appear both above and below the soil line, with a purplish-black color. Crop lodging and yield loss result from foot and stem lesions, which girdle and weaken the stem. When a crop has lodged or there is extended moisture, disease sores appear on the pods. Pod lesions start out small and black, but they have the potential to spread and cause early pod senescence. A severe pod infection can cause the seed to be undersized, shrunken, or discolored; on the other hand, the seed may not exhibit any signs at all.

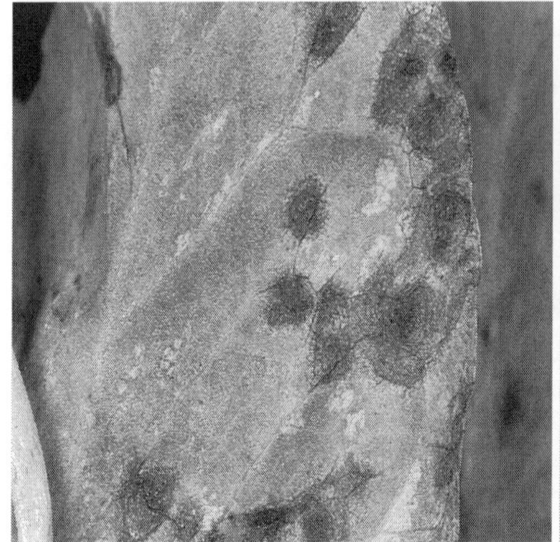

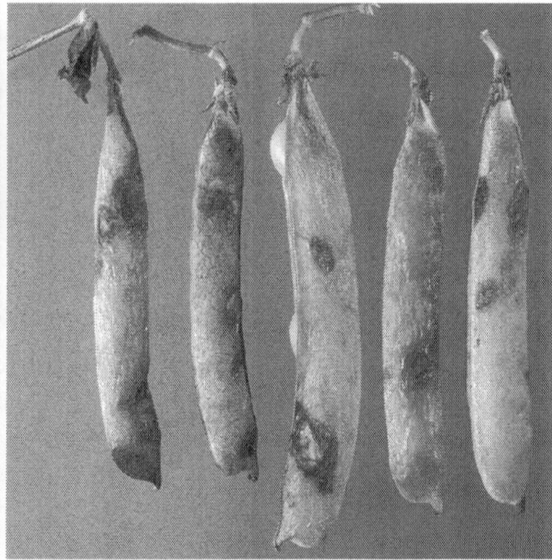

Figure 14.1: Ascochyta Blight of Pea (*Source*: CABI, Plantwiseplusknowledgebank.org).

Favorable Condition

The germination and dispersal of Ascochyta spores are facilitated by atmospheric moisture. Warmer temperatures, between 20 and 25 °C (68 and 77 °F), are ideal for the growth of fungi and the onset of diseases. Rain or irrigation on a regular basis supplies the moisture required for spore propagation and infection. A conductive environment can be produced by dense populations of plants; certain plants are more vulnerable to Ascochyta blight than others. Different crops are impacted by different Ascochyta fungal species. Inoculums from past growing seasons may linger in plant waste, soil, or seeds, creating a re-infection environment that facilitates the spread of ascochyta blight.

Disease Cycle

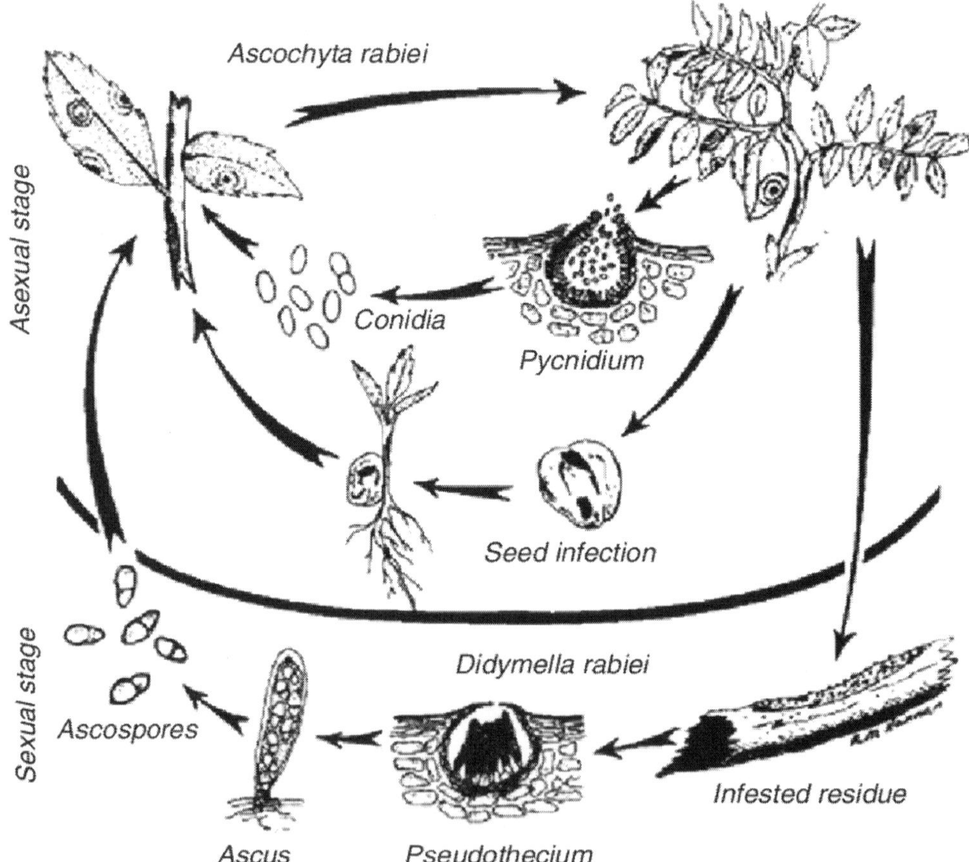

Figure 14.2: Disease Cycle of Ascochyta Blight of Pea (*Source*: Mishra *et al.*, 2016).

Management

1. Sow disease free seed.
2. Intercrop with wheat, barley and mustard.
3. Seed treatment with carbendazim @ 1g/Kg of seed.
4. Hot water seed treatment 52 °C for 10 min to lower the infestation.
5. Spray the crop with Mancozeb @ 2.5g/lit if notice during the growth period.

2. Downey Mildew

Economic Importance and Occurrence

It has been demonstrated that downy mildew significantly reduces field pea yields. Seed output was lowered by 75 per cent and pod numbers by 65 per cent even in a moderately severe infection. The range of 3-7 cm in seeding depth had no effect on downy mildew. Early, mid, and late May sowing dates did not consistently affect disease levels either. Downy mildew was occasionally lessened by late sowing, but low seed production was always the outcome.

Causal Organism

Disease caused by *Peronospora viciae*. The fungus belongs to the Kingdom - *Chromista*, Phylum - *Oomycota*, Order - *Peronosporales*, Family - *Peronosporaceae*, Genus - *Peronospora*, and Species -*viciae*.

Disease Symptom

On one side of the leaf, there is a yellowish patch, and on the bottom leaf surface, there is a grayish white, moldy growth. If the weather is cool and damp, infected leaves may turn yellow and eventually die. Stunted and deformed stems are possible. Pods get brown spots, and mold could develop inside pods.

Figure 14.3: Downey Mildew of Pea.

Favorable Conditions

The growth of downy mildew is encouraged by chilly, damp, and humid weather. The pathogen may grow at temperatures ranging from 5 °C to 30 °C, producing small sac-like structures called sporangia. However, the best conditions for reproduction are found between 15 °C and 20 °C. Unlike powdery mildews, downy mildews need extremely moist or humid conditions to thrive. Infection requires water, while sporulation requires humidity levels above 90 per cent. Cool temperatures promote the growth of the infections.

Disease Cycle

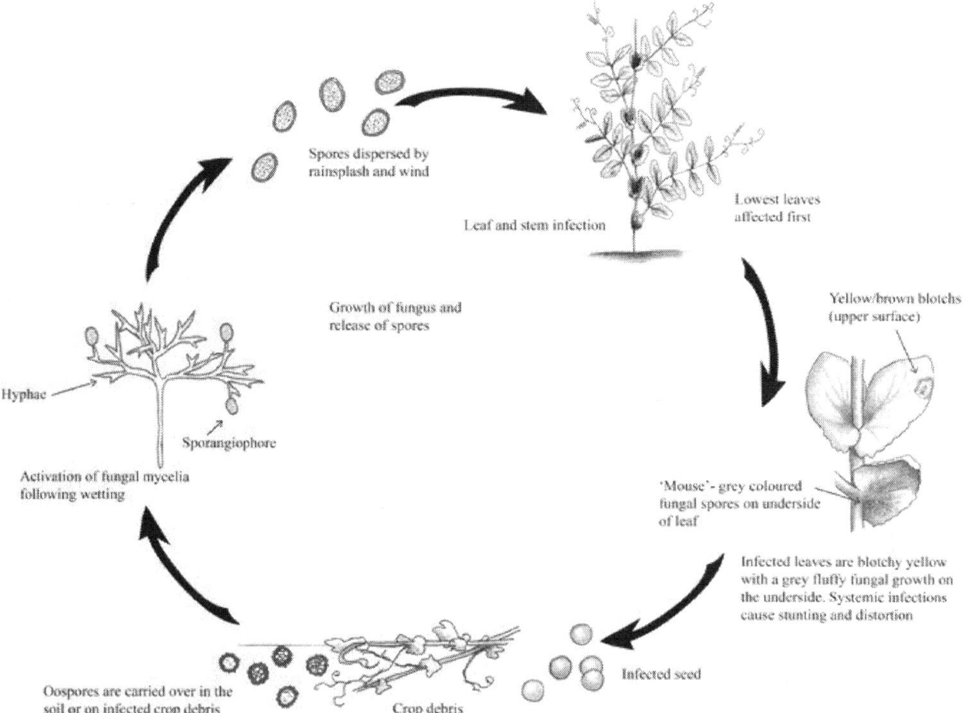

Figure 14.4: Disease Cycle of Downey Mildew of Pea (*Source*: extensionaus.com.au).

The fungus that causes downy mildew thrives in the ground, on decaying pea debris, and may spread by seeds. One potential main source of systemic illnesses is contaminated seed. Under favorable circumstances, late-season disease development might result from systemic infection of plants. In four to five days, when the weather is damp and cold (5–15°C), the sickness can spread swiftly. When seedlings are in the early vegetative stage, this frequently occurs.

Management

1. Sowing of resistant or tolerant varieties.
2. Rotate out of legumes for 2 or 3 years.
3. Burn or bale infested pea straw after harvest.
4. Seed treatment with 2 g/kg with metalaxyl and azoxystrobin.
5. Foliar spray of pyraclostrobin or azoxystrobin or metalaxyl @ 0.2 per cent at 15 days of intervals.
6. Penflufen + Trifloxystrobin, Fosetyl Al spray on standing crop @ 0.2 per cent at 15 days of intervals.

3. Powdery Mildew

Economic Importance and Occurrence

Pea powdery mildew is an airborne disease that is prevalent around the world and is especially significant in regions with warm, dry days and cool nights. It is caused by the obligate biotrophic fungus Erysiphe pisi DC. According to Gritton and Ebert (1975), the disease can result in yield losses of 25–50 per cent by lowering plant height, number of nodes, number of pods per plant, and number of seeds per pod. Yield can be lowered by 10 per cent to 20 per cent due to severe illnesses. In the latter part of the season, powdery mildew is most common. Powdery mildew is more likely to harm late-planted crops than early-planted ones. Low-quality seeds might result from severe pod infection.

Causal Organism

The causal organism *Erysiphe pisi* is a fungi belongs to Kingdom - *Fungi*, Phylum- *Ascomycota*, Class - *Leotiomycetes*, Order - *Erysiphales*, Family- *Erysiphaceae*, Species- *pisi* (Kirk *et al.*, 2008).

Disease Symptom

Figure 14.5: Powdery Mildew of Pea (*Source*: Cornell.edu).

It begins attacking leaves by forming tiny, slightly discolored dots, from which mycelium grows in a powdery, grayish white form. Particle-like growth covering the leaf, stem, and pod. The leaves wither and turn yellow. The fruits may stay little or fail to set. The result is defoliation. Later stages: the pods are also covered in powdery growth. Patches of whitish to grayish powdery growth cover the leaves. New growth may turn yellow and eventually drop, appearing twisted, curled, or dwarfed. Pods are twisted and small. The majority of this issue is with fall beans. Rain and wind spread powdery mildew.

Disease Cycle

The fungus creates spores that are carried by the wind into new crops while it overwinters on infected pea waste. When conditions are right, the disease can infect a plant completely in 5–6 days, and if a few plants are affected, it spreads quickly to other areas. 4-5 days towards the end of the growing season, during flowering and pod filling, warm (15–25 °C) and humid (over 70 per cent) circumstances are conducive to the development of disease. Rainfall hinders the disease's progress because it washes spores off plants. It only takes a few moist nights for illness to develop.

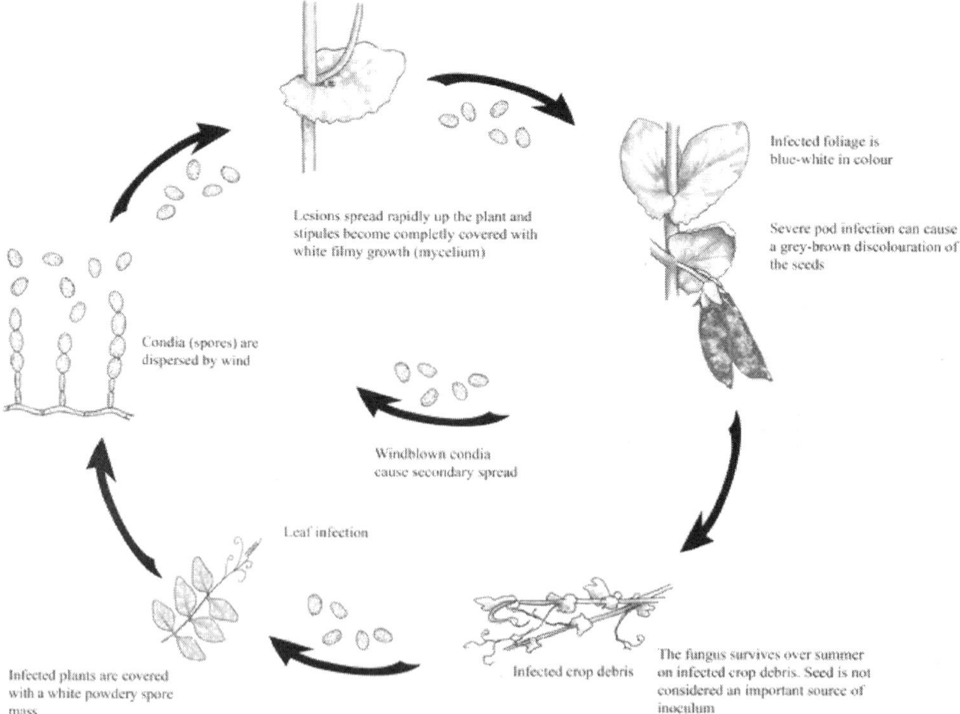

Figure 14.6: Disease Cycle of Powdery Mildew of Pea (*Source*: extensionaus.com.au).

Management

1. Plant early maturity varieties.
2. Resistant varieties should be planted.
3. Rotate plantings at least 1 year.
4. NSKE 5 per cent spray on standing crop after appearance of symptoms.
5. Bicarbonate is used for control such as Milstop at 2 to 5 lb/A on 7 to 14 day interval.
6. Dinocap 48 EC or Tridomorph 80 EC 0.1 per cent at 10 days interval.
7. Hexaconazole 0.05 per cent, penconazole 0.05 per cent spray at 15 days intervals on standing crop.

4. Bacterial Blight

Economic Importance

Although bacterial blight in field peas is common in Victoria, the severity of the disease varies from crop to crop and from season to season.Crop failure due to severe epidemics is possible, but losses are typically less than 20 per cent.

Causal Organism

The causal agent of this disease is *Pseudomonas syringae* p.v. *pisi* belongs to Domain – *Bacteria*, Phylum - Pseudomonadota, Class - Gammaproteo bacteria, Order - Pseudomonadales, Family - Pseudomonadaceae, Species – *syringae* and Pathovar. – *pisi.*

Disease Symptoms

Figure 14.7: Bacterial Blight (*Source*: extensionaus.com.au).

Lesions on the pods are deep and turn an olive-brown color; lesions on the leaflets first turn yellowish and then turn brown and papery. Lesions may also appear on stems that are close to the ground. These start off as wet patches that eventually transform from olive green to dark brown. Lesions in the stem may combine, resulting in the stem shrivelling and dying. The veins frequently place boundaries on the spots, which grow and merge. The leaf spot shrinks and appears on pods.

Favorable Conditions

Plants can become infected at any point throughout their growth. The pathogens don't have to cause symptoms to persist on the surface of plants. However, symptoms may appear after a plant has been damaged by frost, rain, heavy dew, or another method. Bacteria enter the plant tissue through damage to the field pea. Epidemics may result from early infections, although yield losses may also be caused by later illnesses. The ideal conditions for the disease to spread within crops are an abundance of rainfall combined with strong winds.

Disease Cycle

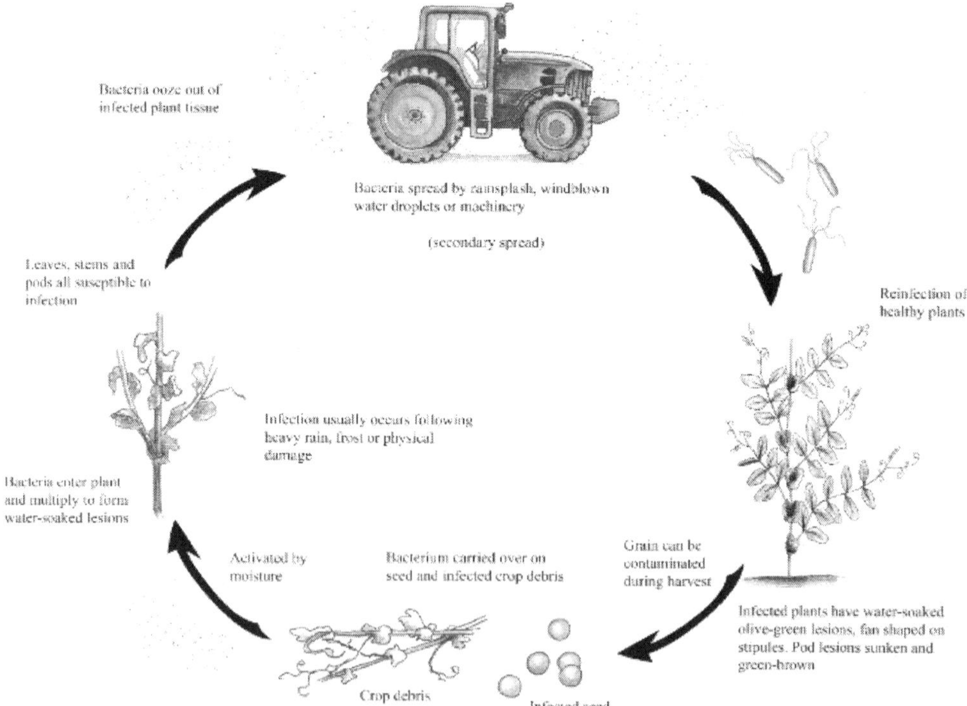

Figure 14.8: Disease Cycle of Bacterial Blight.

While *P. syringae* pv. *syringae* can live on a number of host plants, bacteria can thrive on seed or pea waste. Usually, contaminated seed is sown in a field to spread the disease. In rainy weather, raindrops carried by the wind, plant-to-plant contact, and plant-splash transfer germs from diseased to healthy plants. Plants can become infected at any point throughout their growth cycle, but it is most common after frosts. Frost-damaged plants, as well as any other physical damage, are more prone to infection. Strong winds, chilly temperatures, heavy dews, and rainfall all contribute to the ideal circumstances for disease to spread throughout crops.

Management

1. Use of diseased free seed for sowing.
2. Seed treatment with bioagent *Bacillus subtilis* give effective result to manage bacterial blight.
3. Seed treatment with Blitox 50 0.2 per cent before sowing.
4. When given as a foliar spray at intervals of seven and fifteen days, streptocycline at 300 ppm alone and in combination with 0.2 per cent Blitox 50 WP (copper oxychloride) considerably decreased the percent disease index.

5. Pea Early Browning

Causal Organism

Pea early browning is a viral disease belongs to Kingdom – Orthornavira, Phylum - Kitrinoviricota, Class - Alsuviricetes, Order - Martellivirales, Family - Virgaviridae, Genus - Tobravirus, Species - *Pea early browning virus.*

Symptoms

Figure 14.9: Pea Early Browning (*Source*: Wang *et al.,* 1997).

The Pea Early Browning Virus (PEBV) belongs to the Tobravirus genus. It is spread by trichodorid nematodes, which live in the soil, and by seeds from sick plants. PEBV symptoms on peas, displaying leaf loss and browning. It displays necrosis on the stem and leaves. The pea cv. Rondo seed exhibits significant wrinkling due to PEBV infection when compared to the healthy seed.

Disease Cycle

It is transmitted by soil-inhabiting trichodorid nematodes and through seeds from diseased plants. Some tobraviruses, *e.g.*, pea early browning virus, may be transmitted by 4 to 10 per cent of the seed of infected plants. Tobraviruses overseason in infected perennial cultivated or wild plants from which the nematode vector transmits it to the roots of cultivated plants.

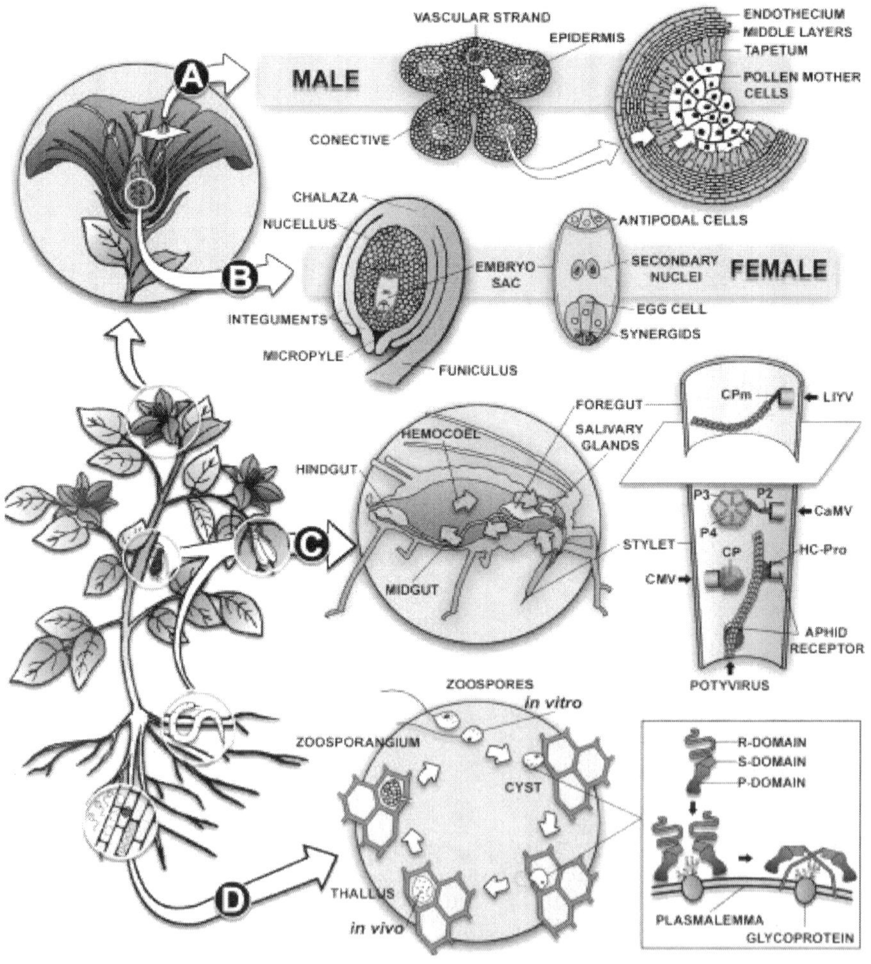

Figure 14.10: Disease Cycle of Pea Early Browning (*Source*: GN Agrios).

Management

1. The use of highly resistant or immune varieties and the avoidance of sources of infection.
2. Identification and exploitation of resistance to the nematode vectors would be another possible approach to crop improvement.
3. Host resistance to *Xiphinema index*, the nematode vector of grapevine fanleaf virus, also provided protection against infection with the virus.
4. Control of volunteer host plants may also help to reduce sources of infection, and the persistence of virus in the vector population.

6. Pea Mosaic Disease

Economic Importance

The primary reason why PSbMV is significant economically in field pea, faba bean, and lathyrus is because of its impact on seed quality. Because it only caused minimal yield loss and moderate symptoms, it was classified as a minor illness. However, research conducted by ICARDA in Syria revealed that yield losses in lentil, chickpea, faba bean, and pea owing to PSbMV were 66 per cent, 41 per cent, 45 per cent, and 49 per cent, respectively.

Causal Organism

Pea seed borne mosaic virus is the causal agent of this disease having ssRNA belongs to the Kingdom - Orthornavirae, Phylum - Pisuviricota, class – Stelpaviricetes, Order – Patatavirales, Family – Potyviridae, and Genus - Potyvirus.

Symptoms

Figure 14.11: Pea Mosaic Disease (*Source*: agriculture.vic.gov.au).

Infected plants may also produce distorted flowers, which often lead to small distorted pods. Ovule development in affected pods may be uneven, with only one or two seeds being produced. Seed coats may split as the seeds mature. The symptoms on seeds result in brown rings and tan spots. Field peas develop various degrees of stunting, downward curling of leaflets as well as a transient clearing and swelling of leaf veins in most cultivars.

Disease Cycle

The virus is believed to have spread worldwide through the exchange of infected seed. Seed transmission rates of up to 100 per cent in field peas and up to 44 per cent in lentils have been reported. PSbMV is also transmitted in a non-persistent manner by more than 20 aphid species, and by mechanical means. The most efficient vector of PSbMV worldwide is the pea aphid (*Acyrthosiphon pisum*). Surveys of Victorian pulse crops have found cowpea aphid (*Aphis craccivora*), foxglove aphid (*Aulacorthum solani*), and green peach aphid (*Myzus persicae*) to be important vectors of the virus.

Management

1. Sown virus free disease plant.
2. Always sown virus free resistance varieties.
3. Control weeds and rotate crops and avoid planting near virus infected plant.
4. Use a floating row cover or aluminum foil mulch to keep these pests away from plants.
5. Seeds should be treated with insecticide such as carbosulfan @1ml per 1 kg of seed.

References

Ben-Ze'ev, N. and Zohary, D. (1973). Species relationship in the genus *Pisum* L. *Israel J. Bot.*, **22**: 73 - 91.

Bretag, T. W., Keane, P. J. and Price, T. V. (2006). The epidemiology and control of ascochyta blight in field peas: a review. *Australian Journal of Agricultural Research*, **57**(8): 883-902.

Chang, K. F., Hwang, S. F., Ahmed, H. U., Strelkov, S. E., Conner, R. L., Gossen, B. D. and Turnbull, G. D. (2013). Yield loss and management of downy mildew on field pea in Alberta, Canada. *Crop Protection*, **46**: 23-28.

Davidson, J. A., Hartley, D., Priest, M., Krysinska-Kaczmarek, M., Herdina, McKay, A. and Scott, E. S. (2009). A new species of Phoma causes ascochyta blight symptoms on field peas (*Pisum sativum*) in South Australia. *Mycologia*, **101**(1): 120-128.

Fondevilla, S. and Rubiales, D. (2012). Powdery mildew control in pea. A review. *Agronomy for sustainable development*, **32:** 401-409.

Jindal, K. K. and Shyam, K. R. (1994). Management of bacterial blight of pea (*Pisum sativum* L.) through chemicals. *COLLOQUES-INRA*, 975-975.

Lesins, K. A. and Lesins, I. (2012). *Genus Medicago (Leguminosae): a taxogenetic study. Springer Science and Business Media.*

Liu, N., Xu, S., Yao, X., Zhang, G., Mao, W., Hu, Q. and Gong, Y. (2016). Studies on the control of Ascochyta blight in field peas (*Pisum sativum* L.) caused by Ascochyta pinodes in Zhejiang Province, China. *Frontiers in microbiology*, **7**: 481.

Misra, J. P., Rao, G. M., Kumar, A., Yadav, A., Kumar, S., Yadav, R. and Srivastava, S. (2016). Molecular assisted breeding for Ascochyta Blight resistance in chickpea (*Cicer arietinum* L.) A review. *Journal of Pure and Applied Microbiology*, **10**(2): 1469-1476.

Mukherjee, A., Sun, Y., Morelius, E., Tamez, C., Bandyopadhyay, S., Niu, G. and Gardea-Torresdey, J. L. (2016). Differential toxicity of bare and hybrid ZnO nanoparticles in green pea (*Pisum sativum* L.): a life cycle study. *Frontiers in plant science*, **6:** 1242.

Patil, N. B., Zacharia, S. and Kumari, M. (2017). Eco-friendly management of powdery mildew of garden pea (Pisum sativum L.). *Int. J. Curr. Microbiol. Appl. Sci*, **6**: 684-689.

Robinson, G. H. J., Balk, J. and Domoney, C. (2019). Improving pulse crops as a source of protein, starch and micronutrients. *Nutrition bulletin*, **44**(3): 202-215.

Singh, R. B. and Singh, R. N. (2003). Management of powdery mildew of mustard. *Indian Phytopathology*, **56**(2): 147-150.

Skoglund, L. G., Harveson, R. M., Chen, W., Dugan, F., Schwartz, H. F., Markell, S. G. and Goswami, R. (2011). Ascochyta blight of peas. *Plant Health Progress*, **12**(1): 29.

Surwase, A. G., Badgire, D. R. and Suryawanshi, A. P. (2009). Management of pea powdery mildew by fungicides, botanicals and bio-agents. *Annals of Plant Protection Sciences*, **17**(2): 384-388.

Wang, D., Macfarlane, S. A. and Maule, A. J. (1997). Viral determinants of pea early browning virus seed transmission in pea. *Virology*, **234**(1): 112-117.

Wardle, P. (1991). *Vegetation of Newzealand*. CUP Archive.

Chapter 15

Diseases of Cowpea Crop

*Arvind Kumar[1] *, Girijesh Kumar Jaisval[2], Rishi NathPandey[1], Arun Kumar[3] and Rajneesh Kumar Awasthi[4]*

[1]Ph.D. (Scholar) Department of Plant Pathology, Banda University of Agriculture and Technology, Banda, Uttar Pradesh
[2]Ph.D. (Scholar) Department of Plant Pathology,
[3]Ph.D. (Scholar) Department of Entomology,
Chandra Shekhar azad University of Agriculture and Technology, Kanpur, Uttar Pradesh
[4]Ph.D. (Scholar) Department of Plant Pathology, P.G. College Ghazipur, VBS Purvanchal University, Jaunpur, Uttar Pradesh
**e-mail: arvindkumarak638419@gmail.com*

Introduction

Cowpea commonly called as Lobia, southern pea, Lobia blackeye pea, crowder pea, yard long bean. The chromosome number are 2n=22. It is rich in protein contain 25 per cent, several vitamins, minerals and Integrates well with maize, sorghum, pearl millet, and teosinte to increase milk production. The presence of vitamins in cowpea helps in boosting immunity. It is one of the oldest human food sources and a short-duration multipurpose pulse crop grown extensively in tropical and subtropical countries, primarily in Africa, with Nigeria and Niger dominating, but also in Brazil, the West Indies, Myanmar, Sri Lanka, India, Australia, and other subtropical regions of the world due to its high quality dietary protein content (23-29 per cent), acceptable palatability, and low production cost. (Pandey and Joshi, 1995; Singh *et al.*, 1997; Etana *et al.*, 2013). This crop is resilient to heat and drought stresses and therefore finds place in several cropping systems across the world. Cowpea plants are typically erect, with ribbed stems and smooth trifoliate leaves grouped alternately on the stems. In India it is grown in central and peninsular regions, Uttar Pradesh, Haryana, Punjab, Gujrat, Karnataka, Rajasthan and Madhya Pradesh The total world pulse requirement for consumption is projected to be around 23 million tonnes, whereas cowpea is cultivated on 10.73 million hectares with a productivity of 387 kg/ha and a yield of 3.84 million tonnes. Cowpea is planted on approximately 1.5 million hectares in India, with a yield of 567 kg per ha and a production of 0.5 million tonnes. The crop's yield potential in Karnataka is 420 kg per ha. (Anon, 2011). Cowpea is more drought resistant than soybean or mung bean due to its thick taproot. It has a competitive niche in sandy soils, does not survive extremely wet circumstances, and should not be planted in poorly drained soils. Cowpea production is influenced by a variety of biotic and abiotic stressors. Among biotic stresses southern leaf blight, cowpea rust, cowpea scab, cowpea mosaic, bacterial blight fusarium wilt, powdery mildew, Anthracnose and root knot nematode are majordiseases that hampers crop establishment, impair forage quality and reduces green fodder and seed yield. In addition to directly reducing yield, they also inhibit nodulation, which negates the maximum nitrogen fixation. Estimates placed the losses in green fodder and seed production at 28.8 and 39.7 percent, respectively. Controlling plant diseases is crucial for producing a sufficient and high-quality output. In this chapter brief information on most of the pathogens attacking and their management of cowpea is given.

Table 15.1: List of Major Cowpea Diseases

Sl.No.	*Name of Disease*	*Causal Agent*
	A. Fungal diseases	
1.	Southern blight cowpea	*Sclerotium rolfsii*
2.	Cowpea scab	*Sphaceloma* spp.
3.	Fusarium wilt of cowpea	*F. oxysporum* f. sp. *tracheiphilum*
4.	Powdery mildew of cowpea	*Erysiphe polygoni*
5.	Cowpea rust	*Uromyces phaseolis* var. *vignae*
6.	Anthracnose	*Colletotrichum lindemuthianum*
	Bacterial disease	
7.	Bacterial blight of cowpea	*X. axonopodis*pv. *Vignicola*
	Nematode disease	
8.	Root knot nematode	*Meloidogyne incognita*
	Viral disease	
9.	Cowpea Mosaic	virus transmitted by aphids

1. Southern Blight of Cowpea

It is fungal disease caused by *Sclerotium rolfsii* Sacc. that attack roots and stem of cowpea. The disease occurs in worldwide, but mostly in warm climates.

Occurrence and Economic Importance

S. rolfsii is a polyphagous soil borne plant pathogens which cause southern blight disease in cowpea. The United States Department of Agriculture stated in 1928 that S. rolfsii and root knot nematode were the pathogens that caused the most damage in the southern USA (Hagan, 1999). A novel tomato disease that caused greater than 70 per cent loss in some Florida fields was originally described by Peter Henry Rolfs in 1892. Saccardo gave the fungus the name S. rolfsii in 1911. (Grecher, 1995).

Host

In *S. rolfsii* has many host range and more than five hundred plant species including lentil, tomato, chickpea, and many perennials plants including petunia and zinnia *etc.*

Symptoms

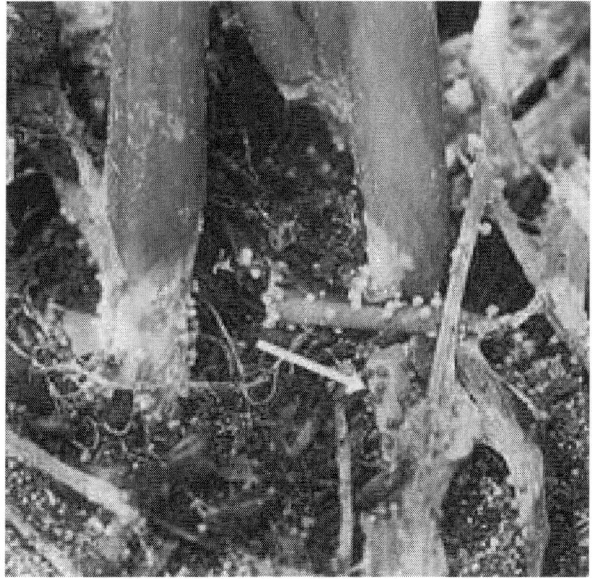

Figure 15.1: Southern Blight of Cowpea.

Sclerotium rolfsii infected Seedlings, herbaceous plants, woody plants, and fleshy roots. This fungus typically infects lower stems close or at the soil surface. The first noticeable sign of southern blight is a progressive yellowing and wilting of the foliage, which starts on the lower leaves. Within a few days, the plant dies. During warm, damp conditions, the fungus' coarse, white mycelium forms distinctive fan-shaped patterns of development on the stem at the soil line. Sclerotia are smooth, spherical, light-tan to dark-brown mustard seed-like entities that develop in the fungus' white mat.

Disease Cycle

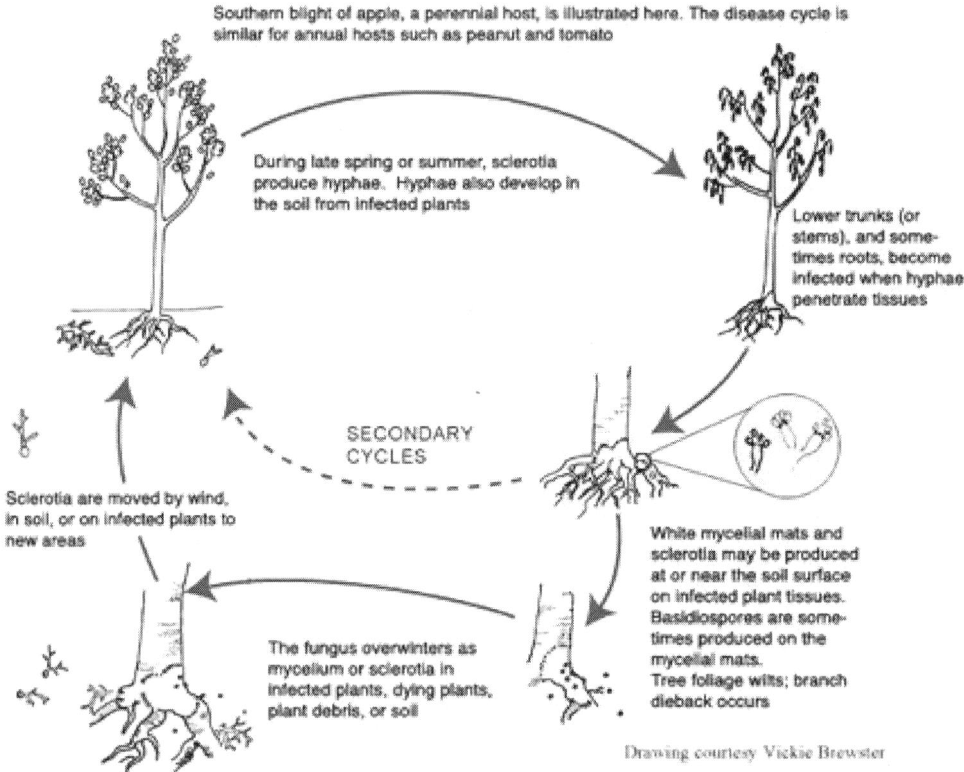

Figure 15.2: Disase Cycle of Southern Blight of Cowpea.

Infected tissues and emerging sclerotia serve as a beginning for hyphal development in the spring. Direct penetration takes place when hyphae come into touch with the tissues of a sensitive crown, root, bulb, fruit, or leaf, although wounds make infection easier. Lower stems, roots, fruits, and leaves may become infected if the tissue is vulnerable and the temperature, humidity, and other environmental conditions are favorable. Both intracellular and intercellular hyphae are possible. Cell division and death are caused by the fungus' production of cellulase, polygalacturonase, and oxalic acid. Soft rot symptoms typically become visible 2-4 days after infection. Foliage wilt, branch dieback, and complete plant mortality frequently occur when the soft rot canker encircles the lower trunk or stem. Secondary cycles occur and disease spreads as hyphae contact new healthy susceptible tissues when warm to hot temperatures and humid conditions prevail during the growing season. Secondary cycles could be influenced by basidiospore formation. Basidiospores enlarge when they come into touch with the plant surface and create one to three germination tubes. The fungus over winters diseased plants, plant debris, and occasionally as emerging hymenial layers as sclerotia and mycelium. Most sclerotia are produced at or near the soil surface and survive longer in well-drained soil.

Management

1. Removed of plant debris from the infected fields, plant treatment, crop rotation, use of resistant varieties, or a combination of these practices. The specific management practices used depends upon the crop and cropping situation.
2. You should only purchased plants, cuttings, or seeds from a trustworthy seller if they are pathogen-free. Every pot, piece of planting equipment, and planting medium used in a greenhouse must be sterile and pathogen-free. If field planting is necessary, locations free of *S. rolfsii* history should be chosen.
3. Sclerotia survive in the soil for 3-4 years. Localized soil removal and replacement may help to control *S. rolfsii*.
4. Fungicide treatments on plants typically protective in nature and used to control southern blight. The majority of fungicides are labelled for use on particular field crops, vegetables, or ornamentals. Azoxystrobin, flutolanil, or tebuconazole foliar treatments have been successful in controlling disease on peanuts. For many years, S. rolfsii was controlled with the fungicide PCNB on some ornamentals, peanuts, various crops, and as a seed treatment.

2. Cowpea Scab

Coepea Scab disease caused by *Sphaceloma* spp. The disease occurs in all cowpea growing areas of Eastern, Northern and West Nile regions and a few districts (Nakasongola, Mubende, and Hoima) in central region. However, the eastern region has the highest incidence followed by West Nile and then Northern. Tororo, Kumi and Serere districts are hotspots for the disease.

Occurrence and Economic Importance

Its seed-borne disease is one of the country's biggest restrictions to cowpea production, capable of generating yield losses of up to 100 per cent due to severe infections have been reported from Nigeria (Mbong*et al.*, 2010).

Host

Host of the cowpea are Cowpea (*Vigna unguiculata*), Lima bean (*Phaseolus lunatus*) and Common beans (*Phaseolus vulgaris*) are to be hosts of the cowpea scab disease (Singh and Allen, 1979).

Symptoms

Cowpea scab are affected most of all plant parts on above the soil cause up to the 100 per cent yield loss. The pathogen's first symptom, which appears 3 to 6 days after inoculation, is puckering of the lamina. On mature leaves, white dots with or without brown edges can be visible. Typical scab lesions on the petiole, stem, peduncle, and pod are white, turning dark brown when chlamydospores form, and oblong-elongate, with the exception of pod lesions, which are oval. Plants remain small and lose leaves, blooms, and pods before they mature when infection levels are high. Leaves, stems, and pods are all extensively marked. The flowering axis scab, which stops flower formation entirely or causes pod and/or flower abortion, is the most devastating phase.

Figure 15.3: Cowpea Scab.

Management

1. Use certified seeds or seeds from disease-free fields.
2. Use resistant/tolerant varieties such as SECOW 2W and SECOW 3B.
3. Avoid movement of sick plant materials between fields.
4. Do not plant new crop in fields with history of the disease for the last 2-3 years.
5. Crop-rotate with sorghum, millet, wheat, rice, and not beans, greengram, groundnuts.
6. Remove and destroy remains of cowpea, beans, greengram from previous cropping season.
7. Treat seed with mancozeb @ 80g/kg of seeds.
8. Monitor cowpea fields regularly, at least once a week, for signs of disease from 3 weeks after emergence to maturity.
9. Apply Mancozeb at 2.5Kg per ha at 7-10 day intervals from the time the first sign of disease was seen.

3. Bacterial Blight of Cowpea

The causal organism of cowpea bacterial blight is *Xanthomonas axonopodis* pv. *vignicola*, which occurs in humid and sub-tropical climates where cowpea is produced. It is a soil borne disease that annually reduced yield and lower quality of cowpea. Bacterial blight of cowpea mostly occurs in poor drainage soils.

Occurrence and Economic Importance

Yield losses due to cowpea bacterial blight are 2.7 - 92.2 per cent depending on susceptibility of the variety (Kishun, 1989). Bacterial blight has the potential to diminish production by up to 92 per cent and emergence in infected seeds by 67 per cent, depending on the cultivar and stage of plant growth at which infection takes place. Plant death might increase by 81 per cent due to secondary infection.

Host

Major host of bacterial blight of cowpea is sunhemp, hyacinth bean, common bean black night shade, purple tephrosia and black gram

Symptoms

Figure 15.4: Bacterial Blight of Cowpea.

The first signs of infection develop on the cotyledons of seedlings sprouting from infected seed and appear reddish and wrinkled. First, necrotic lesions grow on the leaves, and then the stem is attacked. The infection infiltrates vascular bundles, causing the disease to becomes systemic. The afflicted plant loses its growing tip, which causes the plant to eventually die. On the stem, close to where the initial leaves and cotyledons converge, cankers frequently form. Such stems are weak and quickly break in severe winds because they cannot support the weight of the plant. The growing seedling turns brown - red and eventually dies. Irregular to spherical brown patches with chlorotic halos emerge on leaves and spread to the stem. The stem may split, and the pods may become infected, resulting in shrivelled seeds.

Disease Cycle

Infected pods generate smaller, wrinkled, and infected seeds. These contaminated seeds serve as a source of primary inoculum. Secondary spread is caused by rain, insects, and equipment. Secondary infection on leaves exhibited as light yellow circular spots 4 to 10 mm in diameter spread across the lamina.

Management

1. Used resistant varieties.
2. Use healthy and disease free seeds.
3. Seed treatment with copper oxychloride for 15-60 minutes is more effective (Thammaiah and Khan, 1995).
4. Spraying copper and copper plus Sikirou (1999) reported that seed treatment in hot water at 60 °C for 30 minutes or 70 °C for 10 minutes or treatment in hot air (65 °C for 120-144h or 70 °C for 96h) eliminated the bacterium from infected seeds without inhibiting seed germinations.

4. Fusarium Wilt

Fusarium wilt of cowpea is a devastating disease in the southern and eastern United States caused by *Fusarium oxysporum* f. sp. *tracheiphilum*. Fusarium is a fungus genus that includes major plant pathogens as well as mycotoxin-producing pollutants in human and animal diets. This genus causes plant diseases such as wilt, blight, root rots, and cankers, and its range includes soils and organic substrates all over the world.

Occurrence and Economic Importance

Fusarium wilting disease is responsible of more than 50 per cent of cowpea destruction on the field which classified this disease as the main cause of this legume wilting. In India, the yield loss caused by the fungus was 26.8-64.5 per cent in 1954 (Singh, 1954) and 74.6 per cent (Singh and Sinha, 1955).

Host

The major host is cowpea and some minor host such as soyabean, common bean

Symptoms

Many crops are infected by this disease. Acidic soils and high temperatures promote the spread of this disease. This is a soil-borne disease that commonly infiltrates plants via roots or wounds. Fusarium wilt typically causes yellowing of the lower leaves on one side of the plant. As the organism develops in the food and water conducting tissues, infected plants normally become stunted and wilted. When the stem is broken longitudinally, brick red tissue can be seen. Seedlings may rot, and the leaves may become yellow and wilt.

Figure 15.5: Fusarium Wilt of Cowpea.

Disease Cycle

The morphology of *F.oxysporum* f.sp. *tracheiphilum* on oat meal agar has a cotton wool look, with an aerial mycelium that ranges in colour from white to purple. Ellipsoidal, hyaline, and aseptate describe the microconidia. The macroconidia are primarily septate and have a little curvature. The aseptate, terminal or intercalary, ellipsoidal to globose chlamydospores are. On PDA, the fungus also forms small, many sporodochia in 8–12 days at 26–30 °C.At higher temperatures of 27°C, the signs of vascular wilt become more severe (Swanson and Van Gundy, 1985). Furthermore, according to Harris and Ferris (1991), the wilt symptom is especially bad when the fungus is associated with the worm *Meloidogyne javanica*. *F. oxysporum* f. sp. *tracheiphilum* is transferred through seeds and is seed borne. For four years, wilted plant seeds kept in the refrigerator were still viable (Armstrong and Armstrong, 1950).

Management

1. Use an integrated crop management approach.
2. A 4 or 5 year rotation with other crops. Plant in disease free soil and use clean planting material.
3. Burning of crop debris and the planting of resistant cultivars is recommended.
4. Treating high quality seed with fungicides labeled for cowpeas.
5. Planting certified seed of resistant varieties.
6. Spray with approved fungicides –Cuprasan and Kocide.

5. Cowpea Mosaic

Cowpea mosaic virus is a Comovirus-type plant virus. It is an RNA virus with isometric particles 28nm in diameter. The primary source of inoculum is either seed or infected plants of cowpea, French bean or sunhemp. It is transmitted by various beetles with biting mouthparts, grasshopper, aphid and thrips also reported to be vector.

Occurrence and Economic Importance

Cowpea mosaic virus cause a yield loss of 13-87 per cent under field conditions depending upon crop susceptibility, virus strain and the environmental conditions. The type member of the genus Comovirus, one of the three genera that make up the family Comoviridae, is the cowpea mosaic virus. The genome of the cowpea mosaic virus is made up of two positive-strand RNA molecules that are each individually encapsulated in isometric particles with 60 copies of each coat protein type.

Symptoms

The virus that causes it is spread by aphids. The afflicted leaves turn a pale yellow colour and display vein banding and mosaic symptoms. Reduced size and puckering are visible on the damaged leaves. Pods shrink and twist in addition. Tap root discoloration, stem longitudinal cracks, stunting, wilting, and low yields are further symptoms. Stunted plants might not be able to make typical pods. There won't be any pods if the disease affects plants when they are still in the early stages of growth.

Figure 15.6: Cowpea Mosaic.

Disease Cycle

The virus particles are flexious rod measuring 750x15nm. The nucleic acid is a single-stranded RNA. Seed transmission is common and most importance source of initial infection in the crop. From 30-80 per cent seed from infected crop may carry the virus, which is present in the embryo and cotyledons but not in the seed coat. Transmission in the field occurs through the aphid vectors. The aphid transmits the virus more from chlorotic areas than from the green areas of the leaf.

Management

1. Use healthy seed from healthy crop.
2. Disinfect tools and farm implements with chlorine before moving from diseased areas to healthy areas.
3. Remove infected plants as early as symptoms are observed to minimize spread.
4. Practice crop rotation with non-legumes *e.g.* cereals.
5. For controlling aphids spray Oxydemeton methyl 25 EC (Metasystox) @ 1 ml/liter or Imidacloprid 17. 8 SL @ 0.2 ml/liter of water and repeat the spray after 10 days of first spray.
6. Roguing out of cowpea mosaic virus diseased plants in the early stage of growth up to 30 days and spraying twice at fortnightly intervals with Methyldemeton 25 EC 500 ml/ha.

6. Powdery Mildew

It is a fungal disease that affects a wide range of plants and depletes their nutrients. As a result, the plant blooms less and becomes weaker. Its symptoms are distinct and easy to recognise. The disease is commonly known to present from early flowering to pod maturity, and its progression is dependent on cultivars, sowing dates, and climatic conditions.

Occurrence and Economic Importance

It is a particularly important in climates with warm, dry days and cool nights (Sillero *et al.*, 2006). In severe infection may cause 25-50 per cent yield losses (Munjal *et al.*, 1963; Warkentin *et al.*, 1996). Reddy *et al.* (1994) and Shambarkar *et al.* (1996) reported up to 40 per cent loss in Mung bean and 45 per cent in sesame respectively. Emechebe and Florini (1997) reported that the damage due to powdery mildew on cowpea in the Sudan Savanna of Nigeria were moderate.

Symptoms

All of the damaged plants' aerial portions are covered in powdery mildew. Leaf symptoms appear first, then stem, branches, and pod symptoms follow. a thin, powdery growth that occasionally appears on the stems, pods, and leaves. This growth's powdery texture is simple to remove. Plants get yellow and lose their leaves when the illness is severe. The fungus and its spores are contained in this white growth. In general, early-planted cowpeas are unaffected by powdery mildew. Affected leaves shrink in size and twist.

Figure 15.7: Powdery Mildew of Cowpea.

Disease Cycle

The fungus is able to attack different leguminous hosts and survives in conidial or perithecial form. The fungus is ectophytic, in which mycelium spread on the surface of leaves and takes nutrients from the host by making haustoria that enter into the epidermal cells. The hyphae is thin-walled elliptical or barrel or cyclindrical in shape. The conidia are arranged in basipetal manners on the conidiophores. Each cleistothecium has 4-8 asci and each ascus has 3-8 ascospores, which are elliptical hyaline and single celled and can live from season to season.

Management

1. Collect the residue of plants from field and burn them.
2. Spray NSKE 5 per cent or Neem oil 3 per cent twice at 10 days interval from initial disease appearance.
3. Spray Eucalyptus leaf extract 10 per cent at initiation of the disease.
4. Spraying wettable sulphur at 3g/liter or carbendazim at 1 g/liter of water or spraying carbendazim 500 g or propiconazole 500 ml/ha at illness onset and 10 days later can manage the disease.

7. Cowpea Rust

The causal organisms of cowpea rust are *Uromyces phaseolis* var. *vignae.* It is widely distributed throughout the country and is responsible for heavy reduction in yield (Chandra shekar *et al.*, 1988). During the southwest monsoon, the disease is seen in severe form in July seeded crops, resulting in significant production loss. (Chandrasekhar *et al.*, 1989).

Occurrence and Economic Importance

This disease is common in cowpea growing areas around the world, with reports from at least 20 nations claiming 10-15 per cent annual losses. (Anil kumar*et al.*, 1989). Rust is a harmful disease that has a significant economic impact on much of Karnataka and has a negative impact on the production of cowpeas. Premature defoliation caused by the disease, which results in smaller seeds and a considerable yield loss, has been noticed. (Honnur *et al.*, 2016).

Symptoms

Leaf flecking is one of the early signs of rust infestation. These flecks quickly grow into reddish-brown pustules that frequently converge before finally bursting to reveal a mound of brown spores. Reduced seed size and a significant loss in yield are the results of the disease's interference with normal root development and nutrient uptake by plant roots.

Figure 15.8: Cowpea Rust.

Disease Cycle

In cooler region it can survive through teliospores in crop refuse. In other localities, it survives from season to season through continuously grown bean group crop and on other hosts. Intensive cultivation of beans in areas with high atmospheric humidity is one of the conditions favorable for disease development. Under favorable conditions, the spores complete the infection cycle within 5 days and a new crop of spores is produced within 5-10 days. Heavy rain washes down the spores, hence there are less chances of infection and spread. If the daily temperature reaches 34°C there is no disease development.

The repeating spores germinate best at 15-24°C, whereas the teliospores in plant debris germinate best at 10-15°C. Long day hours favors the spread of the disease.

Management

1. Destruction of disease crop debris.
2. Follow wide plant spacing.
3. Fungicidal application using Tridemorph 0.1 per cent or Mancozeb 0.25 per cent effectively control the disease.

8. Anthracnose

It is fungal disease caused by *Colletotrichum lindemuthianum* affecting stem, leaf, branches *etc.* It was previously believed to be caused by the same fungus that affects Phaseolus beans, but it is now recognised as a distinct species, *Colletotrichum destructivum*. When diseased seed is planted, an epidemic begins.

Occurrence and Economic Importance

Economic losses are caused by the disease in tropical parts of Latin America and Asia. Losses of up to 50 per cent have been documented in southwest Nigeria's rainforest zone, particularly where cowpea is planted as a monoculture.

Symptoms

This fungus, which is seed-borne, affects the entire plant's above-ground structure. Dark, deep lesions that penetrate the seed coat identify infected seed. Oval and sunken stem lesions are present. The lesion has a dark brown core and purplish to crimson edges. In its early stages, the fungus grows along the veins and changes colour from purplish to red. The leaves start to look ragged as they age. The pods become infected and develop tiny, elongated reddish patches. Older spots have borders that range from brown to reddish brown and are sunken. Pods typically become infected through wounds. Contaminated pods decay. This causes significant agricultural losses.

Figure 15.9: Anthracnose of Cowpea.

Disease Cycle

Acervuli, which can be found on pods, leaves, stems, and branches, bear conidia that are either spherical or elongated. Hyaline, branching, and septate describe the mycelium. Under ideal climatic conditions, a conidium can germinate in 6–9 hours. The pathogen enters the cuticle and epidermis mechanically. (Leach, 1923). When temperatures are favourable, infectious hyphae expand and proliferate between the cell wall and protoplast for 2-4 days without causing visible damage to host cells). Mordue, (1971) It has been stated that the fungus can persist in seed for at least two years. Infected pods and seeds can live for a long time depending on the climate. The disease might survive for at least five years on air-dried pods and seeds stored at four degrees Celsius, or on dry, contaminated plant materials left in the field in sealed plastic bags. (Tu, 1983).

Management

1. Implement an integrated crop management strategy.
2. Rotate crops and remove and eliminate all crop residues after final harvesting. Seeds from disease-affected areas should not be planted because the agent is seed-borne.
3. To reduce damage, use clean seeds, rotate crops, and eliminate sick plants.
4. Spraying with a fungicide containing copper is the sole approved chemical for anthracnose management.

9. Root-knot Nematode

The most important root knot species are *Meloidogyne incognita, Meloidogyne javanica, Meloidogyne incognita* have been found to be predominant and cause severe crop loss (Adegbite *et al.*, 2005). The infection begins with root penetration of second stage juveniles developed in soil from egg masses produced by females on infected roots. Invaded nematodes cause galling of roots by initiating the formation of large cells in the root's meristematic cortical and xylem tissues. It adversely affects nodulation in cowpea at low population while nodulation may be completely prevented in cowpea plants under high nematode population.

Symptoms

The flowering, podding, seedling, and vegetative growth stages are all impacted by the nematode. The nematode also attacks the leaves, roots, and entire plant, causing various symptoms on the roots, corms, and tubers. Affected plants experience early death due to severe root system damage, which may be severely galled. Because the root system is unable to absorb enough water and nutrients, above-ground nematode symptoms include nutrient shortages, stunting, and frequently withering.

Figure 15.10: Root Knot Nematode of Cowpea.

Disease Cycle

Root knot nematodes can survive the winter as eggs or juveniles in the soil. The juveniles can infect roots, and the eggs hatch anytime the earth warms up, regardless of the season. Male and female larvae enter a tiny root of a vulnerable plant and begin feeding on the vascular system.

Management

1. Due to M. incognita's large host range, cultural control strategies are less well researched, and crop rotations are harder to establish.
2. Olabiyi *et al.* (2007) The addition of compost to *M. incognita*-infested soil in a cowpea cropping system resulted in a considerable reduction in soil nematode population and root gall index. Application of *Bacillus* spp. significantly reduces hatching of larvae of *M. javanica*.
3. Carbosulfan@0.05 or 0.1 per cent soak for 6 hours or Monocrotophos@o.1 per cent soak for 6 hours are effective for *M. incognita* control.
4. Fenamiphos 10 per cent granule is more effective for the control of *Meloidogyne* spp.

Conclusion

From this study it can be concluded that normally in cowpea crops are affected by different types of diseases and cause severe losses to yield. The management of cowpea diseases is not possible by a single method of plant disease management. It can be managed by the integration of all available control measures such as physical, chemical, biological and growing of resistant varieties have been used for the management of cowpea diseases.

References

Adegbite, A.A. and Amusa, N.A. (2003). The major economics field diseases of cowpea in the humid agro-ecologies of South-Western Nigeria. *African J. f Biotechnology*7: 4707-4712.

Adegbite, A.A., Amusa, N.A., Agbaje, G.O. and Taiwo, L.B. (2005). Screening of cowpea varieties for resistance to Meloidogyne incognita under field conditions *Nematropica,* **35**: 155-159.

Anilkumar, B. T., Chandrashekar, M. and Veerappa, K. B. (1989). Assessment of cowpea genotypes for multiple disease resistance (Abs).*Indian Phytopathol.,* **42**: 334.

Armstrong, G. M. and Armstrong, J. K. (1950). Biological races of the Fusarium causing wilt of cowpeas and soybeans. *Phytopathology,* **40**: 181-193

Chandrashekar, M., Anilkumar, T. B. and Saifulla, M. (1989). Effect of different dates of cowpea on the severity of leaf rust caused by *Uromycesphaseoli* var. *vignae.Trop. Agric.,* **65**: 149-152.

Chandrashekar, M., Anilkumar, T. B. and Saifulla, M., and Salimath, P. M. (1988). Leaf rust epidemics of certain promising cowpea varieties compared by using Logistic and Gompertz models. *Trop. Agric.,* **65**: 37-40.

Emechebe, A. M. and Florini, D.A. (1997). Shoot and pod diseases of cowpea induced by fungi and bacteria. In: Advances in cowpea research. Singh BB. Mohan Raj, DR, Dashiel, KE and Jackai, LEN (eds.). Copublication of OOTA and JIRCAS. IITA Ibadan Nigeria. Pp: 176-192.

Etana, A., Tadosse, E., Mengistu, A. and Hassen, A. (2013). Advances evaluation of cowpea (*Vigna unguiculata*) accessions for fodder production in the central rift valley of Ethiopia. J. *Agril. Ext. and Rural Devt.,* **5**: 55-61.

Grecher, W. J. (1995). *Phytopathology,* **21**: 11-115.

Harris, A. R. and Ferris, H. (1991). Interactions between *Fusarium oxysporum f.sp. tracheiphilum* and Meloidogyne spp. in *Vigna unguiculata*. 1. Effects of different inoculum densities on Fusarium wilt. *Plant Pathology,* **40**(3): 445-456.

Hagan, A. K. (1999). Plant disease, **3**: 73-75.

Honnur, R. B., Yadahalli, K. B. and Jahagirdar, S. (2016). Identification of susceptible stage for rust in cowpea, Biochem. Cell Arch. **16**(1): 141-143.

Kishun, R. (1989). Appraisal of loss in yield of cowpea due to *Xanthomonas campestris* pv. *vignicola. Indian Phytopathol.,* **42**: 241–246.

Leach, J. G. (1923). The parasitism of *Colletotrichum lindemuthianum*. Minnesota Agricultural Experiment Station Technical Bulletin, 14

Mordue, J.E.M. (1971). *Colletotrichum lindemuthianum*. Descriptions of Pathogenic Fungi and Bacteria Set 32, Sheet No. 316. Wallingford, UK: CAB International.

Mullen, J. (2001). Southern blight, Southern stem blight, White mold. *The Plant Health Instructor*. DOI: 10.1094/PHI-I-2001-0104-01.

Olabiyi, T.I., Akanbi, W.B. and Adepuju, I.O. (2007). Control of certain nematode pests with different organic manure on cowpea. *American-Euracian J. Agric. Envion. Sci.,* **2**: 523-527.

Pandey, S. and Joshi, P.K. (1995). Constraints and prospectus of legumes in the rice-wheat based cropping system in Terai region of Nepal. Trip report 7 Dec. 31 Dec., 1995 Patancheru, 502, 324, Andhra Pradesh, India, ICRISAT, pp.5.

Reddy, K.S., Pawar, S.E. and Bhatia, C.R. (1994). Inheritance of powdery mildew (*Erysiphe polygoni*) resistance in Mungbean (*Vigna radiate* L Wilczek). Theor. Appl. Genet. **88**: 945-948.

Sillero, J.C., Fondevilla, S., Davidson, J., Vaz, Patto, M.C., Warkentin, T.D., Thomas, J. and Rubiales, D. (2006). Screening techniques and sources of resistance to rusts and mildews in grain legumes. Euphytica**147**: 255-272.

Shambarker, D. A., Shinde, Y.M. and Baviskar, A.P. (1996). Genetic resource evaluation against major diseases of sesame. Sesame safflower Newsl. **12**: 6-10.

Sikirou, R. (1999). Epidemiological investigation and development of integrated control method of bacterial blight of cowpea caused by *Xanthomonas campestris* pv. *Vignicola*. Ph.D. thesis University of Gottingen, Germany, p. 218

Singh, R.S. and Sinha, R.P. (1955). Studies on the wilt disease of cowpea in Uttar Pradesh. *J. India Bot.Soc.,* **34**: 375-381.

Singh, B.B., Raj, D., Dashiel, K.E. and Jackai, L.E.N. (1997). Advances in cowpea research, co-publication of International Institute of Tropical Agriculture (IITA) and Japan International Research Centre for Agriculture Sciences. IITA, Ibadan and Nigeria Devon, pp.230-320

Singh, S.R. and Allen, D.J. (1979.) Cowpea Pests and Diseases. International Institute of Tropical Agriculture, Ibadan, Nigeria 108pp.

Swanson, T.A., Gundy, S. D. V. (1985). Influences of temperature and plant age on differentiation of races of *Fusarium oxysporum* f.sp. *tracheiphilum* on cowpea. Plant Disease, **69**(9): 779-781

Thammaiah, N. and Khan, A.N.A. (1995) Effect of seed treatment on *Xanthomonas campestris* pv. *vignicola* causal agent of bacterial blight in cowpea. *Adv. Agric. Res.* India, **4**: 93-102

Tu, J.C. (1983). Epidemiology of anthracnose caused by *Colletotrichum lindemuthianum* on white bean (*Phaseolus vulgaris*) in southern Ontario: survival of the pathogen. Plant Disease, **67**(4): 402-404

Chapter 16

Diseases of Cluster Bean Crop

Kavita Kansotia[1], Shivam Maurya[2], Pinki Sharma[3], Kiran Kumawat[3] and Lalita Lakhran[4]*

[1]Ph.D. Scholar, Department of Plant Pathology, Sri Karan Narendra Agriculture University, Jobner, Jaipur, Rajasthan
[2]Assistant Professor, Department of Plant Pathology, School of Agriculture, Lovely Professional University, Phagwara, Punjab
[3]Ph.D. Scholar, Department of Plant Pathology, MPUAT, Udaipur, Rajasthan
[4]Assistant Professor, Agriculture University, Jodhpur, Rajasthan
**e-mail: kavitakansotia6@gmail.com*

Introduction

Cluster bean [*Cyamopsis tetragonaloba* (L.) Taub], often recognized as guar, is a plant in the *Leguminaceae* family generally used for food, animal feed, green manure and industrial 'Guar gum' (Reddy *et al.*, 2014). In addition to being grown for its use as animal fodder, guar gum (also known as galactomannan) is also produced from it and used in the paper, pharmaceutical, petroleum, and cosmetic industries. Additionally, the guar crop enhances the soil by storing nitrogen there (Yogi *et al.*, 2016). It is believed to have originated in Africa and has been grown in India since ancient times. A resilient crop that can withstand drought is guar. It is the country's annual *Kharif* crop, which is largely farmed in the country's dry and semi-desert zones. Guar can also be cultivated well in irrigated locations during the *Zaid* season. Countries account for 80 per cent of worldwide guar output and 75 per cent of global guar commerce (Swamy and Naveena, 2015). The crop has gained recognition for its capacity to supply industry with high-quality gum. Several cultivars that can easily be grown in arid and semi-arid environments have been developed by various Indian research institutes, but some of these varieties are vulnerable to a number of phytopathogenic bacterial and fungal infections. The productivity and quality of grain and fodder production are significantly impacted by these biotic diseases (Sharma *et al.*, 2020). Kishnawat *et al.* (2021) observed various fungal diseases which infect and cause huge damage to the crop are *viz.*, Alternaria blight (*Alternaria cucumarina* var. *cyamopsidis*), Anthracnose (*Colletotrichum capsici* f. sp. *cyamopsidis*), *Curvularia lunata*, dry root rot/leaf blight (*Fusarium solani* and *Rhizoctonia solani*) and vascular wilt (*Fusarium moniliformae*). During favourable weather conditions for disease development very meager work has been carried out on weather parameters, which are responsible for disease development. There is a positive correlation between weather parameters favourable for progressive development of disease in the crop. These diseases cause huge loss on quantitative and qualitative aspects of the crop. Disease suppression by biocontrol agents such as *Trichoderma harzianum*, *Trichoderma viride*, and *Bacillus subtilis* is the result of long-term interactions between the plant, the pathogen, the biocontrol agent, the microbial community on and around the plant, and the physical environment, with pathogen growth being inhibited. Essential oils and plant extracts include a slew of bioactive compounds that are toxic to fungus, bacteria, and nematodes. Various researchers discovered that neem oil, turmeric, and garlic were effective against clusterbean infections. Chemical control, such as carbendazim, topsin M-70, captan, thiram, mancozeb, copper oxychloride, and foliar sprays of carbendazim, topsin M-70, captan, thiram, mancozeb, and copper oxychloride, is an effective method for treating several clusterbean diseases.

A. Fungal Disease

1. Alternaria blight of cluster bean
2. Dry root rot of cluster bean
3. Powdery mildew of cluster bean
4. Anthracnose of cluster bean
5. Wilt disease of clusterbean

B. Bacterial Disease

1. Bacterial blight of cluster bean

A. Fungal Disease

1. Alternaria Blight of Cluster Bean

Economic Importance and Occurrence

Alternaria is an extremely damaging pathogen that attack on vegetables and other economically valuable crops. Alternaria blight, caused by *Alternaria cucumarina* var. *cyamopsidis*, is a significant disease of cluster beans that is found across Northern India's guar-growing regions. In all of the clusterbean emerging places, primarily in rainfed conditions, the climatic components favour disease sprouting and propagation without delays and indirectly, hampered production. Areas where guar is grown are now very concerned about Alternaria blight because it poses a serious threat to crop productivity, especially in rainfed areas (Anupriya *et al.*, 2020).

Causal Organism

Disease caused by Alternaria blight, caused by *Alternaria cucumarina* var. *cyamopsidi*. Fungus belongs to the Kingdom – *Fungi*, Phylum – *Ascomycota*, Class – *Dothideomycetes*, Order –*Pleosporales*, Family – *Pleosporaceae*, Genus – *Alternaria* and Species – *cucumarina*.

Symptomatology

The disease manifests itself in practically every aerial component of the plant, including the cotyledons, leaves, petioles, stems, and pods. Water-soaked sores on the leaf blades are the first sign of infection. The lesions grow from greyish brown to dark brown, with concentric zonation delineated by light brown lines. In certain lesions, prominent darker border ridges with or without chlorotic halo can be detected. If the plants are infected prematurely in their development cycle, blooming and pod production may not occur.

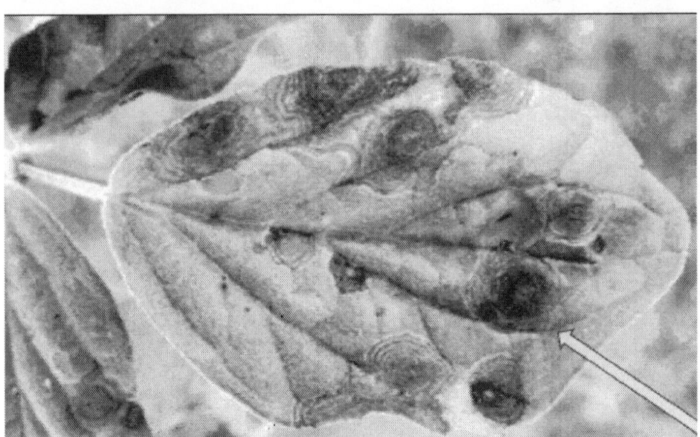

Figure 16.1: Alternaria Affected Leaves (*Source*: krishisewa.com).

Disease Cycle

Alternaria cucumarina var. *cyamopsidis*, causes guar blight are seed borne as well as soil borne in nature. The pathogen persists on seeds and infected plant debris in the form of conidia and mycelium in the soil. Various agents, such as wind and rain, disperse conidia and mycelia cause secondary infection.

Disease Management

1. To prevent soil-borne main inoculum, prior crop plant debris should be burned and seed treatment with carbendazim @ 2 g/kg seed is required to eliminate seed-borne disease.
2. The seed must be collected from a disease-free crop, the land must be kept clean and heavy summer plowing must be followed.
3. Garlic bulb extract 10 per cent and the fungicidal combo azoxystrobin + difenoconazole 0.29 Kg a.i/ha.
4. Spraying with a fungicidal mixture of carbendazim (12 per cent) + mancozeb (63 per cent) @ 0.2 per cent and garlic bulb extract @ 10 per cent twice at 10 days interval to control Alternaria blight.

2. Dry Root Rot of Cluster Bean

Economic Importance and Occurrence

It is a soil and seedborne polyphagous pathogen that affects over 500 crop species, causes dry root rot, also known as charcoal rot, a devastating disease of cluster beans. In South and Southeast Asia, *Macrophomina* species infect a variety of field crops, particularly common bean (*Phaseolus vulgaris* L.), cowpea (*Vigna unguiculata* L.), Walp, urdbean (*Vigna mungo* L.) Hepper, soybean (*Glycine max* L.) Merr. and potato (*Phaseolus tuberosus* L.).

Causal Organism

Rhizoctonia bataticola (*Macrophomina phaseolina*), it belongs to the Kingdom – *Fungi*, Pzzhylum-*Ascomycota*, Class – *Dothideomycetes*, Order – *Botryosphaeriales*, Family- *Botryosphaeriaceae*, Genus – *Macrophomina*, and Species - *phaseolina*.

Symptomatology

The characteristics symptoms are browning of the leaves, which lasts for two or three days before shrinking. After the onset of the first adverse effect, the plant may wither within seven days. When the stem is examined closely, dull lesions on the bark at the ground level may be visible. Sclerotial bodies can be seen on infected tissues. When a pathogen infects a plant's roots, it causes root rot, plant wilt and eventually death when the disease progresses. *M. phaseolina* is a soil-borne pathogen that may infect chickpeas at any stage of growth, although it prefers to infect them after they have finished reproducing. Infection on the host, the fungus produces cell wall disintegrating enzymes and phytotoxins such as phaseolinone and botryodiplodin by rotting the roots of plants.

Disease Cycle

M. phaseolina is soil and seedborne pathogen. *M. phaseolina* grows anamorphically (sclerotial) in soil and on plant debris. On medium culture, the fungus produces brown to grey mycelium that develops into darker, thin, hyaline, septate, dichotomously branched hyphae that eventually generate black sclerotia. A distinguishing feature of the fungus is right-angle branching of the mycelium with constriction of the hyphal branch at the place of origin. The sclerotia are dark in colour.

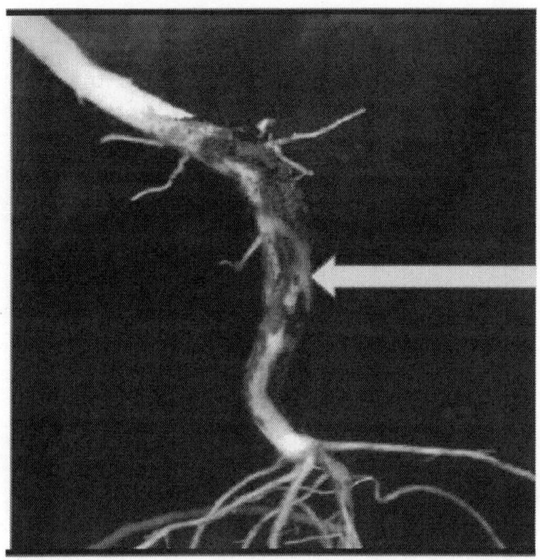

Figure 16.2: Dry Root Rot Affected Root.

Management

1. Soil solarization is a cost-effective method for reducing pathogen populations in the soil, although it takes a long time and is dependent on weather conditions.
2. Soil application and seed treatment with *Trichoderma* spp. Botanical extracts include a diverse range of bioactive chemicals, including fungus, bacteria and fungi are effective against dry root rot causing pathogen.
3. To avoid the disease, it's also important to use cultivars that have ability to reduce risk of infection.
4. The severity of the disease is limited by changes in social behaviors and seed treatment. Soil drenching with mancozeb @ 250 ppm concentration and seed treatment with carbendazim + mancozeb @ 2 g/kg of seed.

3. Powdery Mildew of Cluster Bean

Economic Importance and Occurrence

Powdery mildew is most commonly occurring disease of cluster bean crop. Powdery mildew under favorable condition causes yield losses up to 50-60 per cent (Richard *et al.*, 1999). Alternaria blight cause yield losses 43-87 per cent was recorded when leaves are infected at seedling stage than at older stages (Sharma, 1983).

Causal Organism

The disease caused by *Leveillula taurica* (Lev.). The fungus belongs to *Ascomycota*, Class - *Leotiomycetes*, Order - *Erysiphales*, Family- *Leveillula*, and Species- *taurica* (Kirk *et al.*, 2008).

Symptomatology

The disease occurs almost all guar growing regions of the world. Under severe conditions entire plant can cover by white cottony mycelial growth of the fungus. At final stage small black or brown cleistothecia (Sexual fruiting bodies) appears on the mycelial mats.

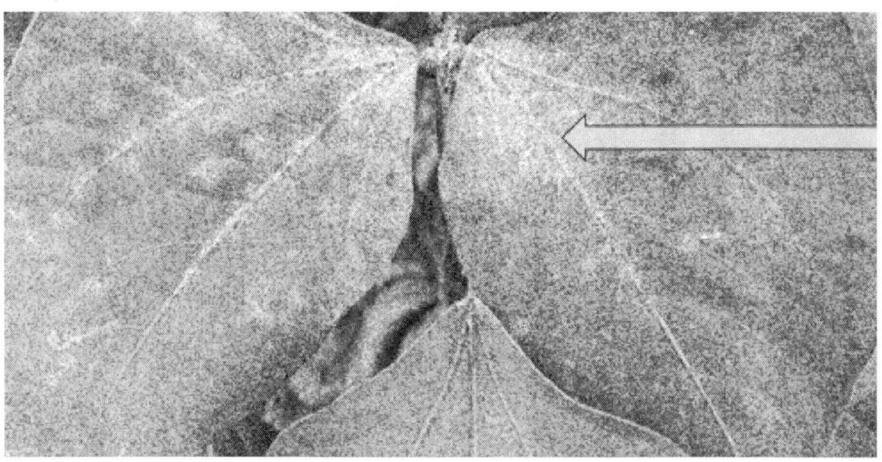

Figure 16.3: Powdery Mildew Affected Leaves.

Disease Cycle

Because the pathogen is an obligate parasite, it could only persist in cleistothecial form without even a living host. The primary inoculum is disseminated via the air, possibly across considerable distances from the primary source of inoculum. Powdery mildew is a fungus that spreads by air regionally and over large distances, and it can be caused by both conidia and ascospores. When circumstances are good, the virus infects fast, causing symptoms to manifest within three to seven days and producing a significant number of conidia in a short amount of time. Powdery mildew in beans requires a temperature range of 22-27°C, as well as high humidity at night and low humidity during the day (Agrios, 1988). In heavy dews, disease incidence rises, allowing the germ tube to penetrate the host epidermal surface, where the fungus produces haustoria to absorb nutrients from the host cells.

Disease Management

1. Plant early maturity varieties.
2. Resistant varieties should be planted.
3. Rotate plantings at least 1 year.
4. NSKE 5 per cent spray on standing crop after appearance of symptoms.
5. Dinocap 48 EC or Tridomorph 80 EC 0.1 per cent at 10 days interval.

4. Anthracnose of Cluster Bean

Economic Importance and Occurrence

Colletotrichum is one of the most common plant pathogenic fungi that cause anthracnose in a variety of plant types (Dean *et al.*, 2012). Anthracnose of cluster beans is caused by *Colletotrichum capsici* f. sp. *cyamopsicola,* a major disease that affects pod quality, grain quality and production. Desai and Prasad initially reported it in 1955 at the Agriculture Institute in Anand, Gujarat. Diseases may be found all throughout the country, wreaking havoc on crops. In nature, pathogens are both soil and seed carried. Corda characterized the *Colletotrichum* fungus for the first time in 1831.

Causal Organism

Anthracnose of cluster beans is caused by *Colletotrichum capsici* f. sp. *cyamopsicola. Colletotrichum* conidia are single celled, hyaline, oval to elongate, falcate conidia that form on conidiophores in acervuli with setae. Septate, branching at the base, hyaline to pale brown, and smooth conidiophores. Butler (1918) investigated the saltation of *C. capsici* and proposed that it is a polymorphic species that produces morphologically similar Colletotrichum under the same species. *Vermicularia capsici* was renamed *C. capsici* by Butler and Bisby in 1931.

Symptomatology

The infection infects seedlings initially, leaving necrotic, black patches on the cotyledon and main leaves. Necrotic signs appear on the leaves and stems, which later become brown with a black border. Later, these spots expand to cover a vast area of plant parts and consolidate, causing blighting, withering, and drying of diseased tissues. On stem lesions, fungus produces black acervuli. The pods turn dark, with brown to black patches on the surface, and when the pods are split open, the seeds seem to be diseased, tiny, malformed, and purple in colour.

Disease Cycle

Pathogens need warm, humid weather to infect a variety of host plants, including fruits, ornamentals, vegetables, field crops, and even grasses. *Colletotrichum* is a fungus that affects the majority of crops farmed across the world.

Figure 16.4: Anthracnose of Affected Pods of Cluster Bean (*Source*: Fitsum *et al.,* 2014).

Disease Management

Anthracnose of cluster beans may not be manageable with a single technique. Because the fingus is seed-borne, wind-borne, and water-borne rather than soil-borne, methods to restrict its spread should focus on three key areas in the field.

1. Use disease free seeds.
2. Crop rotation and evacuation of any tainted plant sections.
3. To manage the disease in an environmentally responsible manner, plant extracts and bio-agents can be used instead of chemicals. The application of extracts from Neem (*Azadirachta indica*) @ 10 per cent, Mahogany (*Swietenia mahagoni*) @ 10 per cent, and Garlic (*Allium sativum*) @ 5 per cent reduced the occurrence of anthracnose.
4. Traditionally uses of combination fungicide are more effective against anthracnose diseases than single one. Tebuconazole or Azoxystrobin @ 0.1 per cent foliar application at 15 days of interval.

5. Bacterial Blight of Cluster Bean

Economic Importance and Occurrence

Bacterial blight of cluster beans is the most serious disease caused by *Xanthomonas axonopodis* p.v. *cyamopsidis*, which causes yield reductions of 50-70 per cent in extreme cases.

Causal Organism

Bacterium caused the disease belongs to Phylum – Pseudomonadota, Class –

Gammaproteo bacteria, Order – Xanthomonadales, Family – Xanthomonadaceae, Genus – *Xanthomonas*, and Species – *axonopodis*.

Symptomatology

Seedling mortality is the principal cause of harm caused by the disease. Infection of above-ground plant components is caused by secondary disease transmission. The disease symptoms include wilt of seedling plants at later stages of plant growth, the infected plants exhibit rotting near the soil which results in wilting of the plant. The infection from blighted leaves spreads to stem through petiole and in advance stages, stem gets cracked, wilting which causes defoliation and in severe cases vascular necrosis, dieback and black streaking of the stem along with continuous oozing of exudates. The pods also show heavy spotting.

Figure 16.5: Bacterial Blight of Cluster Bean (*Source*: Mohamed *et al.*, 2020).

Epidemiology

According to Singh (2014), disease propagation is aided by sporadic rainfall, high temperatures (20-30 °C), and greater humidity. Higher nitrogen dosages and a plant's early development stage (20-30 days) made it more susceptible to stem rot, causing it to dry out.

Management

Infected seeds are major source of primary inoculum; therefore, generally seed treatment is advocated to prevent the disease. Diseased free seeds from the summer grown irrigated crop should be used and the seed from the monsoon should be hot water treatment. Singh (2014) asserted that bacterial blight of cluster bean can be managed by

1. Use disease free Seeds.
2. Bioagent *Pseudomonas fluoresense* 10 g/kg as seed treatment found effective.

3. Seed dip in Agrimycine (250 ppm) or Streptocycline (200 ppm) solution for 3 hrs.
4. Spray on standing crop of 35-40 days with Streptocycline 5 gm or Plantomycine 50 gm in 100 litres of per hectare can also effective to control the disease.
5. Seed therapy with streptocycline 500 ppm + 2 sprays of copper hydroxide (0.2 per cent) over a 15-day period produced the highest yield and disease intensity, followed by seed therapy with streptocycline 500 ppm + 2 sprays of copper hydroxide (0.2 per cent) + streptocycline 250 ppm over a 15-day period.

6. Cluster Bean Wilt

Economic Importance and Occurrence

It is grown for green pods used as vegetable and dry seeds. It is also grown as forage and green manure crop. Some of the varieties are used for extraction of gum (Guar gum). Fusarium solani is serious pathogen which causes wilt, root rot and damping off in cluster bean (Satyaprasad and Ramaroa, 1981; Dwivedi, Dubey and Dwivedi, 1991). The pathogen is seed borne and transmitted in all plant parts of cluster bean (Mathur and Shekhawat, 1988; Dwivedi, Dubey and Dwivedi, 1991). The heavy infection of pathogen causes severe losses in quality and yield of crop which reduce the medicinal and industrial value of the crop.

Causal Organism

Fusarium solani is polyphagous fungi which causes important soil borne as well as seed borne diseases like wilt, root rot, damping off in legumes, oilseeds, medicinal and ornamental crops and cause severe economic losses. The pathogen *F. solani* which incited the wilt and root rot in cluster bean crop and cause severe damage to the crop. The ungus belongs to the Division – *Ascomycota*, Class – *Sordariomycetes*, Order – *Hypocreales*, Family – *Nectriaceae*, Genus – *Fusarium*, and Species – *solani*.

Etiology

The pathogen is seed borne and transmitted to all parts of cluster bean plant. Cluster bean wilt was first reported by Singh (1951) from Kanpur. Among the diverse pathogens infecting the crop, *F. solani* (Mart.) Sacc.is the major pathogen, which is responsible to cause the wilt of cluster bean. The seed borne inoculum of *F. solani* transmitted from seed to seedling give rise to pre and post emergence damage to the crop. Microconidia, macroconidia and chlamydospores produced from the mycelium. Microconidia are hyaline ova to cylindrical, slight allantoidal with blunt end and macroconidia are hyaline, fusoid and septate.

Disease Cycle

F. solani can be found in soils worldwide, where its chlamydospores overwinter on plant tissue/seed or as mycelium in the soil. The pathogen enters hosts through developing roots, where it can infect the host. After infection, *F. solani* produces asexual macro and microconidia which are dispersed through wind and rain. The pathogen can persist in the soil for a decade, and if left unchecked can cause complete crop loss.

Management

Controlling soil-borne diseases like *Fusarium* wilt has always been a problematic task. To remove the pathogen, soil treatment is challenging. As a result, wilt control options include seeding resistant cultivars and other cultural measures.

1. Crop rotation is the most effective approach to eliminate soil-borne pathogens.
2. By increasing soil temperature and covering the soil with mulching materials, *F. oxysporum* inoculum can be reduced.
3. Seed and soil treatment of *T. viride* + soil application of Neem cake (150 kg/ha.) + mancozeb (0.3 per cent) soil soaking effectively reduced the disease.
4. Seed treatment with carbendazim (2 g/kg) + Neem oil (15 ml/kg).

Conclusion

Various fungal and bacterial diseases damage the clusterbean crop. Which is a huge manufacturing restriction. Non-chemical opportunity procedures are time-consuming and ineffective against soil-borne plant infections. Chemical management is an effective method of disease control in agricultural crops. Disease suppression by biocontrol agents such as *Trichoderma harzianum, Trichoderma viride,* and *Bacillus subtilis* is the result of long-term interactions between the plant, the pathogen, the biocontrol agent, the microbial community on and around the plant, and the physical environment, with pathogen growth being inhibited. Essential oils and plant extracts include a slew of bioactive compounds that are toxic to fungus, bacteria, and nematodes. Various researchers discovered that neem oil, turmeric, and garlic were effective against clusterbean infections. Chemical control, such as carbendazim, topsin M-70, captan, thiram, mancozeb, copper oxychloride, and foliar sprays of carbendazim, topsin M-70, captan, thiram, mancozeb, and copper oxychloride, is an effective method for treating several clusterbean diseases. Carbendazim and biocontrol agents *Trichoderma viride, T. harizianum* have been found to be effective against numerous clusterbean diseases, and compatibility of biocontrol agents with fungicides has now been discovered.

References

Agrios, G.N. (1988). Plant Pathology, 3rd Ed.; *Academic Press*, Inc.: New York, NY, USA.

Anupriya, Singh, R. S. and Prahlad (2020). Management of *Alternaria cucumarina* var. *cyamopsidis* through plant extracts, bio products and fungicides *in vitro* and *in vivo*. *International Journal of Current Microbiology and Applied Sciences.* **9**(3): 2595-2605.

Dean, R., Van Kan, J.A., Pretorius, Z.A., Hammond-Kosack, K.E. and Di Pietro, A. (2012). The Top 10 fungal pathogens in molecular plant pathology. *Mol. Pl. Pathol.,* **13**: 414-430.

Kishnawat, P., Singh, J. and Singh, M. (2021). Efficacy of different Fungicides, Plant Extracts and Bio-agents for the Management of Root rot of Cluster bean incited by *Macrophonina phaseolina. The Pharama Innovatio,* . **10**(3): 675-678.

Massoud, O. N. and Kamel, S. M. (2015). The inhibitory effects of free and encapsulated arbuscular mycorrhizal fungi and *Trichoderma viride* against charcoal rot (*Macrophomina phaseolina)* on common bean (*Phaseolus vulgaris* L.) *Egyptian Journal of Biological Pest Control Pest Control,* **25**(2): 489-97.

Reddy, D. S., Nagare, P. K., Reddaiah, K. and Reddy, B. R. (2014). Effect of integrated nutrient management on growth, yield, yield attributing characters and quality characters in clusterbean (*Cyamopsis tetragonoloba* L. Taub.). *The Ecoscan.,* **4**: 329-332.

Satish, S., Singh, R. and Kumar, A. (2020). Field evaluation of selected botanicals and fungicides for management of Alternaria blight of clusterbean. *Int. J. Curr. Microbiol. App. Sci.,* **9**(7): 1045-1053.

Sharma, S., Singh, R. and Pandiya, R. K. (2020). Epidemilology and Alternaria blight of clusterbean caused by *Alternaria cucumarina* var. *cyamopsidis. Journal of Pharmocology and Phytochemistry,* **9**(3) : 1886-1890.

Singh, R. (2014). Improved Cultivation Practices for Clusterbean in *Kharif* and Summer Season. Central Arid Zone Research Institute (ICAR).

Swamy, D. and Naveena, K. P. (2015). An analysis of production performance and yield variability of guar (Cluster bean) in India. *J. Prog. Agric.,* **6**: 1.

Yogi, R., Sharma, O. P., Verma, H. P., Kanwar, K. and Shivran, A. C. (2016). Response of fertility and agro-chemical on growth and yield of different varieties of clusterbean. *Ann. Agric. Res.,* **37**(3): 295-299.

Sharma, A. K. (1983). Occurrence of Alternaria blight of pea in Kumaun hills. *Indian Phytopathology.* **36**: 760-761.

Sileshi, Fitsum., Mohammed, Amin., Selvaraj, Thangavel, Alemayehu and Adugna (2014). In vitro evaluation of some fungicides and bioagents against common bean anthracnose (*Colletotrichum lindemuthianum* Sacc. and Magnus) Briosi and Cavara. *African Journal of Microbiology Research,* **8**: 1.

Mohamed, E. M., Suliman, M. E. and Ismail, M. I. (2020). Identification and Evaluation of Bacterial Blight Occurrence on Guar Crop in Gadarif State

(Sudan). *International Journal of Trend in Scientific Research and Development,* **4**(3): 809-814.

Satyaprasad, K. and Ramarao, P. (1981). Root rot of guar caused by *Fusarium solani. Indian Phytopath.,* **34:** 523-524.

Dwivedi S. K., Dubey N. K. and Dwivedi R. S. (1991). Damping off of *Cyamopsis tetragonoloba* (L.) Taub. Due to seed-form inoculums. *Nat. Acad. Sci. Lett.* **14:** 371-373.

Mathur K. and Shekhawat K. S. (1988). Fusarium root rot of guar. *Indian J. Mycol. Pl.Pathol.* **17**: 237.

Chapter 17

Diseases of Rapeseed and Mustard Crop

Harshit C. Chavda and Mayur R. Thesiya*

College of Agriculture, Parul University, Vadodara – 397 015, Gujarat
****e-mail: harshitchavda9328@gmail.com***

Introduction

India grows mustard (*Brassica compestris* L.), rapeseed, and other oilseed crops as one of its key Rabi crops. Representing 32 percent of India's total oilseed production, mustard is the country's second-biggest native oilseed crop. India, one of the world's top producers of oilseeds, cannot supply its enormous population's need for edible oil. An important part of India's agricultural economy is its oilseed crop industry. In the states of Assam, Bihar, Gujarat, Haryana, Himachal Pradesh, Jammu and Kashmir, Madhya Pradesh, Orissa, Punjab, Rajasthan, Uttar Pradesh, and West Bengal, Indian mustard is grown as a Rabi crop. The Brassicaceae family includes oilseed Brassicas, sometimes known as rapeseed-mustard. In terms of acreage and output, oilseed brassicas rank second in India after groundnut and are the third most significant oil seed crop globally, behind soybean and palm (Kumar and Chopra, 2014). After oil palm and soybeans, rapeseed is regarded as the third-largest edible oil source in the world. Since ancient times, people have used every aspect of the mustard plant which is essential to human survival for flavoring, medicine, and preservation. After China and Canada, India is the world's third-largest producer of rapeseed-mustard, accounting for about 11 per cent of global output.

Indian mustard makes up almost 80 per cent of all the hectares planted to these crops in the nation, according to ICAR-DRMR. Rapeseed-mustard makes for 23.2 per cent of the total acreage and 26.2 per cent of the total production of the nine oilseed crops farmed in India (2014–15 to 2018–2019). In India, the average yield of rapeseed and mustard is 1499 kg/ha. Worldwide, 26.09 million hectares are used for the cultivation of mustard and rapeseed, yielding 571,880 tons of output (FAO, 2021). 6.42 million hectares of land and 6.33 million tons of mustard are produced in India (Anonymous, 2017–2018). India's Punjab region grows rapeseed and mustard during the Rabi season, which runs from mid-September to November. The harvest occurs in late February or early March. A mustard crop needs temperatures between 10°C and 25°C as well as 25–40 cm of annual rainfall. Mainly planted during the Rabi season, mustard seeds are sown in October and harvested in March. Among growers in Rajasthan, Haryana, Madhya Pradesh, Uttar Pradesh, and other states, it is a significant crop. A significant rise in mustard production is anticipated in 2022–2023; this information comes from a recent Economic Times report. There is a big discrepancy between the potential output and the yield that is actually realized in the farmer's field because as mustard production rises, so do the attacks of several illnesses. The poor production of the crop that India produces must be addressed in order to meet the increased demand for mustard. Numerous diseases that harm mustard, including as white rust, alternaria blight, powdery mildew, and downy mildew, present serious issues for the nation. Seed yields and oil content are negatively impacted by these diseases, but Alternaria leaf spot disease causing the most harm. On the other hand, illnesses including downy mildew, powdery mildew, and white rust also require treatment. It is possible to maximize the mustard crop's economic production by controlling these diseases well. The causes, symptomatology, etiology, spread, and management strategies for significant mustard illnesses are the main topics of this chapter. In order to enable the effective implementation of disease management methods, the goal is to give thorough information and comprehension of the numerous illnesses that affect mustard.

1. Alternaria Blight

Occurrence and Economic Importance

Around the world, Alternaria blight is the most prevalent and deadly disease affecting mustard. In Europe and Canada, it is sometimes referred to as Alternaria dark spot or black spot. This disease attacks siliquae, penetrates the seeds, and damages the assimilatory tissues of the leaves and stem, resulting in a significant reduction in seed output. Severe disease infection causes the seed to shrink and its oil content to decrease, which alters the seed's chemical makeup and affects its protein, total carbohydrate content, and ash level. The overall losses of yield estimated up to 70 per cent and about 47 per cent reduction in the seed yield in India. Thus, both qualitative as well as quantitative loses can be occur. Alternaria blight caused by three different species namely – *Alternaria brassicae, A. Raphani, A. Brassicicola.*

Causal organism: *Alternaria brassicae* (Berk) Sacc.

Taxonomic Tree (Kirk I 2008)

Kingdom- Fungi

Fungi Imperfecti (Anamorphic group of fungi)

Genus-*Alternaria*

Pathogen-*Alternaria brassicae, A. brassicicola, A. Raphani*

Table 17.1: Characteristics of *A. brassicae, A. brassicicola, A. Raphani* (Meena *et al.*, 2010)

Fungal Structure	*Alternaria brassicae*	*A. brassicicola*	*A. Raphani*
Mycelium	Brown to brownish-gray septate	A grayish- to olive grey in septate form	Cottony witish to greenish grey or dark olive
Conidiophore	Dark, divided (Septate), arise in fascicles, measuring 114-74µ x 4-8 µ	Olivaceous, septate, branched, measuring 35-45 µ x 5-8 µ	Septate, olive-brown, simple or branched, measuring 29-160 µ x 4-8 µ
Conidia	Obclavate, muriform, brownish black, longer, wider, and with more septations on its beak; produced individually or infrequently in chains of two to three	Dark, cylindrical to oblong, muriform without beak, few septation and smaller than other two species of *Alternaria*. Produce long chains of 8 to 10 spores	Olivaceous brown, dark, muriform spores that can appear alone or in chains of up to six, with a poorly formed or absent beak.

Symptoms

This disease severely reduces photosynthesis and the vigor of the plant by attacking all of the above-ground portions, including the leaves, stem, and silique. The symptoms began as chlorotic specs on the lower leaves and eventually changed into tiny brown to dark brown dots. Eventually, they grew into a circular, irregular lesion that was dark brown and surrounded by a yellow halo. The conidiophore and conidia were organized in concentric rings at the center, which gave the appearance of a target board. When these spots grow and merge together, covering a significant portion of the leaf, favorable weather circumstances cause the leaves to lose their outer layer. Eventually, the illness spreads to the plant's other sections. Long, dark patches devoid of concentric rings were seen on the plant's stem and siliques. Infected slices ripen early, contract, break, or crumble readily. The seed beneath the diseased siliques has dark-brown patches on its coat that have turned shriveled, twisted, and discolored. These contaminated seeds germinate poorly and produce less oil content.

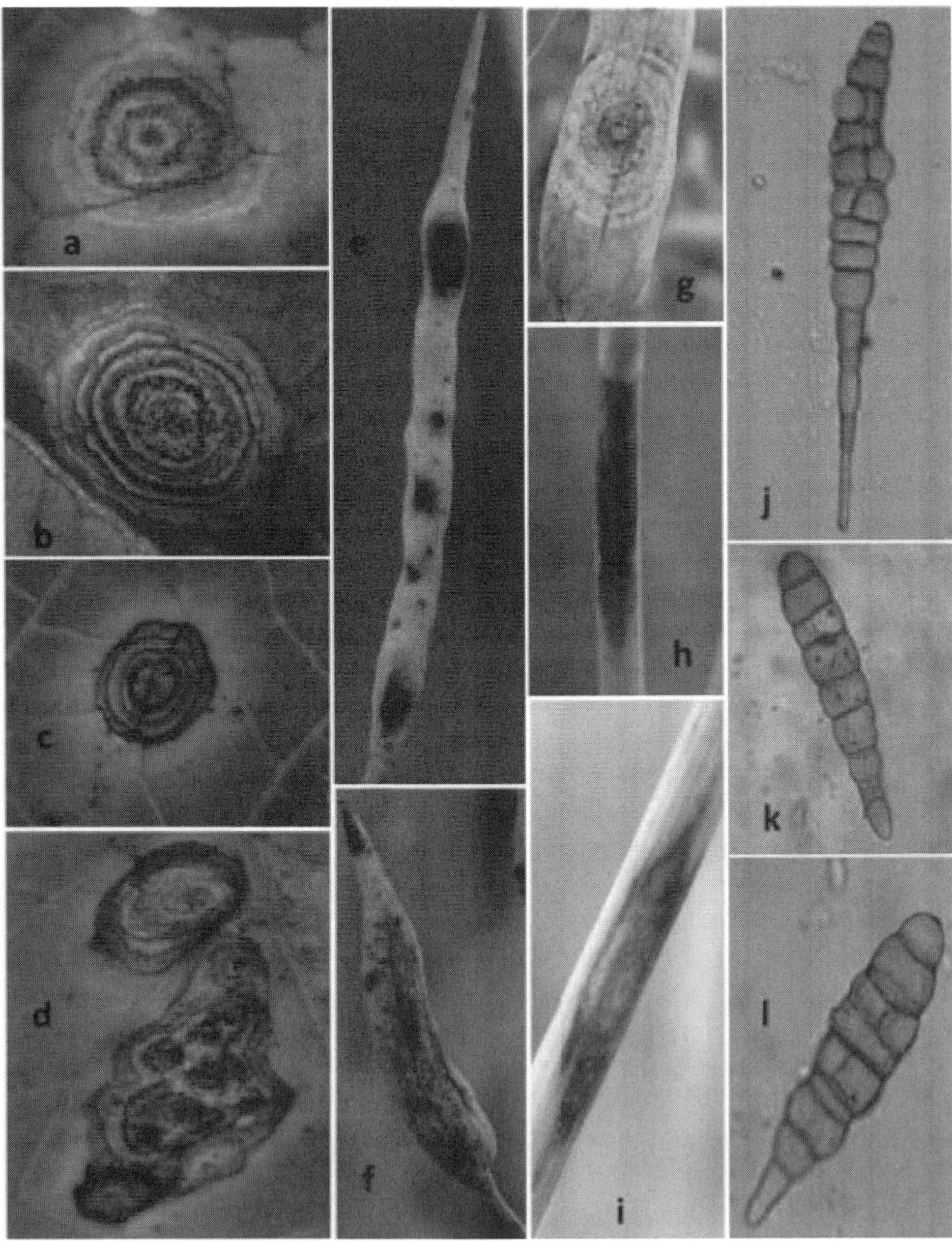

Figure 17.1: Alternaria Blight Symptoms in Mustard (*Source*: Kumar *et al.,* 2014).

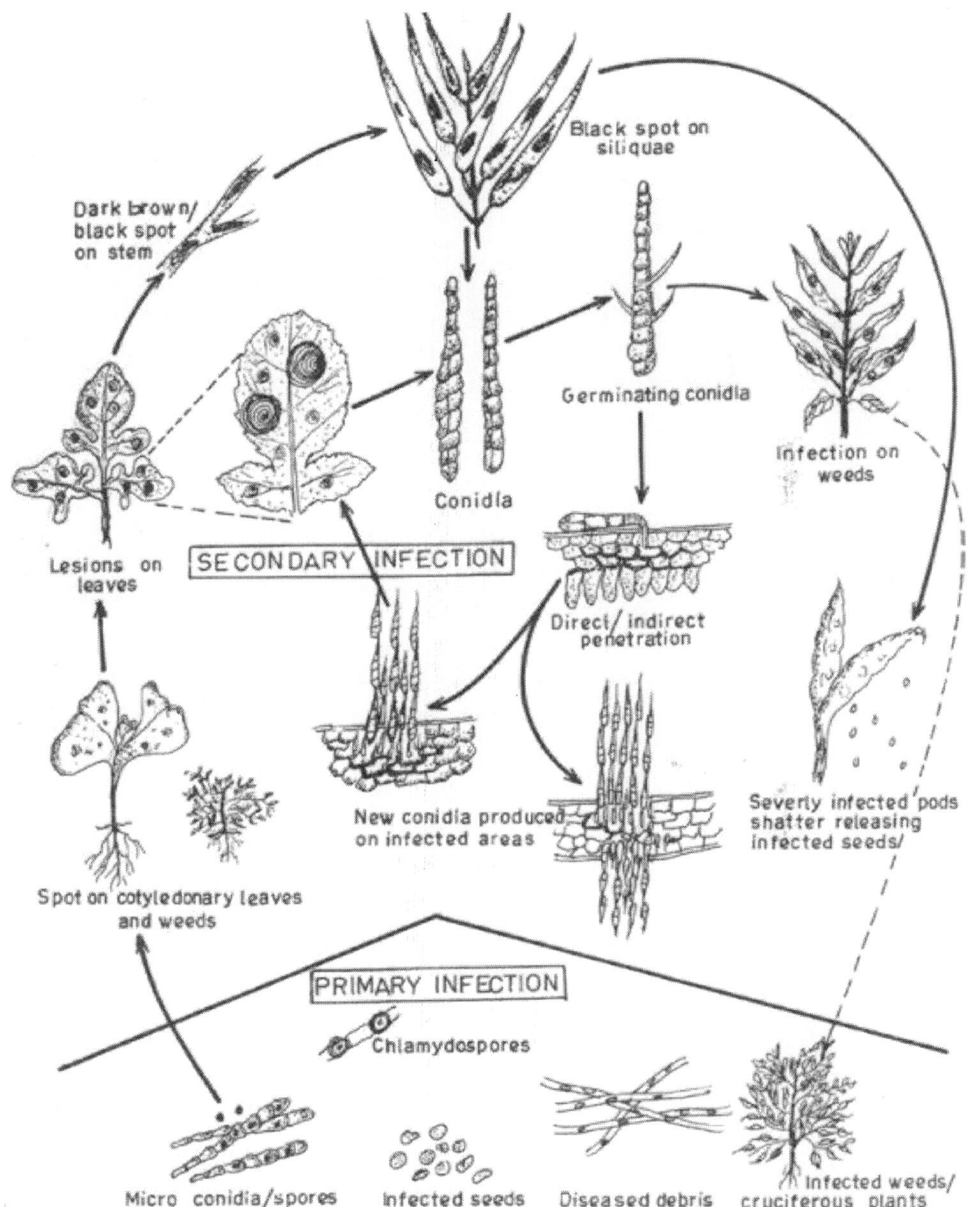

Figure 17.2: Disease Cycle of Alternaria Blight of Mustard (*Source*: Mehta, 2019).

Favourable Conditions

During the crop-growing season, relative humidity levels must be greater than 90 per cent. For *A. brassicae* to sporulate, the ideal temperature is 18–24°C, and for *A. brassicicola*, it is 20–30 °C. In 12 to 14 hours, both pathogens generate spores. A cool, dry climate is also conducive to the development of disease. High infection requires 16–24 hours of moisture in the leaves at the threshold temperature of 25°C. *Alternaria raphani* conidial germination requires a temperature of 31 °C. 98 per cent of the spores sprout after 6 hours of incubation, and 23°C is the ideal temperature for *A. raphani*. Additional environmental elements, such seasonal rainfall or irrigation, wind speeds between 2 and 5 km/h, closer spacing between plants (30 ×15 cm), and high nitrogen dosages (80 kg/ha), are also accountable for disease development.

Disease Cycle

The primary source of food for Alternaria is plant detritus. *A. brassica* generates chlamydospores and microsclerotia, which aid in the pathogen's survival. It is known that *A. brassicae* can spread by seeds, however this means of persistence and spread is not very important. *A. brassicae, A. brassicicola*, and *A. raphani*'s spores and mycelium are harbored by diseased plant materials during the off-season, and infected seeds are the primary means of infection for the disease. When conidia are spread by wind, secondary infection happens.

Management

1. It is advisable to use blight resistant varieties of mustard crop. Otherwise sound and bold, clean seed should be used for the sowing.
2. To minimize the inoculum of pathogen, collection and destructions of infected plant material should be destroy outside the field after harvesting of the crop.
3. Early sowing (Last week of September to early October) of the crop usually escape the disease.
4. Balanced use of the nitrogenous fertilizers.
5. Spraying of Eucalyptus leaves (2 per cent) or Neem leaves (15 per cent) effectively manage the disease.
6. Seed treatment with Captafol @ 4g, Iprodione @ 1.25g and Carbendazim @ 2.5 g or Mancozeb @ 2 g/kg of seed, eradicate the seed borne infection.
7. Use of fungicides during the severe infection of the disease, such as Azoxystrobin 23 per cent SC and Propiconazole 25 per cent EC effectively manage the disease.

2. White Rust

Occurrence and Economic Importance

White rust is a very serious disease of mustard, it is also known as a White blister and staghead disease and it is responsible for considerable yield losses in mustard crop. Disease intensity of white rust increases with the increase in the cultivated area of mustard in India. Disease infection at the flowering stage can cause complete loss of the crop. Due to both local and systemic white rust infection, yield losses were from 23 to 54.5 per cent (Saharan *et al.*, 1984). According to Mehta (2019), a combination of downy mildew and white rust infection can result in a 37–47 per cent decrease in pod formation and a 17–37 per cent decrease in grain production. the approximate decrease in mustard seed oil content from 1 to 90 per cent. Although the disease has been documented from over fifty different countries, given the nature of the virus, it is most likely widespread everywhere mustard is produced.

Causal organism: *Albugo candida.* Kuntze

White rust of mustard is caused by *Albugo candida,* which belongs to Kingdom- *Chromista,* Phylum- *Oomycota,* Class- *Oomycetes,* Order- *Albuginales,* Family- *Albuginaceae,* Genus- *Albugo,* and Species- *candida* (Kirk *et al.*, 2008). An obligate parasitic fungus, produce well-developed profusely branched mycelium which consist of aseptate, coenocyte hyphae. It produces oospore and zoospores, globose sporangia produce in chain.

Symptoms

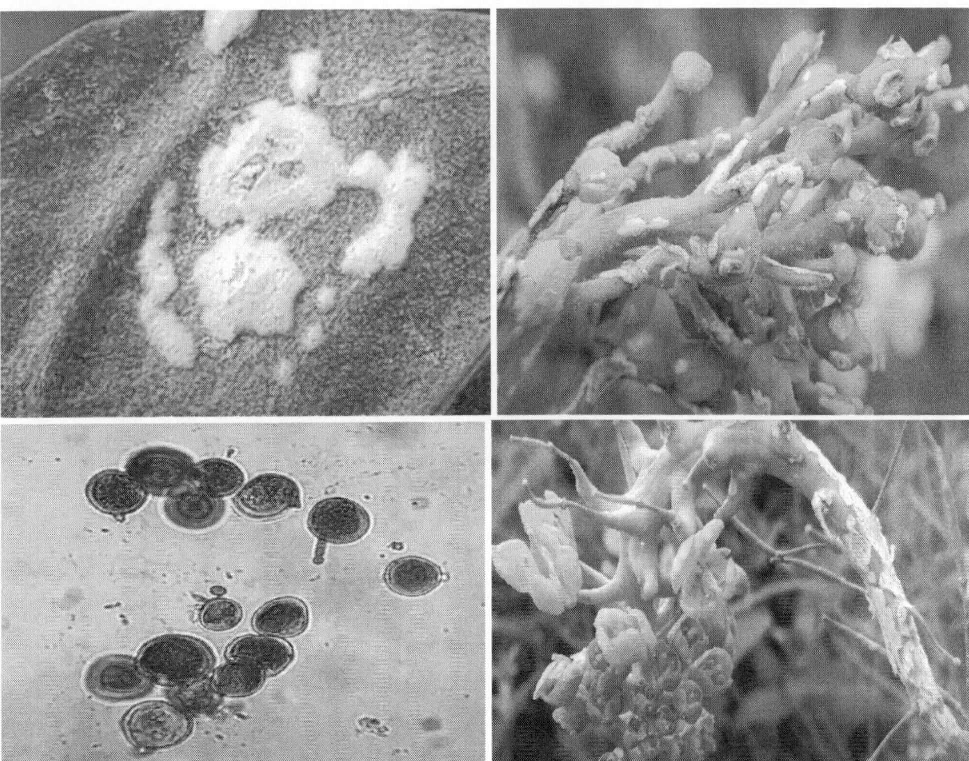

Figure 17.3: White Rust Symptoms and Sporangia (*Source*: B.S. Kharayat).

The leaf, stem, and inflorescence are the primary sites of the symptoms. In crops, the fungus causes both systemic and local diseases. A systemic infection can harm any part of the plant, but a local infection mostly affects the green sections of the plant (Rachel, 2006). In the early stages of the disease, a local infection causes white blisters, which are cream-colored pustules that range in color from white to cream. However, as the infection worsens, circular patches of light green to yellow leaf discoloration are seen on the that corresponds upper surface, and pustules on stems and pods may also form. The disease spreads throughout the plant's tissue in cases of systemic or flower bud infections, causing significant deformation, hypertrophy, hyperplasia, and sterility of the inflorescences. The term "staghead" refers to a spiny, enlarged structure that resembles a deer's antlers when the whole flowering structure is infected. These days, staghead downy mildew and white rust infections are frequently seen together.

Favourable Conditions

The disease exhibits rapid spread in the field under high humidity conditions, particularly when the relative humidity ranges from 60 per cent to 80 per cent. The other factors, such as temperature ranges from 12°C -14°C, cool dew nights, slightly wormer days and intermittent rainfall favours the disease development. Optimum temperature for sporangial germination is about 10°C and oospore germination 10°C - 20°C found most favourable. Late showing of mustard crop is also favours the disease development.

Disease Cycle

Albugo survives by means of oospores in the soil along or moving with infected plant parts. Seed also play important part in the survival of pathogen. The oospores can survive for more than 20 years under storage in infected plant material. Perennial weeds also harbor pathogen in off season. Sporangia and zoospores produced in the infected pustules are the means by which the disease is secondarily transmitted.

Management

1. Cultural practices: choosing healthy seeds, rotating crops because the illness is soil-borne, eliminating weeds and diseased plant waste from the field to lower pathogen inoculums, By the end of the month of October, the crop should be sown because a delay in sowing could result in a serious infection during flowering.
2. White rust resistant varieties should be used in severe infected areas.
3. Botanicals: Application of extracts of various botanicles found effective against white rust disease such as Neem, Tulsi, Datura, Eucalyptus leaf extracts *etc.*

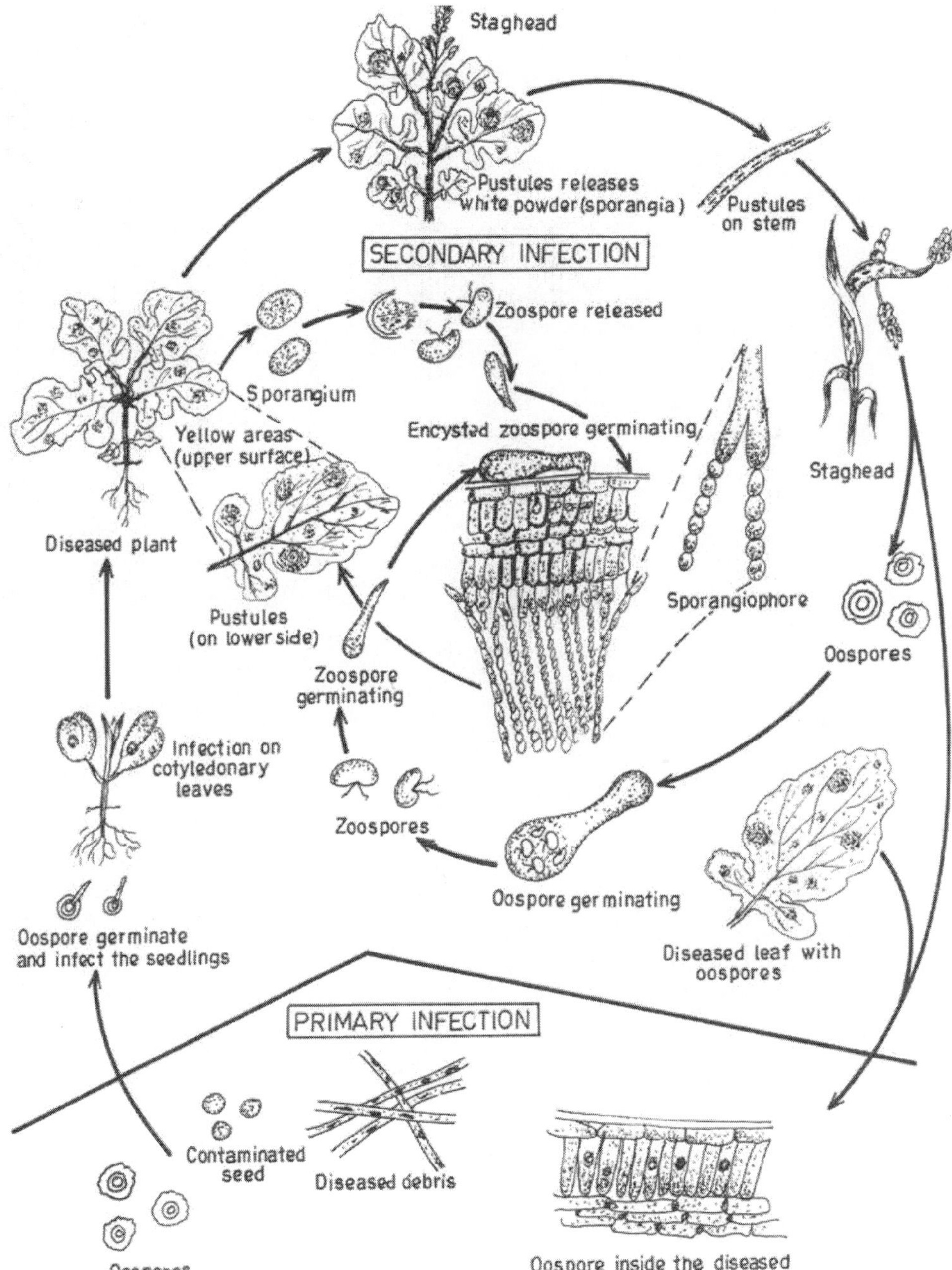

Figure 17.4: Disease Cycle of White Rust of Mustard (*Source*: Mehta, 2019).

4. Chemical treatments: White rust disease can be effectively managed with a variety of fungicides. The best treatments for the disease

were determined to be three sprays of benomyl (0.1 per cent), triphenylstannanol (0.1 per cent), tridemorph (0.1 per cent), mancozeb (0.2 per cent), copper oxychloride (0.3 per cent), and metalaxyl and mancozeb (0.2 per cent) spaced fifteen days apart.

5. Apply Apron SD-35 to the seeds at a rate of 6 gm per kilogram of seeds, and then, at 15-day intervals, spray three times with Dithane M-45 (0.2 per cent). Spritzing after sixty days of seeding will maximize yield and profit. It is imperative that the crop be well covered. At stag head phase, three sprays of metalaxyl (0.2 per cent) followed by mancozeb (0.2 per cent) found very effective, At least two sprays should coincide with flowering.
6. Benzothiadiazole (0.2 per cent) sprayed on mustard plants stimulates their defensive mechanisms and effectively controls illness.

3. Downy Mildew

Occurrence and Economic Importance

Downy mildew generally occurs with the combination of white rust in mustard crop causing severe losses in yield. Estimating yield losses from downy mildew infections alone is particularly challenging because they are almost invariably linked to white rust. On the other hand, systemic infection results in severe yield losses by deforming the stem and inflorescence. When a severe infection strikes a crop early in its growth, up to 75 per cent of the seedlings will die. A combination of downy mildew and white rust infection can reduce pod production by as much as 47 per cent.

Causal Organism

Downy mildew of mustard is caused by *Peronospora parasitica*, which belongs toKingdom-*Chromista*, Phylum- *Oomycota*, Class- *Oomycetes*. Order- *Peronosporales*, Family- *Peronosporaceae*, Genus- *Perenospora*and Species- *parasitica* **(Kirk *et al.*, 2008).** The fungus is an obligate parasite. It causes systemic infection in plant and sends haustoria in plant cell. The sporangiophores are short, stout, determinate, dichotomously branched, hypophyllous or amphigenous, clavate and non-septate

Symptoms

Initial symptoms appear on leaves as a yellowish-light green patches on the upper surface whereas, greyish- whitish, downy fungal growth observed corresponding to the lower surface of leaves. Finally, these converted in to necrotic patches on lower surface of leaves. In severe infection of the leaves, entire leaves, stems as well as siliques also covered by whitish fungal growth of the pathogen. The most conspicuous and pronounced symptom is the infection of inflorescence causing hypertrophy of the peduncle of inflorescence and develop stag head structure. There are no seeds or siliqua produced by the infected inflorescence.

Figure 17.5: Downy Mildew Symptoms in Mustard (*Source*: David B. Langston, University of Georgia).

Favorable Condition

The disease tends to be more prevalent in areas with low-lying and poorly drained soil. It thrives in conditions characterized by high humidity (around 90 per cent) and temperatures ranging between 22-25°C. The presence of water on the leaves further facilitates its development.

Disease Cycle

The oospores can maintain their viability in the soil for a period of five years or more. During this time, they have the potential to cause primary infections in host seedlings. These infections occur through direct penetration of the root hairs and coleoptiles by germ tubes. Secondary spread begins when sporangia (zoospores) become active, particularly during the rainy season. The zoospores enter the leaf stomata to initiate infection. The formation of sporangiosphores is favored by specific conditions, including high humidity (around 90 per cent), presence of water on the leaves, and temperatures ranging from 15 to 25 °C.

Management

1. When sowing, use certified seeds of resistant or tolerant kinds.
2. Cultural actions such as timely crop sowing, crop rotation with non-host crops, adequate field sanitation, *etc.*, should be implemented in order to reduce the disease's incidence.
3. Rough out infected plants carefully during early stage of infection.
4. Late sowing is avoided.
5. Apply 5g of trichomeviride per kg of seed to the seeds.

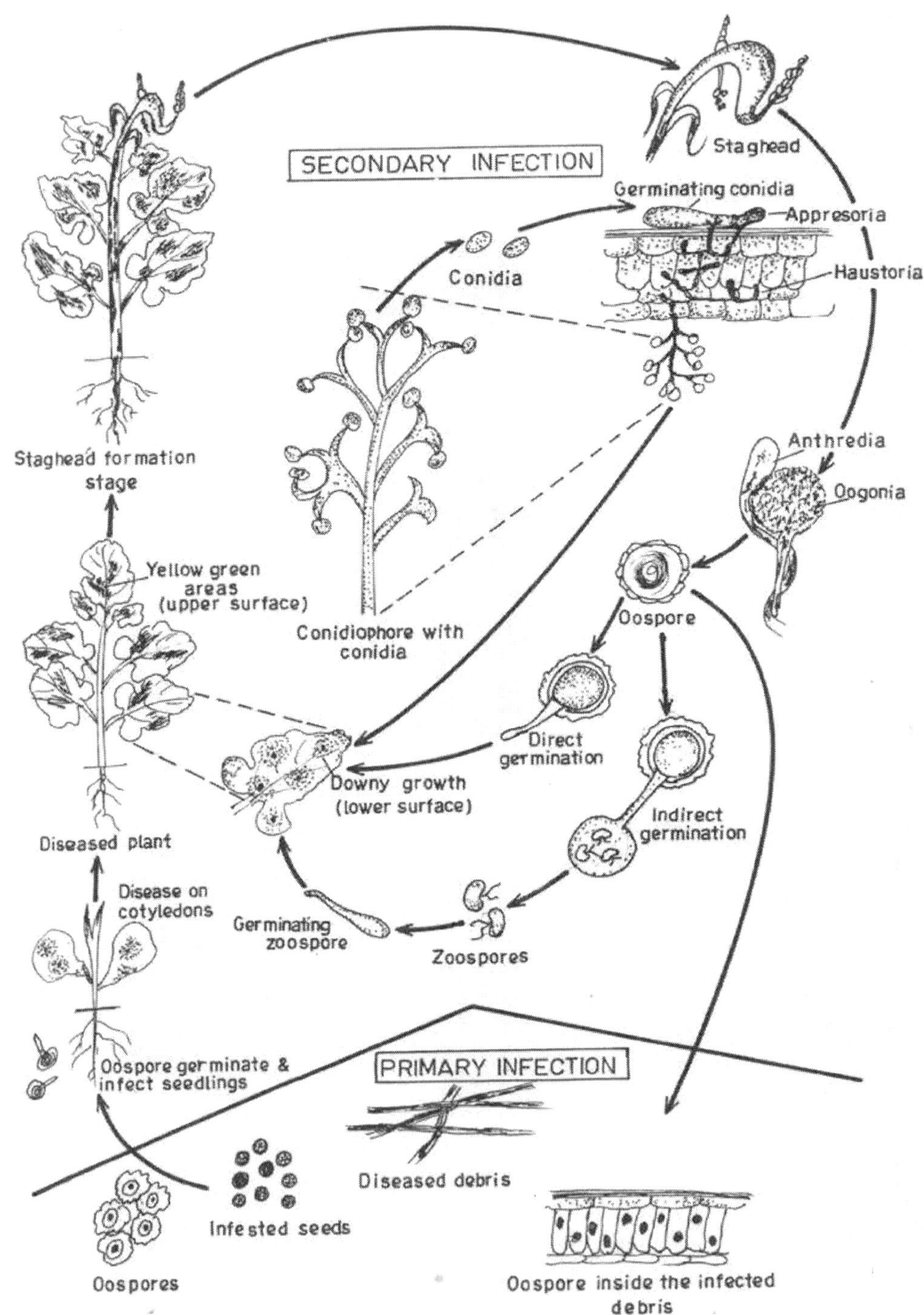

Figure 17.6: Disease Cycle of Downy Mildew of Mustard (*Source*: Mehta, 2019).

Figure 8.5: Small Millet Rust.
(*pg.* 172)

Figure 8.6: Leaf Blight of Small Millet.
(*pg.* 173)

Figure 8.7: Helminthosporium Leaf Spot. (*pg.* 174)

Figure 8.8: Banded Sheath Blight
(*pg.* 175)

Figure 8.9: Sheath Rot.
(*pg.* 176)

Figure 8.10a-b: Foot Rot of Finger Millet. (*pg.* 176)

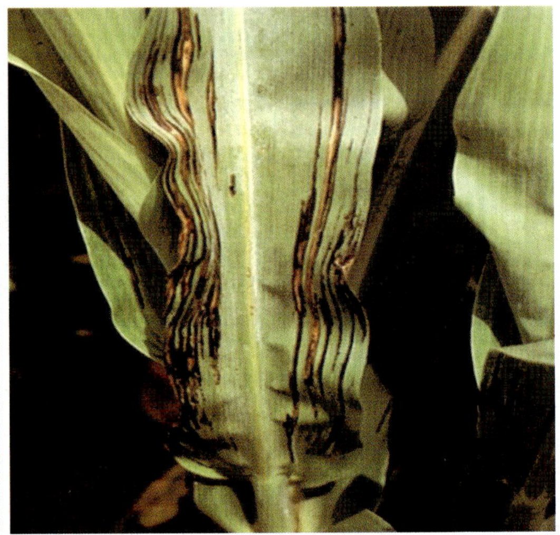

Figure 8.11: Bacterial Leaf Spot. (*pg.* 178)

Figure 8.12: Bacterial Leaf Streak. (*pg.* 179)

Figure 8.13: Maize Stripe Virus (MStV-S). (*pg.* 180)

Figure 8.13: a. Maize Mosaic (MMV-S); b. Red Stripe (SRSD). (*pg.* 181)

Figure 9.1: Disease Symptoms of Chickpea Wilt in Infected Field. (*pg.* 185)

Figure 9.3: Collar Rot in Chickpea (*pg.* 187)

Figure 9.4: Symptoms of Ascochyta Blight (*pg.* 189)

Figure 9.6: Damaging Symptoms of Botrytis Grey Mold. (*pg.* 191)

Figure 9.8: Damaging Symptoms of Chickpea Stunt Disease. (*pg.* 193)

Figure 10.1: Wilt Disease of Pigeon Pea (*pg.* 199)

A **B**

Figure 10.3: Characteristics Symptom of Stem Blight of Pigeon Pea (*pg.* 202)

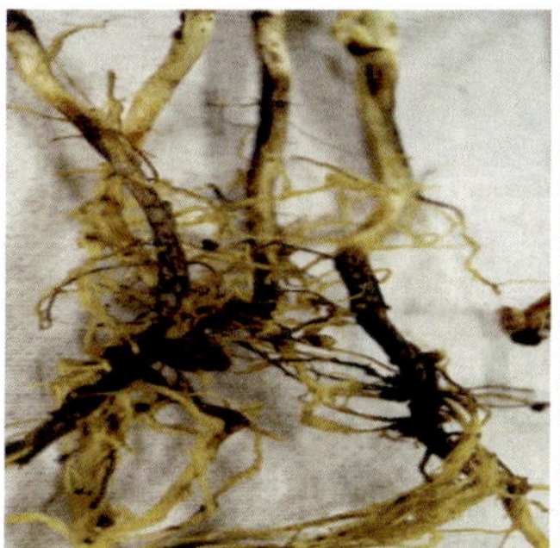

Figure 10.4: Root Rot of Pigeon Pea. (*pg.* 203)

Figure 10.6: Symptoms of Powdery Mildew (*pg.* 205)

Figure 10.12: Alternaria Leaf Spot Symptoms (*pg.* 208)

Figure 10.14: Bacterial Leaf Spot and Stem Canker (*pg.* 210)

Figure 10.15: Symptoms of Sterility Mosaic of Pigeon Pea (*pg.* 212)

Figure 10.17: Yellow Vein Mosaic of Pigeon Pea (*pg.* 216)

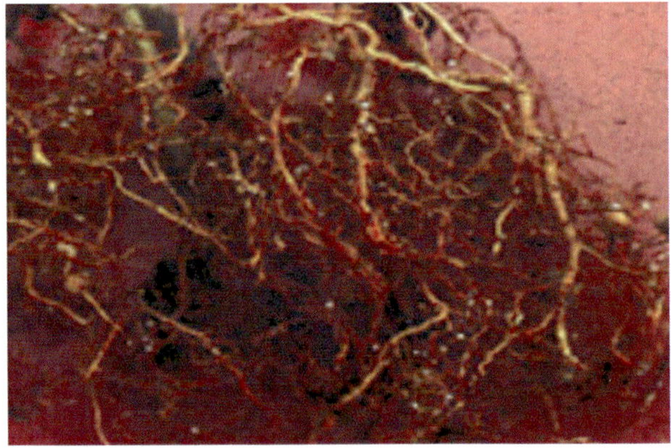

Figure 10.18: Pearly Root Symptoms of Pigeon Pea (*pg.* 217)

Figure 11.1: Wilt Symptoms

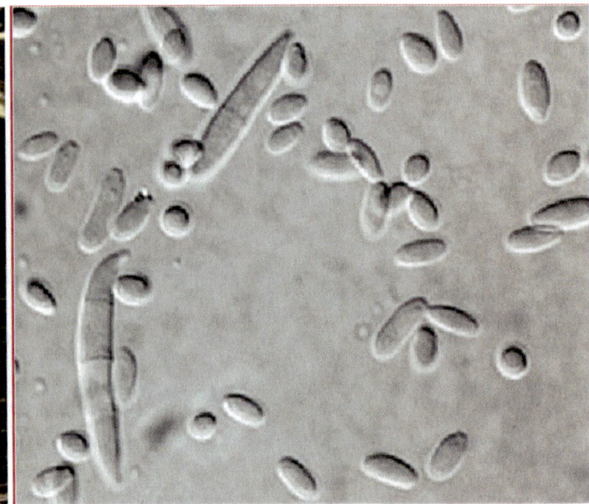

Macroconidia and Microconidia (*pg.* 222)

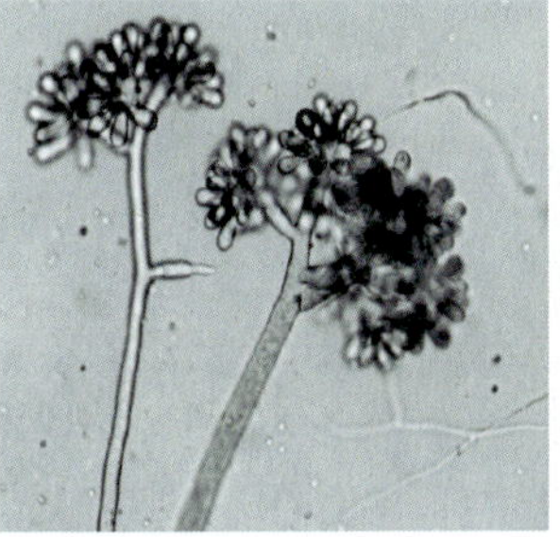

Figure 11.3: Lesions on Diseased Stem Conidiophore of Pathogen (*pg.* 224)

Figure 11.4: Rust Symptoms on Foliage. (*pg.* 226)

Figure 11.5: Lesions on Leaves

Lesions on Pods

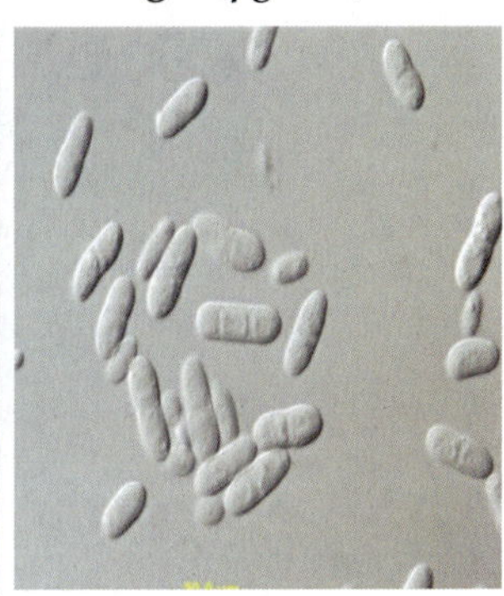

Conidia of Fungus (*pg.* 228)

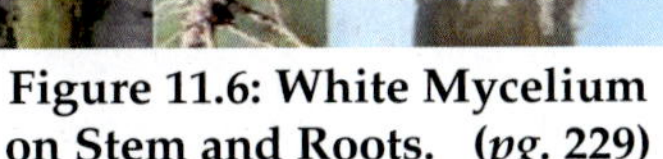

Figure 11.6: White Mycelium on Stem and Roots. (*pg.* 229)

Figure 11.7: A. Blighted Leaves; B. Muriform Conidia of Pathogen. (*pg.* 230)

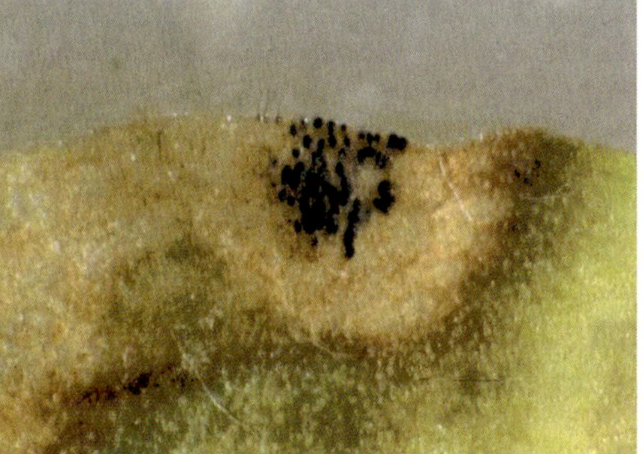

Figure 11.8: A. Stem Lesions; B. Microsclerotia. (*pg.* 231)

Figure 12.1: A. Cercospora Leaf Blight; B. Purple Seed Stain (Pat Miller). (*pg.* 238)

Figure 12.2: Fusarium Wilt of Soybean. (*pg.* 239)

Figure 12.3: Anthracnose Infected Soybean Plants Displaying Rolled Leaves (*pg.* 241)

Figure 12.4: Septoria Leaf Spot of Soybean Leaf. (*pg.* 243)

Figure 12.5: Upper and Lower Leaf Surface Leaves with Downey Mildew (*pg.* 245)

Figure 12.6: White Powder like Patches Observed on Infected Leaf (*pg.* 246)

Figure 12.7: Chlorotic and Necrotic Blotches between Central Leaf Veins on Plants Infected with Sudden Death Syndrome (*pg.* 248)

Figure 12.8: Small Necrotic Lesion with Yellow Halos (*pg.* 249)

Figure 12.9: Small Reddish Brown Lesions (*pg.* 251)

Figure 12.10: A. Mosaic Pattern on Infected Leaf (Source: J. Zhou) B. Leaf Yellowing at Late Infection Stage (Source: J. Zhou). (*pg.* 252)

Figure 13.1: Web Blight (*pg.* 257)

Figure 13.2: Cercospora Leaf Spot (*pg.* 259)

Figure 13.4: Anthracnose Disease (*pg.* 261)

Figure 13.5: Powdery Mildew (*pg.* 263)

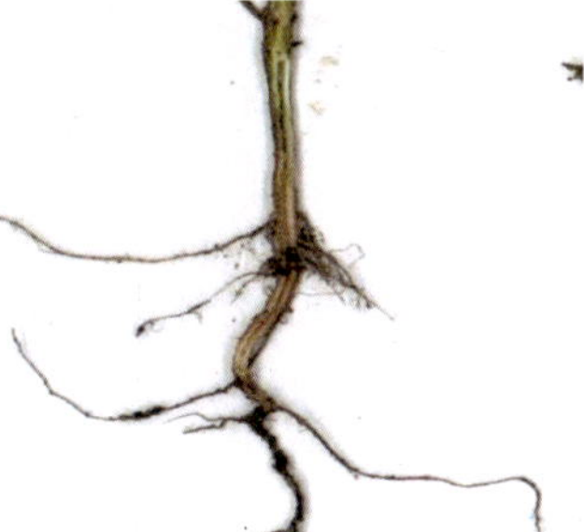

Figure 13.7: Root Rot and Leaf Blight (*pg.* 265)

Figure 13.9: Rust of Urd Bean.(*pg.* 267)

Figure 13.11: Dry Root Rot (*pg.* 270)

Figure 13.12: Mung Bean Yellow Mosaic Virus. (*pg.* 271)

Figure 13.14: Leaf Crinkle (*pg.* 273)

Figure 13.15: Leaf Curl/Necrosis (*pg.* 274)

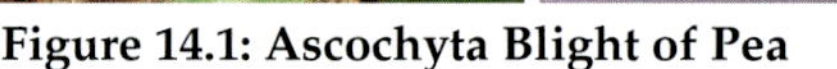

Figure 14.1: Ascochyta Blight of Pea (*pg.* 281)

Figure 14.3: Downey Mildew of Pea. (*pg.* 283)

Figure 14.5: Powdery Mildew of Pea (*pg.* 285)

Figure 14.7: Bacterial Blight (*pg.* 287)

Figure 14.9: Pea Early Browning (*pg.* 289)

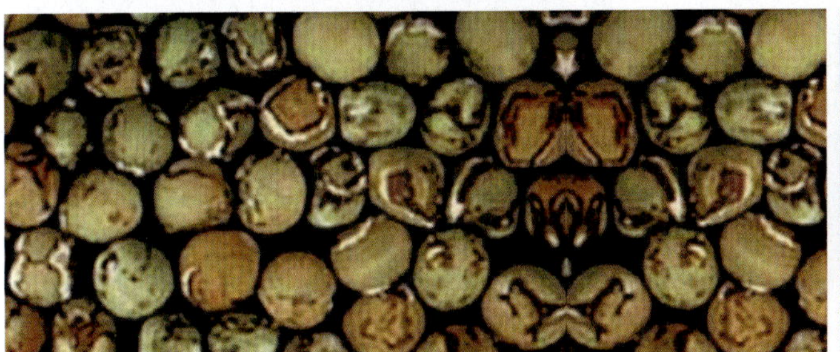

Figure 14.11: Pea Mosaic Disease (*pg.* 291)

Figure 15.1: Southern Blight of Cowpea. (*pg.* 297)

Figure 15.3: Cowpea Scab. (*pg.* 300)

Figure 15.4: Bacterial Blight of Cowpea. (*pg.* 301)

Figure 15.5: Fusarium Wilt of Cowpea. (*pg.* 303)

Figure 15.6: Cowpea Mosaic. (*pg.* 304)

Figure 15.7: Powdery Mildew of Cowpea. (*pg.* 306)

Figure 15.8: Cowpea Rust. (*pg.* 307)

Figure 15.9: Anthracnose of Cowpea. (*pg.* 308)

Figure 15.10: Root Knot Nematode of Cowpea. (*pg.* 310)

Figure 16.1: Alternaria Affected Leaves (*pg.* 317)

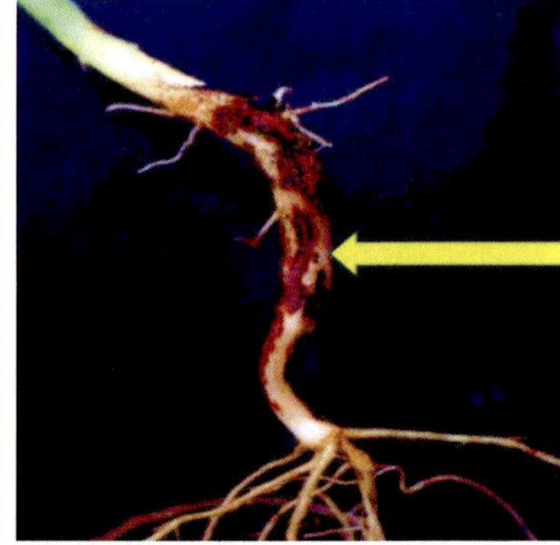

Figure 16.2: Dry Root Rot Affected Root. (*pg.* 319)

Figure 16.3: Powdery Mildew Affected Leaves. (*pg.* 320)

Figure 16.4: Anthracnose of Affected Pods of Cluster Bean (*pg.* 322)

Figure 16.5: Bacterial Blight of Cluster Bean (*pg.* 323)

Figure 17.1: Alternaria Blight Symptoms in Mustard (*pg.* 332)

Figure 17.3: White Rust Symptoms and Sporangia (*pg.* 335)

Figure 17.5: Downy Mildew Symptoms in Mustard (*pg.* 339)

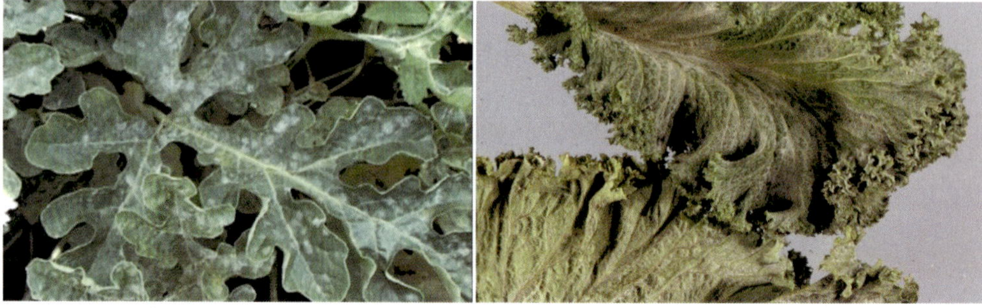

Figure 17.7: Powdery Mildew Symptoms in Mustard (*pg.* 341)

6. Treat the seed with *Pseudomonas fluorescens* at 5 g per kg seed. Spraying is also benefiacial to reduce the disease incidence at 10g per 1 lit. of water.

Ridomil Gold @ 6g/Kg seed treatment is followed by two sprays at 25 and 35 days after sowing with either Fluopicolide 5.56 per cent w/w + Propamocarb Hydrochloride 55.6 per cent SC (Infinito) or Cyazofamid 34.5 per cent SC (Ranman) or Metalaxyl 8 per cent +Mancozeb 64 per cent WP (Ridomil Gold).

4. Powdery Mildew

Occurrence and Economic Importance

Powdery mildew of mustard was first time reported by Butler in 1907 at Chenab canal of Punjab. At that time the disease was considered as minor but recently it is become very severe and destructive disease in some parts of the mustard growing area in India, especially in cool and dry conditions the disease become devastating if the immediate measures not taken place. In India, Gujarat, UP, Rajasthan and Haryana are the major state, where the diseases prevalent in sever conditions. Yield losses up to 45 per cent and quality and quantity of the seed gets affected and damaged up to 17 per cent. The disease causes crop losses that range from 19.5 per cent to 29.5 per cent on average, depending on the variety that is utilized.

Causal Organism

Powdery mildew of mustard is caused by *Erysiphecruciferarum/E. polygoni,* which belongs to Kingdom- *Fungi,* Phylum- *Ascomycota,* Class- *Leotiomycetes.* Order- *Erysiphales,* Family- *Erysiphaceae,* Genus- *Erusiphe*and Species- *cruciferarum* (Kirk *et al.,* 2008).

Symptoms

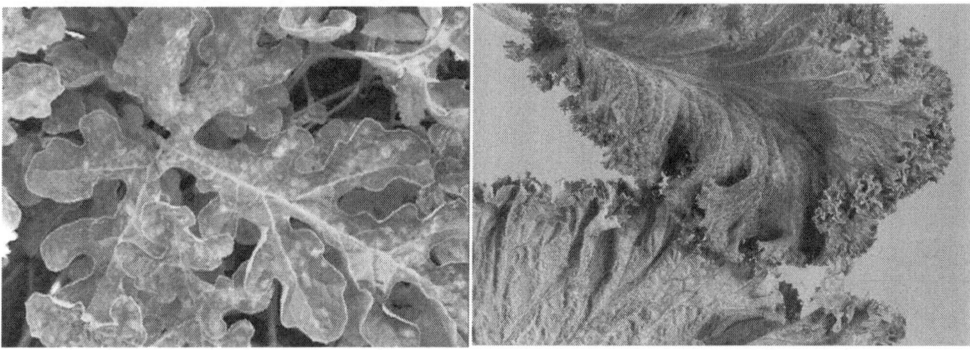

Figure 17.7: Powdery Mildew Symptoms in Mustard (*Source*: Gerald Holmes, Strawberry Center, Cal Poly San Luis Obispo, Bugwood.org).

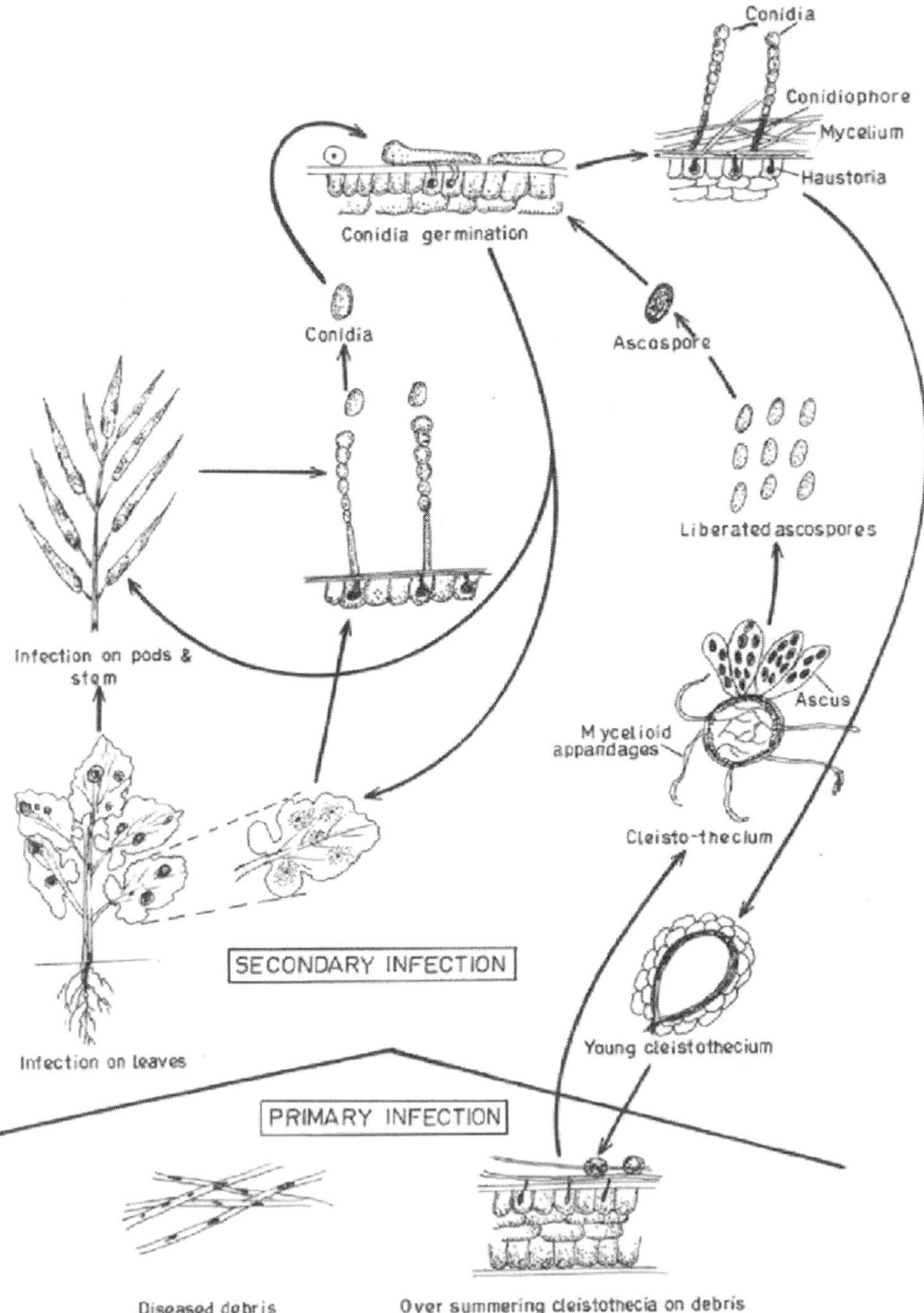

Figure 17.8: Disease Cycle of Powdery Mildew of Mustard (*Source*: Mehta, 2019).

Initially the symptoms observed on the leaves, irregular whitish powdery patches observed on the either sides of the leaves. Later, the patches enlarge and cover entire leaves as well as the stem, inflorescence and siliques of the mustard plant. Infected leaves may curl at the margins, soon become discoloured and defoliation of the leaves takes place. In severe infection, necrotic spots also observed. The powdery masses contain mycelium, conidiophores and conidia of the fungus. Severe infection can be observed on pods; pods reduced in size and remain empty. Infected seeds become shriveled, deformed and having less oil content.

Favorable Conditions

Conditions that are dry and cool are often ideal for the powdery mildew fungus. The ideal temperature range for conidia germination, germ-tube formation, and appressorium production is 20–25 °C. Conidia do not germinate at temperatures below 15 °C or above 30°C. At 50 per cent relative humidity, conidia germinate to their maximum extent. Sporangulation is increased by factors like the host's aging, dry soil, low relative humidity, and poor nutrition.

Disease Cycle

Being a biotroph, the pathogen lives on undesirable plants as mycelium or cleistothecia. Fungal mycelium and conidia seen in dry, infected plant sections make up the primary inoculum. Later on in the infection process, the fungus that formed clestothecium on diseased plant tissues also served as the main source of infection. Conidia carried by the air cause the disease's secondary spread.

Management

1. Timely sowing of the crop matures early and escapes the disease.
2. Early removal of contaminated plant debris decreased the disease's incidence; prudent application of nitrogenous fertilizers is also crucial.
3. Treat the seeds with Iprodione 25 per cent + Carbendazim 25 per cent WP at 2g per kg seed followed by the removal of three lower leaves found effective.
4. Foliar spray of Tebuconazole 50 per cent + Trifloxystrobin 25 per cent WG (0.1 per cent) or mycobutanil (0.1 per cent) found effective in management of the disease
5. Apart from these, application of $ZnSO_4$ + Sulphur found most effective.
6. Foliar spray of Dinocap, Tridemorph, Tridimefon and wettablesulpure successfully manage the powedery mildew disease.
7. Growing of resistant/Tolerant cultivars or breeding progrmme should incorporate powdery mildew resistant accessions to develop resistant cultivars

REFERENCES

FAOSTAT (2021). Food and Agriculture Organization of the United Nations Statistical Database; Statistical Division; FAO: Rome, Italy.

Holmes, G. (2023). Gerald Holmes, Strawberry Center, Cal Poly San Luis Obispo, Bugwood.org. (Accessed on 31.07.2023)

Kharayat, B. S. (2023) ePlantHealth, https://www.youtube.com/watch?v=j0yfbFa3Zik,

Kirk, P. M., Cannon, P. F., Minter, D. W., and Stalpers, J. A. (2008).*Dictionary of the Fungi* (10[th]edn). *Wallingford, UK.*

Kumar, V., and Chopra, A.K. (2014). Fertiirrigational response of hybrid cultivar of Indian mustard (Brassica junceaL.) to distillery effluent in two seasons. Analitical Chemical Letters, 4 (3): 190- 206.

Kumar. D., Maurya, N., Bharti, Y.K., Kumar, A., Kumar, K., Srivastava, K., Chand, G., Kushwaha, C., Singh, S.K., Mishra, R.K. and Kumar, A. (2014). Alternaria blight of oilseed Brassicas: A comprehensive review. Afr. J. Microbiol. Res., 8(30): 2819-2829.

Langston, D. B. (2023). University of Georgia, Bugwood.org https://www.ipmimages.org/browse/detail.cfm?imgnum=5077029

Mehta, N. (2019). Development of prediction models for the management of rapeseed-mustard diseases-Current scenario, Pl. Dis. Res. 34 (2): 81-112.

Rachel, L. (2006). White blister control in vegetable brassica crops. Farm note.112.

Saharan, G.S., Kaushik, C.D., Gupta, P.P. and Tripathi, N.N. (1984). Assessment of losses and control of white rust of mustard. Indian Phytopath. 37: 397 (Abstr.).

ORCID ID: https://orcid.org/0000-0002-6960-940X

Chapter 18

Diseases of Groundnut Crop

Mayur R. Thesiya and Harshit Chavda*

Assistant Professor, Department of Plant Pathology, College of Agriculture, Faculty of Agriculture, Parul University, Vadodara – 391 760, Gujarat
****e-mail: mayur.thesiya30257@paruluniversity.ac.in***

Introduction

Groundnut or peanut (*Arachis hypogaea*) is an annual warm-season legume plant that originated in South America. It is primarily grown as a rainfed crop during the rainy season, covering approximately 85 per cent of the total area and contributing to 82.5 per cent of overall production. Globally, groundnut is cultivated across nearly 30.4 million hectares, resulting in a total production of about 51.5 million tonnes and an average productivity of around 1696 kg/ ha from 2018 to 2020 (FAOSTAT, 2022). Groundnut is now the world's fourth most important source of edible oil and the third most important source of vegetable protein. Groundnut seeds are valuable, containing roughly 40-50 per cent oil and 20-30 per cent proteins, along with being an excellent source of B vitamins. Varieties of groundnut differ based on seed size; large-seeded types are used for roasting and confections, while small-seeded varieties are used for oil extraction. The remaining groundnut meal after oil extraction is protein-rich and can be utilized for either food or feed (Pardee, 2002).

In India, groundnut cultivation is widespread in several states, with Gujarat, Andhra Pradesh, Tamil Nadu, Karnataka, Rajasthan, and Maharashtra being major contributors, collectively accounting for about 80 per cent of the total area and production in the country. The lower productivity in India can be attributed to the reliance on rainfed farming methods, exposing the crop to various diseases and insect pests. Over 55 pathogens, including viruses, have been reported to affect groundnut crops. Seed and seedling diseases, such as collar rot, stem rot, and charcoal root rot, are recognized as economically significant issues in groundnut cultivation. These diseases result in severe seedling mortality, leading to uneven crop stands, especially in sandy loam soil, and can cause yield reductions of 25-50 per cent (Ghewande, 1985). Among the foliar fungal diseases, early leaf spot and late leaf spot, commonly known as "tikka disease," are economically important. These diseases affect groundnut crops wherever they are grown, but their incidence and severity vary depending on location and seasons, showing short and long-term fluctuations in relative proportions (McDonald *et al.*, 1985). Yield losses caused by these diseases can range from 10 to 70 per cent worldwide but vary considerably between different regions and seasons (Subrahmaniyam and McDonald, 1983). Among viral diseases, groundnut bud necrosis, peanut mottle, Indian peanut clump, and stem necrosis diseases are economically significant. It has been reported to cause yield losses of up to 50 per cent in early-infected groundnut crops in India (Ghanekar, 1980). Groundnut is also susceptible to various nematode-induced diseases. Root knot disease, caused by four species of Meloidogyne (*M. arenaria, M. hapla, M. incognita,* and *M. javanica*), is the most significant nematode disease affecting groundnut. *M. arenaria* is the most widespread, and a locally known nematode-induced disease called "Kalahasti malady" has been observed in Chittoor and Nellore districts of Andhra Pradesh, India. In addition to these challenges, groundnut growers, processors, and exporters face aflatoxin contamination caused by *Aspergillus flavus* and its processed products, posing another major concern in groundnut production.

This chapter aims to discuss the causes, symptoms, etiology, spread, and management practices for diseases of groundnut such as collar rot, stem rot, rust, charcoal root rot, and root knot disease. The goal is to provide comprehensive knowledge and understanding of these diseases, enabling the implementation of proper disease management strategies.

A) Fungal Diseases

1) Tikka Diseases or Leaf Spot

Occurrence and Economic Importance

Leaf spots damage the plant by reducing the available photosynthetic area, by lesion formation, and by stimulating leaflet abscission. Worldwide, yield losses range from 10 per cent to over 50 per cent, but vary considerably from place to place and between seasons. Eight states *viz.*, Gujarat, Andhra Pradesh, Rajasthan, Karnataka, Madhya Pradesh, Maharashtra and Tamil Nadu and Uttar Pradesh were identified to have acreages greater than one lakh hectares and these states jointly accounted for 95.5 per cent of the national acreage.

Causal Organism

Early leaf spot: *Cercospora arachidicola* (Perfect stage-*Mycosphaerella arachidis*) and Late leaf spot: *Cercospora personatum*/*Phaeoisariopsis personata* (Perfect stage- *Mycosphaerella berkeleyi*). The pathogen belongs to Kingdom: *Fungi*, Division: *Ascomycota*, Class: *Dothideomycetes*, Order: *Capnodiales*, Family: *Mycosphaerellaceae*, Genus: *Cercospora*, and Species: *arachidicola*/*personata* **(Kirk *et al.*, 2008)**. The *C. arachidicola* appears about month earlier than the *P. personatum*. The fungal mycelium is confined to the leaf spot and is present both inter/intra cellular. The hyphae develop into the hymenial layer in the sub epidermal region and short conidiophores are produced in bunches from the hymenial layer on which hyaline, whip like conidia are born we singly. The perfect stages of the two fungi have been frond on the dead host tissues.

Symptoms

Early Leaf Spot

Lesions are sub-circular in shape and from 1 to over 10 mm in diameter. They are dark brown in colour on the upper leaflet surface where most sporulation occurs and a lighter shade of brown on the lower leaflet surface. The early leaf spot usually has a light to dark brown center and a yellow halo. Distribution of fruiting structures is random on upper leaflet surface. Lesions are also produced on petioles, stems and pegs. These are oval to elongate in shape and have more distinct margins than the leaflet lesions. On severity of the diseases, leaflets become chlorotic, then necrotic, lesions coalesce and leaflets are shed.

Figure 18.1: A. Early Leaf Spot (Left); B. Late Leaf Spot (Right).

Late Leaf Spot

Lesions circular to sub-circular in shape with 1 to 6 mm diameter. All lesions are dark brown to black in colour. On the lower leaflet surface where most sporulation occurs the lesions are black in colour and slightly rough in appearance. Fruiting structures are in concentric rings on the lower leaflet surface. Lesions on other parts and effects on disease severity are similar to that of early leaf spot.

Favourable Conditions

Warm and moist weather, Optimum temperature, 24-28°C.

1. High relative humidity, 90 per cent for a period of 3 days.
2. High doses of nitrogen N_2 and phosphorus fertilizers.
3. Magnesium deficiency increases susceptibility to the disease.

Disease Cycle

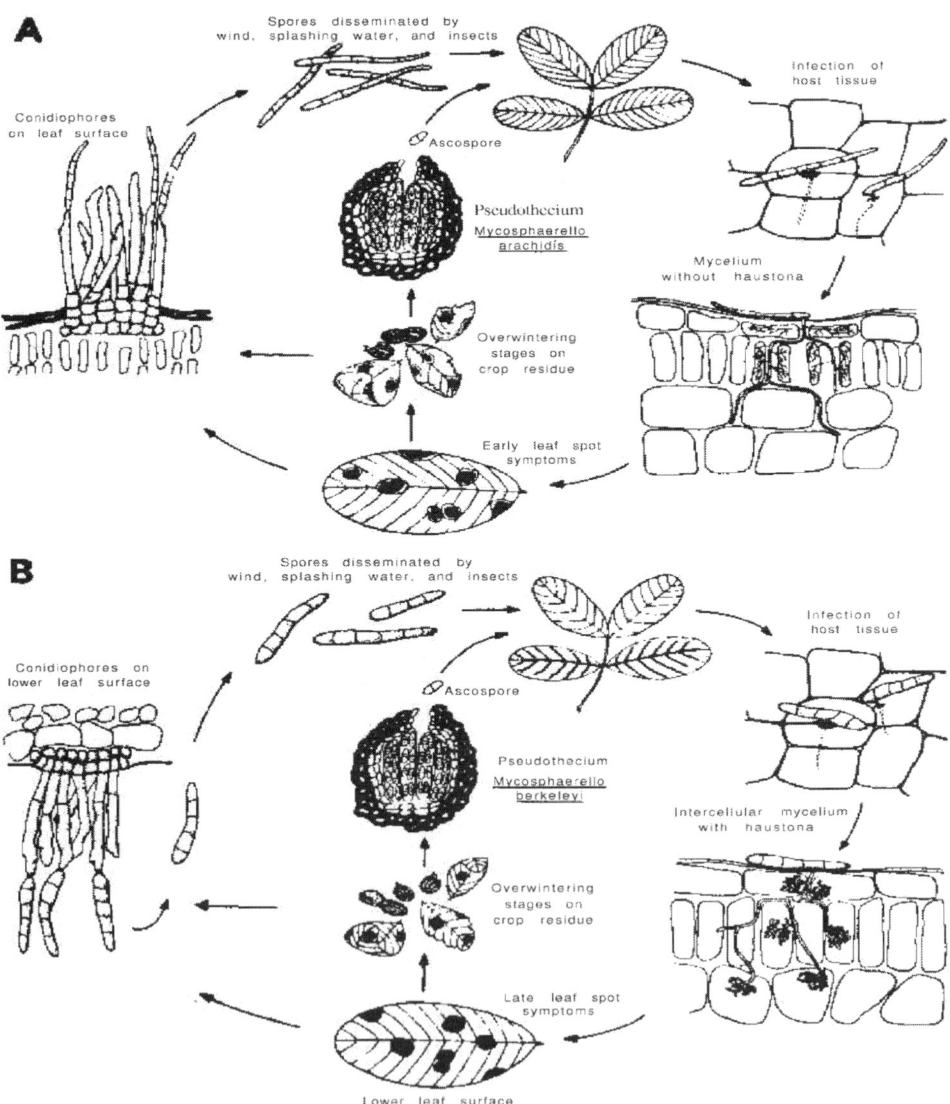

Figure 18.2: Disease Cycle of Leaf Spot caused by *Cercospora* spp. (*Source*: McDonald, 1985).

The perpetuate of leaf spot fungi occurs through conidia present in the soil and on diseased plant debris. Additionally, conidia can be found on the surface of the nuts. The primary source of infection is attributed to seed-borne conidia and the presence of conidia in plant debris within the soil. Secondary spread is facilitated by the dispersal of airborne conidia.

Management

1. Grow tolerant varieties *i.e.*, GJG 32.
2. Implementing a planting strategy of intercropping pearl millet or sorghum with groundnut at a 1:3 ratio.
3. Utilizing crop rotation with non-host crops, preferably cereals.
4. Ensuring the deep burial of crop residues in the soil and the removal of any volunteer groundnut plants.
5. Reducing the primary source of infection by burning previous year's diseased plant debris to a significant extent.
6. Applying seed dressing with Carboxin 37.5 per cent + Thiram 37.5 per cent DS (Vitavax Power) at a rate of 2 grams per kilogram of seed.
7. Administering foliar applications of 5 per cent Neem seed kernel extract (NSKE) at two-week intervals, three times, commencing four weeks after planting.
8. Controlling the spread of secondary inoculum through the use of various fungicides such as Propiconazole 25 per cent EC (Tilt) or Chlorothalonil 75 per cent WP (Kavach), which have yielded positive results.

2. Rust

Occurrence and Economic Importance

Groundnut rust is known to affect regions including Central and South America, the West Indies, Venezuela, China, the former USSR, the United States, and India. In India, it was first documented in 1971, with reported occurrences in Punjab, Tamil Nadu, West Bengal, Andhra Pradesh, and Gujarat. In the southern parts of India, rust has been linked to yield losses ranging from 10 per cent to as high as 92.9 per cent. The economic impact, specifically in terms of reduced pod yield due to rust, has been estimated to range from 14 per cent to 52 per cent. Additionally, a loss in oil content ranging from 7 per cent to 10 per cent has been observed.

Causal Organism

It caused by fungus *Puccinia arachidis*, which belongs to Kingdom: Fungi, Division: Basidiomycota, Class: Pucciniomycetes, Order: Pucciniales, Family: Pucciniaceae, Genus: *Puccinia*, and Species: *arachidis* (Kirk *et al.*, 2008). The pathogen exhibits the production of both uredial and telial stages. Abundant production is noted in the uredial stages, while the production of telia is limited. Teliospores are characterized by their dark brown color and consist of two cells. Notably, no records exist regarding the presence of pycnial and aecial stages. Furthermore, information concerning the involvement of an alternate host remains unavailable.

Symptoms

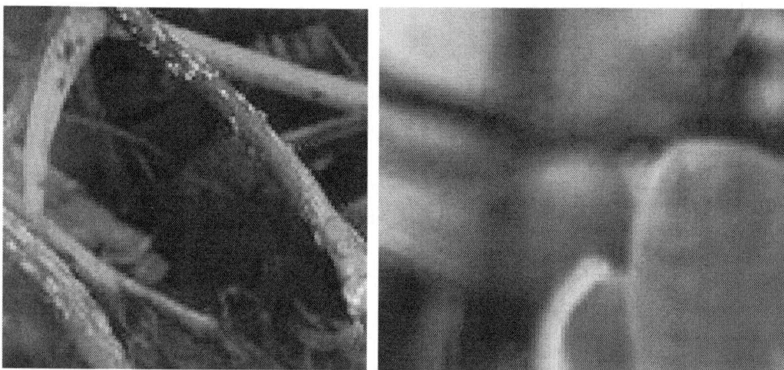

Figure 18.3: Rust Symptoms in Groundnut (*Source*: CABI).

The disease affects all above-ground parts of the plant, typically becoming noticeable when the plants are approximately 6 weeks old. Rust manifests as distinctive orange-coloured pustules, known as uredinia, which emerge on the lower surface of the leaflets and subsequently rupture, revealing clusters of reddish-brown urediniospores. Initially, pustules develop on the lower leaflet surface, and in highly susceptible cultivars, they may be surrounded by colonies of secondary pustules. Additionally, pustules may also appear on the upper surface of the leaflet, typically presenting as circular formations with diameters ranging from 0.5 to 1.4 mm. In cases of severe infection, lower leaves may wither and prematurely detach from the plant. Furthermore, kernels produced on affected plants tend to be small and shrivelled.

Favourable Conditions

1. Low temperatures ranging from 20-25°C accompanied by wet weather.
2. High relative humidity exceeding 85 percent.
3. Pronounced and consistent heavy rainfall.

Disease Cycle

There is no report of existence of collateral/alternate host however due to continuous cultivation of groundnut there is overlapping of harvesting and sowing of the crop. Uredospores of ground nut rust remain viable for about 20-30 days and hence they do not find any difficulty in carrying themselves from one crop season to other crop season. The teleospore do not found in India except in Punjab in 1963.

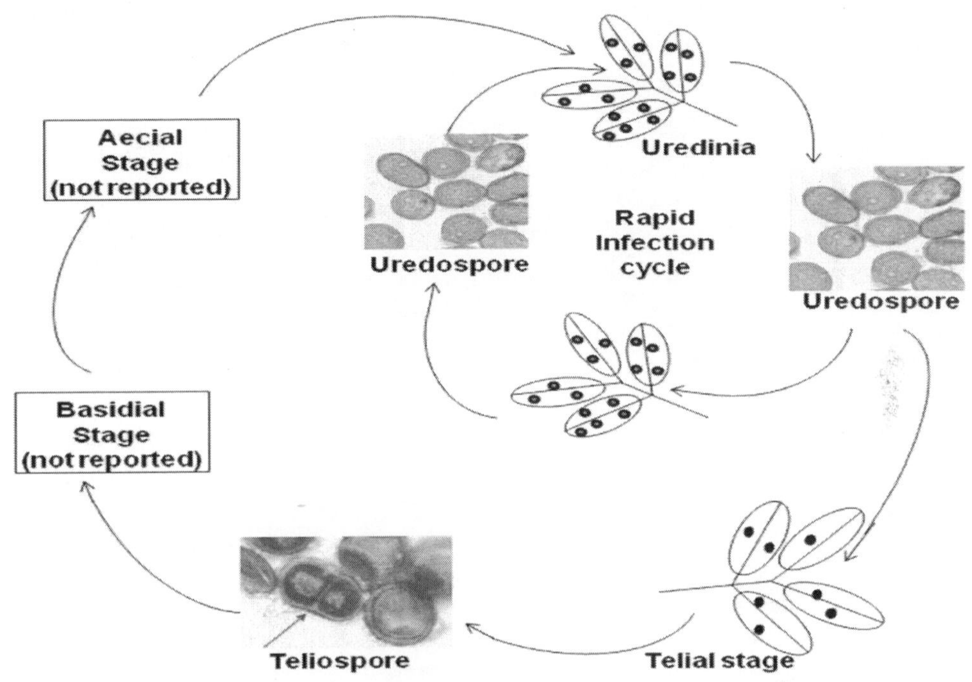

Figure 18.4: Disease Cycle of Groundnut Rust (*Source*: Mondal and Badigannavar, 2015).

Management

1. Use resistant/tolerant varieties *i.e.*, GJG-32.
2. Early sowing in the first fortnight of June to avoid disease incidence.
3. Implementing crop rotation and maintaining field sanitation practices.
4. Avoiding the practice of monoculturing groundnut crops.
5. Thoroughly removing volunteer groundnut plants and any potential reservoir hosts.
6. Employing intercropping techniques, such as combining pearl millet or sorghum with groundnut at a 1:3 ratio.
7. Utilizing biocontrol agents, such as *Verticillium lecanii*, which have been reported to effectively inhibit the germination of *P. arachidis* (Daudi *et al.*, 2019).
8. Ensuring the destruction of volunteer (self-sown) groundnut plants and crop debris.
9. Foliar application of aqueous Neem leaf extract @ 5 per cent.

10. Spray Azoxystrobin 11 per cent + Tebuconazole 18.3 per cent SC (Sukoyaka) or Chlorothalonil 75 per cent WP (Kavach).

3. Collar Rot/Crown Rot and Seedling Blight

Occurrence and Economic Importance

Collar rot was first reported from Saurashtra region (Gujarat, India). It was then reported to occur in Australia, East Africa, Fiji, Island, India, Iran, Israel, Sudan and the US. This disease is most common in Saurash1ra region of Gujarat. As the disease causes considerable seedling mortality in the initial stage of crop growth. Crop yield is directly affected by reductions in the plant stand of the crop. In India under Punjab condition, the losses may about to 40-50 per cent in term of mortality of plants.

Causal Organism

Collar rot/Crown rot of Groundnut is caused by *Aspergillus niger*, a member of the following taxonomic classification: Kingdom: Fungi, Division: Ascomycota, Class: Eurotiomycetes, Order: Eurotiales, Family: Trichocomaceae, Genus: *Aspergillus,* Species: *A. niger* (Kirk *et al.*, 2008). The mycelium of this pathogen is described as hyaline to sub-hyaline. Conidiophores originate directly from the substrate and are characterized as septate, featuring thick walls and exhibiting a hyaline or olive-brown coloration. Vesicles are primarily globose and contain two rows of hyaline phialides, both primary and secondary. The conidial heads are typically dark brown to black, and they produce conidia that are globose, dark brown in colour, and arranged in elongated chains.

Symptoms

Figure 18.5: Collar Rot Symptoms in Groundnut (*Source*: Krishisewa).

The disease presents in two distinct phases: pre-emergence and post-emergence. During the pre-emergence phase, the seeds may deteriorate and become coated with a sooty black mass of spores, known as seed rot. Upon germination, the emerging hypocotyl is swiftly affected by lesions below ground, leading to the rotting of seedlings before they emerge from the soil (pre-emergence phase). In the post-emergence phase, a circular light brown lesion initially appears on the cotyledons. As time progresses, the tissue of the hypocotyl or stem region becomes water-soaked and displays light brown discoloration. Subsequently, the seedlings collapse and perish due to the rotting of their tender hypocotyls. As the disease advances, the collar region becomes fragmented and darkened. Mature plants may also be affected at ground level due to collar rot, resulting in the girdling of the collar region, followed by the leaves turning chlorotic. This progression is subsequently marked by drooping, wilting, and the death of the affected branches. The impacted collar region eventually becomes shredded and is often covered with a profuse growth of conidiophores and conidia.

Favourable Conditions

1. High soil and air temperature predispose the plants to infection.
2. Maximum seed rot occurs at 15 °C and 40 °C.
3. Temp. 30-35 °C favours the disease.
4. Monoculturing increase the inoculum in soil.
5. Infection occurs from seed borne as well as soil-borne inoculums of the fungus.
6. Seed is infected if stored in damp condition.
7. Deep sowing, late germination, heavy soil and high air temperature favours the disease.

Disease Cycle

1. **Primary source of infection:** Seed-contaminated with spores, soil -debris in the soil.
2. **Secondary source of infection:** Conidia dissemination through irrigation/rain water and farm implements.

Management

1. GG 40- tolerant to collar rot.
2. Thorough removal and disposal of plant debris.
3. Elimination and responsible disposal of crop debris from the previous season's infested crops in the field.
4. Application of *Trichoderma harzianum* and *Pseudomonas fluorescence* to the soil at rates of 2.5 to 5 kg/ha, preferably in combination with neem cake or mustard cake at a rate of 500 kg/ha.

5. Treating seeds with Carboxin 37.5 per cent + Thiram 37.5 per cent DS (Vitavax Power) at a rate of 2 grams per kilogram of seed.
6. Drenching the soil with Carbendazim 25 per cent + Mancozeb 50 per cent WS (Sprint) at a concentration of 0.25 per cent.

4. Stem Rot/Sclerotial Rot

Causal Organism

It caused by *Sclerotium rolfsii* (= *Arthelia rolfsii*), which belongs to Kingdom: Fungi, Division: Basidiomycota, Class: Agaricomycetes, Order: Atheliales, Family: Atheliaceae, Genus: *Athelia* and Species: *S. rolfsii* **(Kirk *et al.*, 2008)**. This disease is prevalent across groundnut-growing regions worldwide, spanning tropical areas and the warmer parts of the temperate zone. In Saurashtra, it represents a significant threat due to the practice of mono-culturing this crop. It's worth noting that India has reported yield losses of around 27 per cent or more due to this disease.

Symptoms

Figure 18.6: Symptoms of Stem Rot (*Source*: Woodward *et al.*, 2013).

The initial indication of the issue manifests as the abrupt drooping of a branch, particularly when it makes contact with the ground, impacting either the entire branch or a segment of it. The leaves on the affected branch become discoloured and shrivel, yet they persistently cling to the plant. In close proximity to the ground contact point, a white fungal layer begins to develop in conjunction with the emergence of mycelium. As the ailment advances, a network of white mycelium extends throughout the soil and the lower part of the plant's foliage. With the ongoing progression of the condition, small structures resembling mustard seeds in both size and color, known as sclerotia, emerge in the infected regions.

Favourable Condition

1. Frequent rain/irrigation favours the disease.
2. Moderate to high temperature and Fluctuation in temperature and moisture.
3. In sandy and sandy loam well aerated soil, disease development take place more rapidly.
4. High P_2O_5 increase disease, gypsum increase disease.

Disease Cycle

1. Primary source of inoculums: Soil borne/seed borne/sclerotia present in soil/multiplying in fallen plant debris. Sclerotia remain viable for 8 years in soil.
2. Secondary source of infection: water irrigation/rain and farm equipment

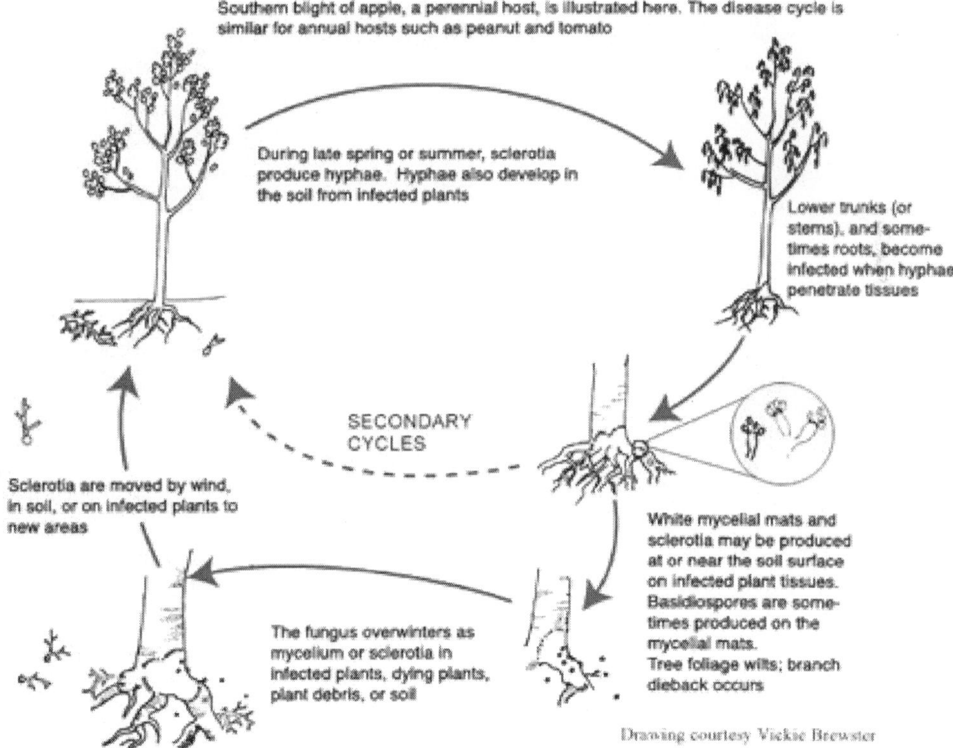

Figure 18.7: Disease Cycle of Stem Rot Disease Cycle (*Source*: APS).

Management

1. GG 40: tolerant to stem rot disease.
2. Crop rotation with maize, wheat, cotton and sorghum.
3. Filed sanitation.
4. Deep ploughing.
5. Utilize T. viride at a rate of 2.5 kilograms per hectare for soil drenching, ideally 30 days after sowing. Alternatively, incorporate *T. viride* at the same rate of 2.5 kilograms per hectare into the furrows at the time of sowing. Enhance the effectiveness of T. viride by combining it with either castor cake or Farm Yard Manure (FYM) at a rate of 100 kilograms per hectare, mixed into the furrows during sowing.
6. Soil solarisation with thickness transparent polyethylene having 100 microns.

5. Dry Root Rot/Charcoal Rot

Causal Organism

It caused by *M. phaseolina* Kingdom: Fungi, Division: Ascomycota, Class: Dothideomycetes, Order: Botryosphaeriales, Family: Botryosphaeriaceae, Genus: *Macrophomina* and Species: *phaseolina*. The pathogenic agent gives rise to mycelium that ranges from hyaline to a subdued brown hue. Sclerotia, which are small, solid structures measuring between 0.1 to 1 millimetre in diameter, appear black, smooth, and are found within various plant parts, including roots, stems, leaves, and fruits. On leaves and stems, conidiomata are formed, taking on a dark-brown coloration and occurring either individually or in clusters. Conidiophores, which are hyaline, are short and cylindrical in shape. The conidia produced by this pathogen are also hyaline and take on an ellipsoid to obovoid form. Notably, *R. bataticola* or *M. phaseolina* has been identified within the seed coat, cotyledons, and embryo of groundnut. This information is provided in a unique and non-plagiarized manner.

Favourable Conditions

1. The prevalence of the disease is favoured by dry and hot weather conditions that persist during and after the flowering stage.
2. The disease is more likely to occur in soil with temperatures ranging from 30 to 42°C, low soil moisture levels, and a slightly acidic soil pH ranging from 5.4 to 6.0.
3. Prolonged periods of rain during the seedling stage, particularly in low-lying areas, can also contribute to the development of the disease.

Symptoms

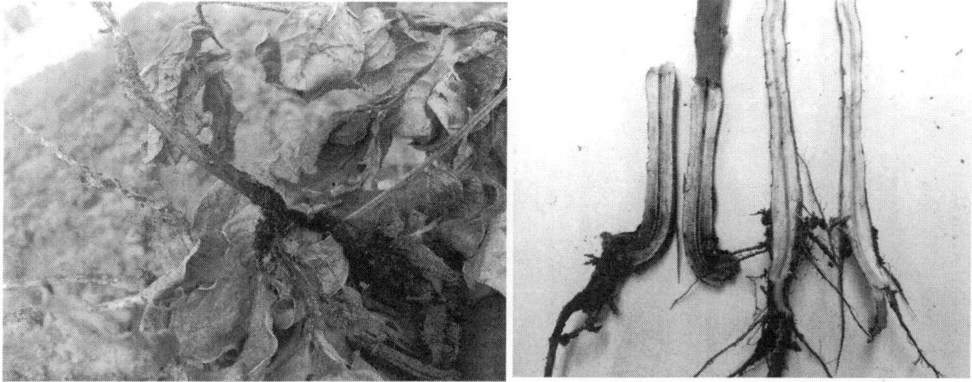

Figure 18.8: Symptoms of Dry Root Rot.

Disease Cycle

The pathogen lies in a dormant state as sclerotia for an extended duration within infected plant remnants and in the ground. Initial contamination transpires via soil-based and seed-based sclerotia. The subsequent dissemination of sclerotia or conidia is facilitated by means such as irrigation water, tools, and human involvement, among other factors.

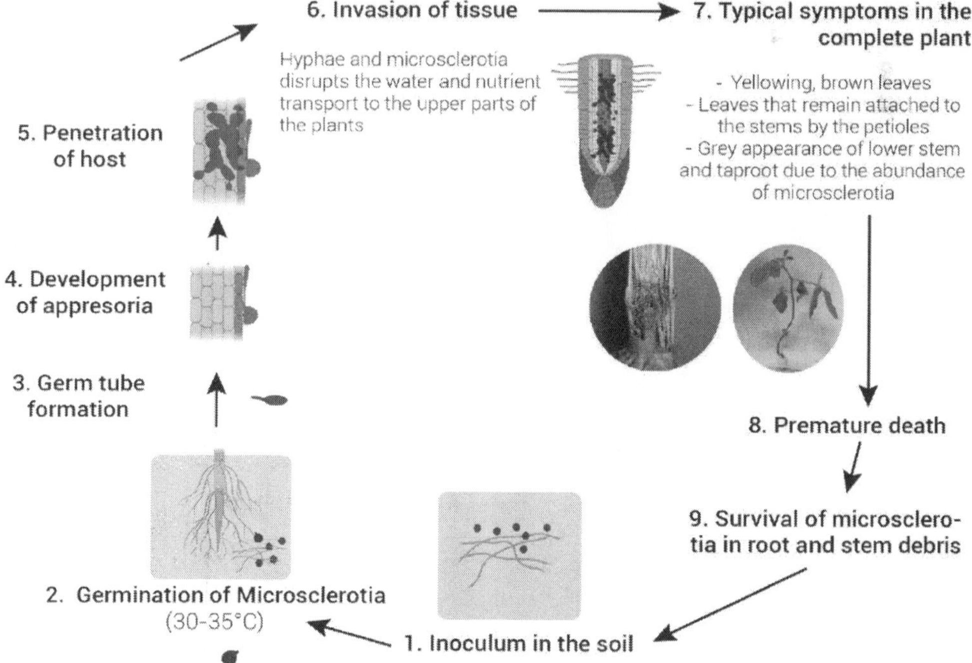

Figure 18.9: Disease Cycle of *Macrophomina phaseolina* (*Source*: Marquez *et al.*, 2021).

Management

1. Adequate fertilization and irrigation.
2. Employ seed treatment with either Carbendazim at a rate of 2 grams per kilogram of seeds, Thiram at 3-5 grams per kilogram of seeds, or Trichoderma viride at 4 grams per kilogram of seeds.
3. At the moment of sowing, distribute a mixture of talc-based Trichoderma viride 1 per cent WP at a rate of 1.5 kilograms per hectare along with Pseudomonas fluorescens 1 per cent WP at the same rate, combined with 500 kilograms per hectare of well-decomposed farmyard manure, directly into the furrows.
4. Spot apply a solution of Carbendazim at a concentration of 0.5 grams per liter.

B) Viral Diseases

6. Bud Necrosis Disease

Family: Tospoviridae

Genus: Orthotospovirus

Pathogen: Peanut bud necrosis virus (PBNV) (ssRNA genome with multipartite genome)

Transmission

The virus is disseminated via thrips, specifically *Thrips palmi*, *T. tabaci*, and *Frankliniella* spp. The virus endures within weed hosts such as *Bidens pilosa*, *Erigon bonariensis*, *Tagetes minuta*, and *Trifolium subterraneum*.

Symptoms

Chlorotic circular patterns appear on the leaves of afflicted plants. One conspicuous sign is the death of the terminal bud. In the early stages of infection, plants exhibit stunted growth, and their shoots produce diminutive, wrinkled leaves with diverse mottling and small ring-like spots. As the plants mature, they tend to exhibit less stunting, but young leaflets still often display chlorotic rings and mottling approximately 40 days after planting. Bud necrosis can extend to affect the petioles and stem, at times resulting in the demise of the plant. Subsequently formed leaves are smaller than usual and display a wide array of symptoms, including distortion, mosaic mottling, and overall chlorosis. Both the size and number of pods are significantly reduced. In cases of later infections, pod size diminishes further. Seeds frequently appear shrunken with mottled and discoloured seed coats.

Figure 18.10: Bud Necrosis Disease: Rings and chlorotic spots, stunting, and necrosis of terminal bud of peanut (Courtesy of B.V. Bhaskara Reddy, Acharya N G Ranga Agricultural University).

Management

1. Early sowing.
2. Tolerant Cultivars: Kadiri 3 and ICGS 11.
3. Remove and destroy infected plants up to 6 weeks after sowing.
4. Plant density can be increased to decrease the proportion of infected plants.
5. Intercropping of cereal crops such as pearl millet.
6. Application of Spirotetramat 11.01 per cent + Imidacloprid 11.01 per cent w/w SC (Movento) 30 days after sowing to manage thrips.

7. Groundnut Rosette Disease (GRD)/Groundnut Chlorotic Rosette

Family: *Tombus viridae*

Genus: *Umbravirus*

Species: *Groundnut rosette virus*

Transmission

Aphid: *Aphis craccivora*

Symptoms

1. Plants infected when young, produce progressively smaller, chlorotic, twisted and distorted leaflets.

Figure 18.11: Symptoms of Groundnut Rosette Disease (*Source*: CABI).

Management

2. Removal of groundkeepers and volunteer plants, as these are primary sources of infection, early sowing at high plant densities
3. Viral disease managed by control vector population by spraying insecticide *i.e.* flonicamid 50 WG or dinotefuran 20 SG or neem based biopesticide.

C) Nematode Diseases

The root knot nematode diseases diseasease caused by *Meloidogyne arenaria*, which belongs to Kingdom: Animalia, Phylum: Nematoda, Class: Secernentea, Order: Tylenchida, Family: Heteroderidae, and Genus: *Meloidogyne.*

Symptoms

Figure 18.12: Symptoms of Root Knot Nematode-Healthy (Left); Infected Plant (Right) (*Source*: Wu *et al.*, 2023).

Visible aboveground indications encompass diminished growth and a reduced number of small, pale green or yellowish leaves, which have a tendency to wilt during warm weather. Blossoms and fruits are infrequent and of subpar quality. Typically, afflicted plants persist throughout the growing season and are rarely prematurely killed. Distinguishing symptoms of this ailment manifest in the subterranean sections of the plants. Infected roots exhibit the characteristic root-knot galls, which are two to several times larger in diameter than healthy roots. Multiple infections along the root give rise to a rugged, clubbed appearance. In instances where certain nematode species are responsible for the infection, roots may develop a bushy root system. However, in most cases, infected roots remain undersized and exhibit necrosis and rot, particularly as the season progresses. When the root-knot nematodes target fleshy underground organs like tubers, carrots, potatoes, peanuts, and yams, they induce small swellings on the organ's surface, which become highly conspicuous and cause deformities or fissures. Trees' roots are also susceptible to attack by root-knot nematodes, resulting in galls that correspond in size to the duration of infection.

Life Cycle of *Meloidoigyne* spp.

1. Each female has the ability to lay approximately 500 eggs encased in a gelatinous substance.
2. The second-stage juvenile emerges from the egg and enters the soil.
3. It's worth noting that the second-stage juvenile represents the singular infective stage of the nematode's life cycle.

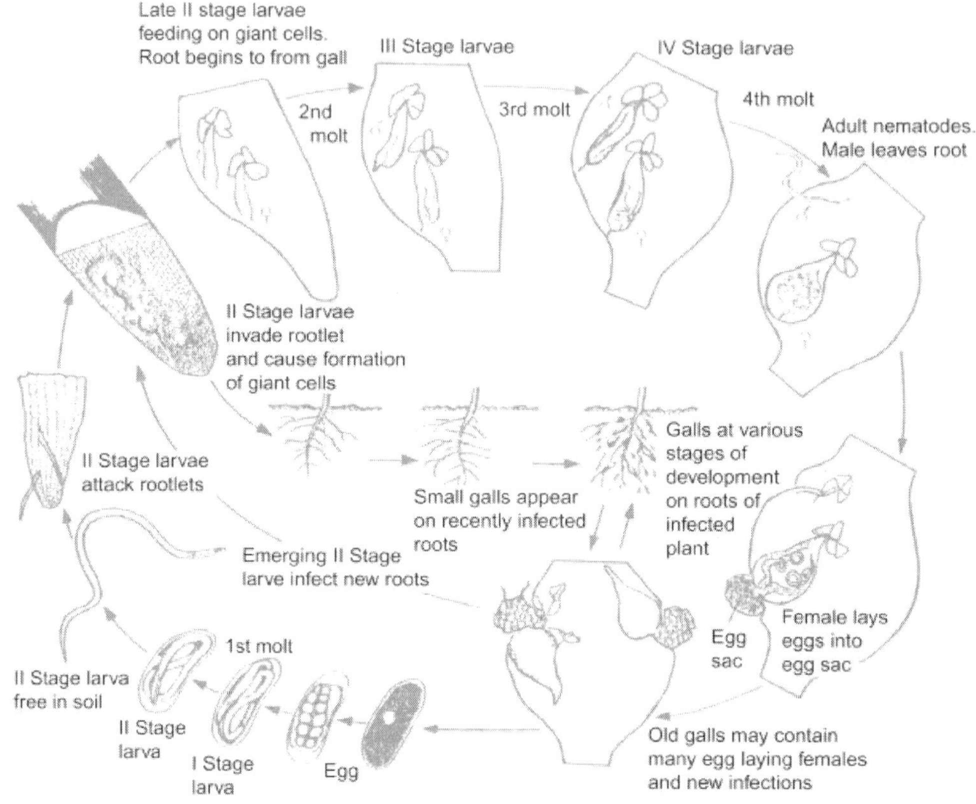

Figure 18.13: Life Cycle of *Meloidoigyne* spp. (*Source*: Jackson 2022).

4. The nematode sustains itself by feeding on the cells surrounding its head. This involves the insertion of its stylet and the secretion of saliva into the cells.
5. Following this, the nematode goes through a second molt, giving rise to the third-stage juvenile. The third-stage juvenile is characterized by a more robust build and subsequently undergoes a third molt. This results in the development of the fourth-stage juvenile, which can be identified as either male or female.
6. These fourth-stage juveniles proceed to undergo the fourth and final molt. The male emerges from the root as a worm-like adult, becoming free-living in the soil. Meanwhile, the female continues to increase in girth and undergoes some elongation, ultimately adopting a pear-shaped appearance.
7. A complete life cycle typically spans 25 days when maintained at a temperature of 27°C. However, it should be noted that the duration

of the life cycle may be extended under lower or higher temperature conditions.

8. Root-knot nematodes are primarily disseminated through water or via soil that adheres to agricultural machinery or is carried on infected plant material introduced into areas that were previously free from infection.

Management of Root-knot Nematode

1. Resistant Cultivars: To thwart the emergence of resistance-breaking biotypes of nematodes, it is imperative to avoid cultivating resistant varieties continuously for extended periods.
2. Fallowing: Implementing fallow periods in crop rotation can help mitigate nematode populations.
3. Root Destruction: Given that root-knot nematodes can endure and reproduce on residual roots post-harvest, it is advisable to uproot and eliminate galled roots from the soil.
4. Neem-Based Solutions: India has extensively researched the use of neem-based oil cakes and related products for root-knot nematode control.
5. Soil Solarization: Elevating soil temperatures to approximately 45°C through solarization is considered lethal for plant-parasitic nematodes.
6. Healthy Planting Material: Opt for nematode-free seedlings when transplanting to minimize the risk of nematode infestation.
7. Non-Host Crops: Rotate highly susceptible vegetable crops with cruciferous crops that display moderate resistance or tolerance to root-knot nematodes.
8. Crop Rotation: Implementing a sequence of sesame, maize, groundnut, sorghum, cabbage, velvet bean, and then resistant sweet potato has proven effective in *Meloidogyne* spp. management.
9. Antagonistic Crops: Certain plants, like marigold (Tagetes), produce compounds toxic to root-knot nematodes, such as terthienyl and bithienyl derivatives.
10. Granular Nematicides: The application of granular nematicides, such as carbofuran at a rate of 3 kg a.i./ha, is effective in controlling root-knot nematodes in vegetable crops under both field and greenhouse conditions.
11. Drip Irrigation: Employing drip irrigation delivers materials directly to the rhizosphere, enabling precise and timely treatment during the growing season.
12. Oviparasitic Fungi: Utilize commercial formulations of oviparasitic fungi like Paecilomyces lilacinus and Pochonia chlamydosporia at a

rate of 50g/m^2, along with an adequate amount of farmyard manure or vermicompost, during nursery preparation.

13. *Pasteuria penetrans*: Manage nematode populations in the field by applying the obligate bacterial parasite *Pasteuria penetrans* at a rate of 50g/m2 in nurseries or 20kg/ha under field conditions.

Plant Disorder

Iron Chlorosis

Figure 18.14: Iron-chlorosis (*Source*: Texas and Agrilife Research).

References

CABI (2023). CABI Compendium. Wallingford, UK: CAB International.

Daudi, H., Shimelis, H., Mwadzingeni, L., Laing, M., and Okori, P. (2019). Breeding groundnut for rust resistance: A review. *Legume Research-An International Journal*, **42**(3), 291-299.

FAOSTAT (2022). Food and Agricultural Organizations of the United Nations, Rome. Statistical database. (Accessed on 31/07/2023)

Ghanekar, A.M. (1980). *Groundnut Virus Research at ICRISAT*. Proceedings of the International Workshop on groundnuts, 13-17 October 1980. Eds. R. W. Gibbons and J.V. Martin, International Crops Research Institute for the Semi- Arid Tropics, Patancheru, India, pp. 211-216.

Ghewande, M.P. (1985). Seed and seedling diseases of groundnut deserve attention for better productivity. *Seeds and Farms* 11: 9-12 and 19.

Jackson, R. S. (2022). *Wine tasting: a professional handbook.* Elsevier.

Marquez, N., Giachero, M. L., Declerck, S., and Ducasse, D. A. (2021). Macrophomina phaseolina: General characteristics of pathogenicity and methods of control. *Frontiers in Plant Science,* 12, 634397.

Mc Donald, D., Subrahmaniyam, P., Gibbons, R.W. and Smith, S.D.H. (1985). Early and late leaf spots of groundnut. *Information Bulletin No.21.* International Crops Research Institute for the Semi-Arid Tropics, Patancheru, A.P., India.

Mondal, S. and Badigannavar, A. M. (2015). Peanut rust (Puccinia arachidis Speg.) disease: its background and recent accomplishments towards disease resistance breeding. *Protoplasma,* 252, 1409-1420.

Pardee, W. D. (2002). Groundnut. *Encarta Encyclopedia. Microsoft Corp. USA.*

Subrahmaniyam, P. and Mc Donald, D. (1983). Rust diseases of groundnut. Information Bulletin No.13. International Crops Research Institute for the Semi-Arid Tropics, Patancheru, A.P., India.

Woodward, J. E., Brenneman, T. B., and Kemerait Jr, R. C. (2013). Chemical control of peanut diseases: targeting leaves, stems, roots, and pods with foliar-applied fungicides. Fungicides–showcases of integrated plant disease management from around the world, 55-76.

Wu, L., Ren, Y., Zhang, X., Chen, G., Wang, C., Wu, Q. and Yuan, M. (2023). Effect of Root-Knot Nematode Disease on Bacterial Community Structure and Diversity in Peanut Fields. Agronomy, 13(7), 1803.

ORCID: https://orcid.org/0000-0003-2027-5055

Chapter 19

Diseases of Sesamum Crop

Kavita Kansotia[1]*, *Shivam Maurya*[2], *Pinki Sharma*[3], *Kiran Kumawat*[3] *and Lalita Lakhran*[4]

[1]*Ph.D. Scholar, Department of Plant Pathology, Sri Karan Narendra Agriculture University, Jobner-Jaipur, Rajasthan*
[2]*Assistant Professor, Department of Plant Pathology, School of Agriculture, Lovely Professional University, Phagwara, Punjab*
[3]*Ph.D. Scholar, Department of Plant Pathology, MPUAT, Udaipur, Rajasthan*
[4]*Assistant Professor, Agriculture University, Jodhpur, Rajasthan*
**e-mail: kavitakansotia6@gmail.com*

Introduction

The oldest oil seed that has been utilized by humans is sesame (*Sesamum indicum* L., 2n=26), which is a member of the *Pedaliaceae* family (Kafiriti and Decheers 2001). Its high-quality polyunsaturated stable fatty acid, which inhibits oxidative rancidity earned the title "Queen of Oil Seeds". Sesamol and sesamolinol, two natural antioxidants that slow down oxidation, contribute to its stability as well (Terefe *et al.*, 2012). It is a popular oilseed crop in India because of its edible oil content of 50 per cent, wholesome protein level of 23 per cent, and adequate carbohydrate content of 15 per cent (Ranganatha *et al.*, 2012). According to Reddy (2006), the plant is adapted to hot temperatures and is a member of the Pedaliaceae family. In rural and urban locations where access to other sources of feed is constrained, the sesame oil cake is a crucial byproduct for animal feed, particularly for the production of milk for human use. With 6335 kcal kg-1 of nutritional energy in seeds, sesame ranks first among oilseed crops in terms of oil content (Kumar and Goel, 1994). The crop has been shown to be afflicted with a number of diseases as a result of intense agricultural practices. There have been reports of 72 fungi, 7 bacterial, 1 phytoplasmal and 1 viral disease from India according to Vyas *et al.* (1984). Many diseases attacks on sesame crop, including Fusarium wilt (*Fusarium oxysporum* f. sp. *sesami*), leaf curl (*Leaf curl virus*), charcoal rot/root rot/damping off (*Macrophomina phaseolina*), bacterial blight (*Xanthomonas campestris* pv. *sesami*), Cercospora leaf spot (*Cercospora sesami*), Sesame phyllody (Phytoplasma), Fusarium wilt (*Fusarium oxysporum* f. sp. *sesami*), leaf curl (*Leaf curl virus*) and mosaic (*Mosaic virus*).

Disease classified as:

Fungal Disease

1. Alternaria leaf blight
2. Cercospora leaf spot
3. Root rot disease/Stem rot disease/Charcoal rot
4. Fusarium wilt disease
5. Stem blight disease
6. Powdery mildew disease
7. Anthracnose disease

Bacterial Disease

1. Bacterial leaf spot (*Xanthomonas* spp.)
2. Phyllody disease (Phytoplasmal disease)

1. Leaf Blight

Economic Importance and Occurrence

The host plant's growth stage determines how much *A. sesami* damages sesame plants (Kolte, 1985). Disease proved to be quite destructive during the seedling stage (Tripathi *et al.*, 1998). In seed samples, *Alternaria sesami* infection ranged from 8.96 to 24.80 per cent reported by Kakamega and Siaya (Ojiambo *et al.*, 1998). In Uttar Pradesh, Alternaria leaf spot caused a 20–40 per cent loss in the sesame crop (Kumar and Mishra, 1992). The Alternaria blight of sesame is a national problem that has persisted in being a serious hindrance to the crop's productivity and production primarily in the state of Rajasthan (Sharma, 2021).

Causal Organism

The disease-causing agent *Alternaria solani*, is a member of the Ascomycota Phylum, Class Dothideomycetes, Order Pleosporales, Family Pleosporaceae, Genus *Alternaria*, Species *solani*. The dark brown septate mycelium of the fungus yields a large number of pale grey-yellow conidiophores that are either straight or curled. Conidiophores of the fungus are simple, erect, cylindrical, 0–3 septate.

They grow singly. Their apex produces conidia. Conidia can form in chains in pairs or alone. They are obclavate, straight or slightly curved, 30-120 9-30 (beak excluded) and range in tint from yellowish brown to dark or olivaceous brown. The conidia have four to twelve transverse and zero to six longitudinal septa. After that, they become somewhat constricted and terminate in a long hyaline. Beak forms vary from basic to branching.

Symptoms

Figure 19.1: Microscopic View and Symptom.

The majority of this disease's symptoms are found on the leaf blades as small, 1–8 mm diameter, brown, rounded to irregular patches. Subsequently, concentric zonations encircled by brown lines surrounding the larger, darker dots on the upper surface. There are lighter-colored brown dots on the lower surface. These spots can damage large sections of the blade and often consolidate, dry up, and eventually shed. There are moist, dark brown lesions that spread throughout the length of the stem. The lesions also damage the midrib and the veins on the petioles, stem, and leaf capsules. The seeds are released from infected capsules when they split open too soon. Less severe infections cause defoliation but they can quickly destroy plants as soon as signs show up. Pre- and postemergence damping-off can occasionally cause young plants and seedlings to perish.

Favourable Conditions

The ideal pH for growth is 4.5, low temperature (20–25°C), high relative humidity, cloudy weather and excessive rainfall encourage the growth of disease.

Disease Cycle

The fungus is soil-borne and seed-borne because it remains dormant in the plant waste of infected plants (Agarwal *et al.*, 2006; Naik *et al.*, 2007). *A. sesami* has the ability to penetrate the seed coat and remain there until the seed sprouts. The fungal spores attached to the seeds or capsule may carry and spread the infection. Ojiambo *et al.* (2008) report that plants i between 8 and 10 weeks of age had the highest seed infection rates, whereas plants inoculated at 1, 6, and 12 weeks of age had the lowest rates.

Management

1. The best approach is to create and use resistant sesame cultivars.
2. Sesame's Alternaria leaf blight can be less severe when intercropped with maize in a single alternating row.
3. Use thiram or carbendazim on the seeds before sowing at the rate of 2 g/kg.
4. Spray 2 kg/ha of mancozeb on standing crop thrice after appearance of symptoms at 15 days of interval.

2. Leaf Spot

Economic Importance and Occurrence

The disease affects every portion of the plant that is above ground, completely defoliating the plant and causing significant financial losses. The disease appears on leaves as little as pinhead-sized cottony patches and can begin to harm leaves as soon as four weeks after planting. Before the plant reaches maturity, extensive infection causes defoliation and capsule destruction, which can lead to yield losses of 20–50 per cent.

Causal Organism

The cause of leaf spot disease is *Cercospora.* According to Kirk *et al.* (2008), the fungus is a member of the phylum Ascomycota, class Dothideomycetes, order Capnodiales, family Mycosphaerellaceae, genus *Cercospora*, and species *sesami.* The fungus's hypha has thick walls, a light brown colour and an uneven septate structure. Conidiophores are 1-3 septate, hyaline at the tip, and light brown at the base. They are formed in clusters. Conidia are broad at the base and taper towards the apex. They are elongated, septate, 7–10, hyaline to pale yellow.

Symptoms

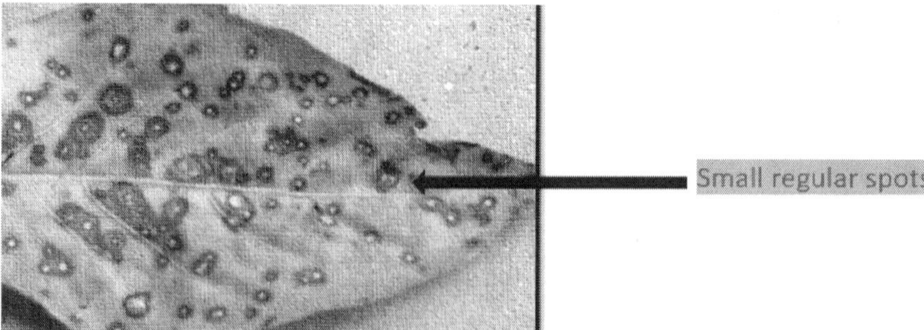

Figure 19.2: Cercospora Leaf Spot of Sesamum.

The disease initially emerges as tiny, water-soaked lesions on the leaves, which grow to become rounded or uneven spots on both surfaces that range in diameter from 5 to 15 mm. Early loss of foliage results from the irregular, variable-sized patches that these spots eventually combine to form. Furthermore, the infection may show up as different-sized spots that develop on the petiole and stem. Additionally, there are black linear patterns on pods that cause drying shedding.

Disease Cycle

The fungus is transmitted by both internal and exterior seeds. In soil, the fungus can also thrive in plant waste. Seeds and contaminated debris may be the source of the initial infection. Conidia that are carried by the wind are the secondary spread.

Disease Management

1. The disease can be effectively controlled by soaking seeds in hot water for 30 minutes at a temperature of 53°C.
2. Cercospora leaf spot has less of an impact on crops that are sown early from the middle of June to the first week of July. The microclimatic effects of intercropping affect the severity of foliar diseases such as sesame leaf spot caused by Cercospora.
3. 10 per cent less disease severity is observed when an aqueous leaf extract of *Aspilia africana, Chromolaena odorata, A. indica,* and *Allium sativum* is sprayed once a week.
4. Plant waste should be burned both before and after threshing.
5. Apply 2g/kg of carbendazin, treinam or mancozeb to the seeds.
6. It is possible to efficiently control Cercospora leaf spot disease with two sprays of 0.2 per cent Mancozeb, Tebuconazole, or Azoxystrobin.

3. Root Rot or Stem Rot or Charcoal Rot

Economic Importance and Occurrence

Farmers' incomes may be impacted by this fungus, which can result in significant yield losses in crops like sorghum and soybeans (Kaur *et al.,* 2012). In the worst situation, diseases that manifested in the pre-emergence stage of groundnut cultivars have been known to cause 100 per cent yield losses (Sharma and Bhowmik, 1986).

Causal Agent

The disease-causing agent is *Macrophomina phaseolina* (*Rhizoctonia bataticola,* Sclerotic stage). The pathogen causes constrictions at the hyphal connections in its dark brown, septate mycelium. The diameter of the tiny, dark-black sclerotia ranges from 110 to 130 µm. The ostiole is clearly visible on the dark brown pycnidia. Hyaline, elliptical, single-celled conidia are present. According to Kirk *et al.* (2008), the kingdom is Fungi, the phylum is Ascomycota, the class is Dothideomycetes, the order is Botryosphaeriales, the family is Botryosphaeriaceae, the genus is *Macrophomina,* and the species is *phaseolina.*

Symptoms

Lower leaf yellowing is the first sign of the disease, which is then followed by falling and defoliation over time. Bark in the collar region is shredded and dark brown lesions are visible in the stem section close to the ground. There are spots where plants have suddenly died. The section of the stem closest to the soil line in adult plants has a high concentration of black pycnidia. It is simple to remove the stem section, leaving the rotting root section in the ground. When the virus spreads to the pods, the young seeds shrivel and turn black, and the pods open early. On the contaminated seeds and capsules tiny pycnidia are also visible. There are a lot of tiny black sclerotia in the stem and rotting root tissues. It's possible that the sclerotia are also on the contaminated seeds and pods.

Figure 19.3: Root Rot or Stem Rot or Charcoal Rot.

Favourable Conditions

30ºC or higher during the day, a protracted drought and abundant irrigation.

Disease Cycle

Latent sclerotia of the fungus can be found in soil and plant debris left over from diseased plants. The diseased plant trash also contains pycnidia. The primary means of fungal transmission are infected seeds, which carry sclerotia and pycnidia. The fungus also spreads through soil-borne sclerotia. The secondary spread is caused by conidia that are dispersed by wind and precipitation.

Management

1. Soil solarization is a cost-effective method for reducing pathogen populations in the soil, although it takes a long time and is dependent on weather conditions.
2. Apply 4g/kg of *Trichoderma viride* to the seedlings.
3. Apply farmyard manure, green leaf manure, or 150 kg/ha of neem cake at a rate of 10 t/ha each.
4. The severity of the disease is limited by changes in social behaviors and seed treatment. Soil drenching with mancozeb @ 250 ppm concentration and seed treatment with Carbendazim + Mancozeb @ 2 g/kg of seed.

4. Wilt

Economic Importance and Occurrence

Among the most significant factors limiting crop yield worldwide is disease infection. Every year, diseases that infect sesame crop cause losses of seven million tonnes globally (Ara *et al.*, 2017). *Fusarium oxysporum* f. sp. *sesami* (FOS) is the cause of the sesame wilt disease, which is a group of diseases that affect sesame crops, particularly those grown in higher Egypt due to the high temperatures. It's one of the factors that contributes to the annual decline in the cultivated area of sesame because farmers lose yield. FOS is a soil-borne fungus, infects root, grows and colonizes xylem vessels and blocking them completely to cause wilt (Bateman *et al.*, 1996) and more than 50 per cent yield losses in sesame (Gaber, *et al.*, 1998; Khaleifa, 2003; El-Bramawy, 2006; El-Shakhess and Khalifa, 2007).

Causal Agent

The disease produced by *Fusarium oxysporum* f. sp. *sesami*, a fungus that grows in soil. Fusarium is a member of the Ascomycota division, the Sordariomycetes class, the Hypocreales order, the Nectriaceae family, the Fusarium genus and the oxysporum species. Chlamydospores, macroconidia and microconidia are all produced by the fungus. Macroconidia are hyaline, falcate-shaped and 5–9-celled. Hyaline, unicellular, with thin walls, and ovoid, microconidia are. Additionally, the thick-walled chlamydospores are generated.

Symptoms

The disease results in yellowing, drooping and withering leaves. The plants' progressive degeneration causes them to wilt and dry out over time. There are long, dark-black lines of vascular necrosis on the afflicted root and stem parts.

Figure 19.4: Sesamum Wilt
(*Source*: infonet-biovision.org).

Disease Cycle

The fungus is able to survive in the soil due to plant trash that is infected. It is also seed-borne with contaminated seeds or soil-borne chlamydospores causing the initial infection. Raindrops and irrigation water-dispersed conidia could be the cause of the secondary infection.

Management

1. Crop rotation is the most effective approach to eliminate soil-borne pathogens from the field.
2. By increasing soil temperature and covering the soil with mulching materials, *F. oxysporum* inoculum can be reduced.
3. Seed and soil treatment of *T. viride* + soil application of Neem cake (150 kg/ha.) + Mancozeb (0.3 per cent) soil soaking effectively reduced the disease.
4. Seed treatment with Carbendazim (2 g/kg) + Neem oil (15 ml/kg).

5. Stem Blight

Economic Importance and Occurrence

The disease incidence (23.84 - 53.72 per cent) recorded during survey in major sesame growing areas of Rajasthan. Disease incidence recorded maximum in Jodhpur district (41.14 per cent) followed by Pali (38.71 per cent), Jaipur (35.90 per cent), Nagaur (28.58 per cent) and lowest incidence was observed in Tonk district (24.41 per cent) (Lakhran, 2021).

Causal Agent

Phytophthora parasitica var. *sesami* is the culprit behind the stem blight of sesamum. The fungus is a member of the following groups: Order: Peronosporales, Family: Peronosporaceae, Genus: Phytophthora, Kingdom: Chromista, Phylum: Oomycota.

Symptoms

Lesions of a black tint develop on the stem close to the soil's surface. The plant dies as a result of the disease spreading and damaging branches and maybe girdling the stem. Additionally, water-soaked areas on leaves may appear and spread until the leaves shrivel. On flowers and capsules, infection may be visible. The seeds inside the poorly developed, shrivelled capsules are infected. The mycelium of the fungus is hyaline and non-septate. The hyaline, sympodially branching sporangiophores are bearing sporangia. The sporangia have a large apical papilla and are spherical and hyaline. The oospores have thick walls, are round, and are smooth.

Favourable Conditions

Prolonged rainfall, low temperature (25°C) and high relative humidity (above 90 per cent).

Figure 19.4: Stem Blight (*Source*: kiosk.nau.in).

Disease Cycle

Dormant mycelium and oospores allow the fungus to persist in seed and soil. The major infection-causing fungus is also carried by the seeds in the form of latent mycelium. Sporangia carried by the wind are the disease's secondary dissemination agent.

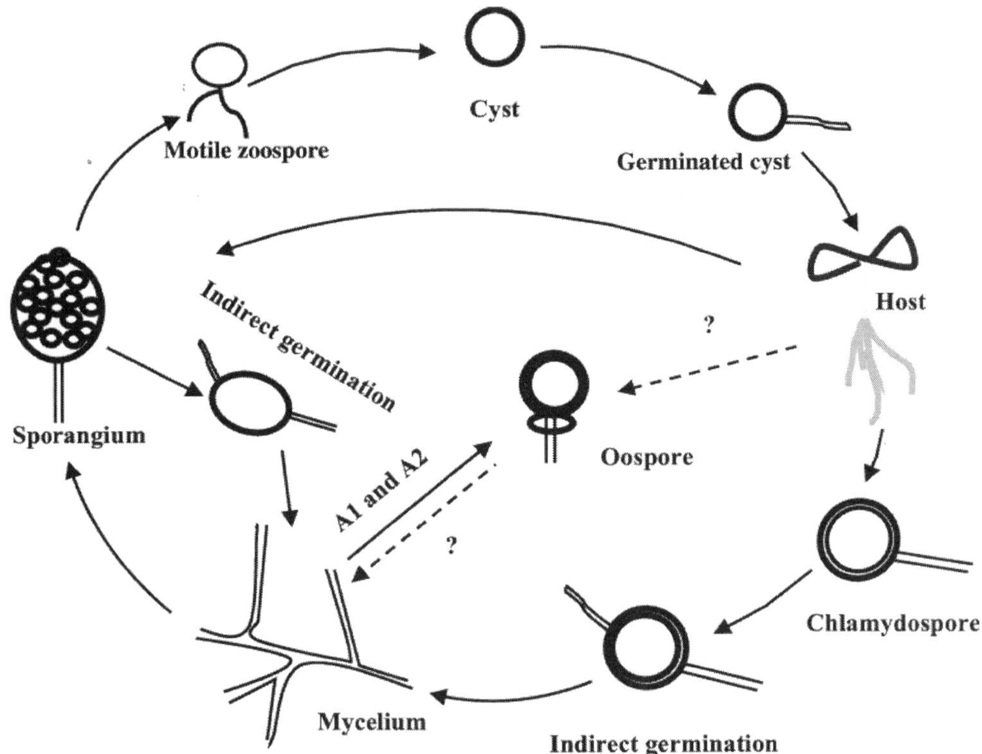

Figure 19.5: Disease Cycle of Stem Blight (*Source*: Meng *et al.*, 2014).

Management

1. Sanitation and clean cultivation and choice of sowing dates depending on the known prevailing locality conditions are taken into practical use in disease management.
2. Avoid continuous cropping of sesamum in the same field.
3. Remove and destroy infected plant debris.
4. Several microbial antagonists such as *T. viride*, *Gliocladim virens*, *Bacillus polymyxa*, *B. subtilis*, *Enterobacter cloacae*, *Pseudomonas aeruginosa*, *P. putida* and *P. fluorescens*, *Streptomyces bikiniensis*, and *S. echinoruber* are inhibitory to the growth of pathogen.

5. Fusarium wilt of sesame can be controlled with application of plant-growth-promoting rhizobacteria, and this practice offers a potential nonchemical means for disease management.
6. Spray metalaxyl 1kg/ha.
7. Treat the seeds with captan or thiram at 2g/kg or metalaxyl @ 4g/kg.

6. Powdery Mildew

Causal Organism

Erysiphe cichoracearum (Syn: *Oidium acanthospermi*)

The pathogen produces a hyaline septate mycelium that is ectophytic and secretes haustoria into the host's epidermis. The basic mycelium gives rise to tiny, non-septate conidiophores that carry conidia in elongated chains. The conidia are ellipsoid or barrel-shaped, single-celled hyaline organisms. The cleistothecia have black, globose appendages that are hyaline, or light brown myceloid. The asci are oblong, and one to two elliptical, pale brown, thin-walled ascospores are produced by each ascus.

Symptoms

Figure 19.6: Powdery Mildew of Sesamum.

Powdery growth that is initially greyish-white first develops on the leaf surface. A powdery coating may cover the entire leaf surface when multiple spots combine. In extreme circumstances, the infection may be seen on the immature capsules and flowers, causing early shedding. Twisted and deformed leaves are possible in multiplely damaged leaves. The development of cleistothecia causes the mycelial growth to turn dark or black in the advanced stages of infection.

Favourable Conditions

Dry humid weather and low relative humidity favourable for development of disease.

Disease Cycle

The pathogen an obligatory parasite, spreads the disease by cleistothecia in the soil-borne plant waste that is affected. Primary infection is brought on by the cleistothecia's ascospores. Conidia that are carried by the wind are the secondary spread.

Management

1. Remove the infected plant debris and destroy.
2. Spray wettable sulphur at 2.5 kg/ha or karathane 1lit/ha repeat after 15 days.

7. Bacterial Leaf Spot

Causal Organism

The causative agent is *Xanthomonas campestris* pv. *sesami.* With a rod-shaped morphology and a monotrichous flagellum, this gram-negative bacterium is a member of the Phylum Pseudomonadota, Class Gammaproteo bacteria, Order Xanthomonadales, Family Xanthomonadaceae, Genus *Xanthomonas* and species *campestris.*

Symptoms

First appearing on the underside of the leaf, water-soaked spots eventually expand to the upper surface. They grow larger, take on an angular form, have constricted veins, and have a dark brown tint. The leaves dry up as a result of several dots coming together to create irregular brown areas. Reddish-brown lesions may also appear on the stem and petioles.

Disease Cycle

The bacterium survives in the infected plant debris and in seeds. The secondary spread is by rain water.

Management

1. Remove and burn infected plant debris.
2. Use healthy seeds for cultivation.
3. Spray Streptomycin sulphate or oxytetracycline hydrochloride or strephocyclin at 100g/ha.

8. Bacterial Leaf Spot

Causal Agent

The bacterium that causes leaf spot on sesamum is *Pseudomonas sesami.* The bacterium has one or more polar flagella and is gram-negative, aerobic and rod-shaped.

Symptoms

On the upper surface of the leaves, the disease manifests as yellow dots that have been soaked in water. As they are constrained by veins and veinlets, they get larger and more angular. The area could have a dark brown tone and glossy bacterial mass exudes.

Disease Cycle

In the tissues of the affected plants, the bacteria are still alive. It is internally seedborne and disperses secondarily by storms and splashing of rain.

Management

1. Keep the field free of infected plant debris.
2. Use of healthy seeds for sowing.
3. Spray with Streptomycin sulphate or oxytetracycline hydrochloride or streptocyclin at 100g/ha.

9. Phyllody

Economic Importance and Occurrence

Phyllody, a phytoplasma-related disease, is one of these serious diseases and is regarded as a major obstacle in many regions, particularly in Asia and Africa, where it significantly reduces the amount and quality of sesame crop. One of the top ten nations in the world for sesame production, India has croplands where the frequency of sesame phyllody disease ranges from 10 to 100 per cent. According to studies by Salehi *et al.* (2017) and Akhtar *et al.* (2013), sesame phyllody is a damaging disease that can reduce seed yield by 80 per cent. The earliest evidence of phytoplasma-associated phyllody disease was found in Pakistan in 1908, according to Vasudeva and Sahambi (1955). Phyllody has long been thought to be a plant pathology connected with viruses.

Causal Organism

It is caused by pleomorphic mycoplasma like bodies present in sieve tube of affected plants, now designated as a phytoplasmal disease.

Symptoms

Depending on the stage of crop growth and the timing of the infection, affected sesame plants exhibit symptoms. Early infection can cause the entire plant to become impacted and the plant will remain stunted to around two-thirds of its typical size. First appearing on the underside of the leaf, water-soaked spots eventually expand to the upper surface. They grow larger, take on an angular form, have constricted veins, and have a dark brown tint. The leaves dry up as a result of several dots coming together to create irregular brown areas. Reddish-brown lesions may also appear on the stem and petioles.

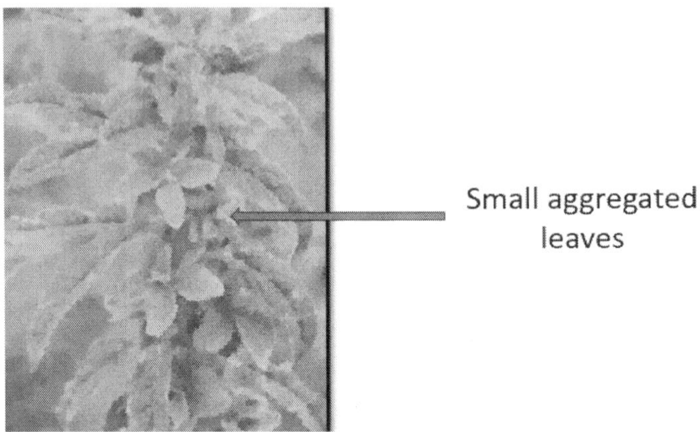

Figure 19.7: Phyllody of Sesamum.

Disease Cycle

The pathogen is able to live on a variety of hosts, including *Brassica campestris* var. *toria, B. rapa, Cicer arietinum, Crotalaria* spp., *Trifolium* spp., and *Arachis hypogaea*, which are used as sources of inoculum. The disease is spread by jassids, *Orosius albicinctus*. The ideal time for a vector to acquire a vector is three to four days, and the feeding period for the inoculation is thirty minutes. The pathogen can incubate for 15–63 days in leaf hoppers and 13–61 days in sesame. The phytoplasma cannot be transmitted by nymphs. The vector population is higher in the summer and lower in the winter.

Management

1. Remove all the reservoir and weed hosts.
2. Avoid growing sesamum near cotton, groundnut and grain legumes.
3. Rogue out the infected plants periodically.
4. An appropriate sowing date may be useful in avoiding severe occurrence of the disease.
5. Spray Monocrotophos or Dimethoate at 500 ml/ha to control the jassids Symptoms.
6. Soil treatment with Thirnet 10 G @ 10 kg/ha or Phorate 10 G @ 11 kg/ha at the time of sowing.

10. Minor Disease

Anthracnose - *Colletotrichum* spp. Dark brown lesions on leaf stem and capsules with black acervuli in the central portion.

References

Agarwal, P.C., U. Dev, B. Singh, I. Rani, and R.K. Khetarpal. (2006). Seed-borne fungi detected in germplasm of *Sesamum indicum* L. introduced into India during last three decades. Indian Journal of Microbiology. **46**: 161–164.

Kafiriti, E. and Deckers, J. (2001). Sesame (*Sesamum indicum* L.). In R. H. Raemaekers, Crop Production in Tropical Africa. Brussels, Belgium: Directorate General for International Co-operation, Pp. 797-804.

Kumar, S. and Goel, P. D. (1994). A great ancient oilseed sesamum. Farmers and Parliament. **12**: 6-7.

Nair, R., Thakur, K.D., Bhasme, S., Vijayan, S. and Tirthakar, S. (2006). Control of root rot (*Rhizoctonia bataticola*) of sesamum by different bioagents. *Journal of Plant Disease Sciences*. **1**: 236–237.

Ojiambo, P.S., Ayiecho, P.O. and Nyabund, J.O. (2008). Severity of Alternaria leaf spot and seed Infection by *Alternaria sesami* (Kawamura) Mohanty and Behera, as affected by plant age of sesame (*Sesamum indicum* L.). *Journal of Phytopathology*. **147**: 403–407.

Ranganatha, A. G. R., Lokesha, R., Tripnthi, A., Asafa, T., Paroha, S. and Srivastava, M. K. (2012). Sesame improvement present status and future strategies. *Journal of Oilseeds Research*, **29**(1): 1-26.

Terefe, G., Wakjira, A., Berhe, M. and Tadesse, H. (2012). Sesame Production Manual. EIAR.

Kolte, S.J. (1985). Disease of annual edible oil seed crops. Volume II: Rapeseed, Mustard, Safflower and Sesame diseases. CRC Press Inc, Boca Raton, FL, USA. Pp. 17-18.

Tripathi, U. K., Singh, S. B. and Singh, P. N. (1998). Management of Alternaria leaf spot of sesame by adjustment of sowing dates. *Ann. of Plant Protection Sci.*, **6**(1): 94-95.

Ojiambo, P. S., Narla, R. D., Ayiecho, P. Q. and Nyabundi, J. O. (1998). Effect of infection level of sesame (*Sesamum indicum* L.) seeds by Alternaria sesami on severity of Alternaira leaf spot. *Trop. Agri. Res. Ext.*, **1**: 125-130

Kumar, P. and Mishra, U. S. (1992). Diseases of Sesamum indicum in Rohilkhand: intensity and yield loss. *Indian Phytopathol.*, **45**(1): 121- 122.

Sharma, J. S., Shekhawat, H. V. S., Yadav, P. C., Patil, P. R. and Bishnoi, J. P. (2021). Occurrence and Distribution of Sesame Alternaria Blight under Environmental Conditions of Transitional Plain of Luni Basin of Rajasthan. *Biological Forum – An International Journal*, **13**(3): 584-588.

Sharma, R. C. and Bhowmik, T. P. (1986). Estimation of yield losses in groundnut due to *Macrophomina phaseolina* (Tassi) Goid. *Indian J. Plant Pathol.* **4**: 108–112.

Kaur, S., Dhillon, G. S., Brar, S. K., Vallad, G. E., Chand, R. and Chauhan, V. B. (2012). Biology, economic importance and current diagnostic trends. *Crit. Rev. Microbiol.,* **38:** 136–151.

Ara, A., Akram, A., Ajmal, M., Akhund, S., Nayyar, B. G.W., Seerat and Chaudhry, S.M. (2017). Histopathological studies of sesame (*Sesamum indicum*) seedlings infected with *Fusarium oxysporum. Plant Pathology and Quarantine,* **1**(7): 82–90

Khaleifa, M. M. A. (2003). Pathological studies on charcoal rot disease of sesame. Ph.D. Thesis, Agron. Dept. Fac. of Agric., Moshtohor, Zagazig Univ., Benha, branch, Egypt, Pp. 295.

Gaber, M. R., Hussein, N. A., Saleh, O. I. and Khalil, M. A. (1998). Susceptibility of certain varieties and genotypes and control of wilt and root rot diseases of sesame attributed to *Fusarium oxysporum* f. sp. *sesami* and *Macrophomina phaseolina. Egypt. J. Microbiol.,* **33**(3): 403-428.

El-Bramawy, M.A.S. and Abd Al-Wahid, O.A. (2007). Identification of genetic resources to Fusarium wilt, charcoal root rot and Rhizoctonia root rot among sesame (*Sesamum indicum* L.) germplasm. African Crop Science Proceedings of African Crop Science Society, El-Minia, Egypt. **8**:1893-1900.

El-Shakhess, S. A. M. and Khalifa, M. M. A. (2007). Combining ability and heterosis for yield, yield components, charcoal-rot and Fusarium wilt diseases in sesame. *Egypt J. Plant Breed.,* **11**(1): 351-371.

Lakhran, L., Ahir, R.R., Kumar, N., Nehra, D and Prajapati, S. (2021). Survey and occurrence of stem and root rot of sesame in different districts of Rajasthan. *The Pharma Innovation Journal,* **10**(9): 428-433.

Youssef, S. A., Safwat, G., Shalaby, A. B. A. and El-Beltagi, H. S. (2018). Effect of phytoplasma infection on plant hormones, enzymes and their role in infected sesame. *Fresenius Environ. Bull.,* **27**: 5727–5735.

Akhtar, K. P., Dickinson, M., Sarwar, G., Jamil, F. F. and Haq, M. A. (2008). First report on the association of a 16SrII phytoplasma with sesame phyllody in Pakistan. *Plant Pathol.,* **57:** 771.

Nagaraju and Muniyappa, V. (2005). Viral and phytoplasma diseases of sesame. In *Diseases of Oilseed Crops*; Saharan, G.S., Mehta, N., Sangwan, M.S., Eds.; Indus Publishing Company: New Delhi, India, Pp. 304–318.

Akhtar, K. P., Sarwar, G., Sarwar, N. and Elahi, M. T. (2013). Field evaluation of sesame germplasm against sesame phyllody disease. *Pak. J. Bot.,* **45:** 1085–1090.

Salehi, M., Esmailzadeh-Hosseini, S. A., Salehi, E. and Bertaccini, A. (2017). Genetic diversity and vector transmission of phytoplasmas associated with sesame phyllody in Iran. *Folia Microbiol.,* **62:** 99–109.

Vasudeva, R. S. and Sahambi, H. S. (1955). Phyllody in sesamum (*Sesamum orientale* L.). *Indian Phytopathol.,* **8:** 124–129.

Chapter 20

Diseases of Linseed Crop

Girijesh Kumar Jaisval[1]*, Arvind Kumar[2], Vishwa Vijay Raghuvanshi[3] and Prabha Siddharth[1]

[1]Department of Plant Pathology, Chandra Shekhar Azad University of Agriculture and Technology, Kanpur – 208 002, Uttar Pradesh
[2]Department of Plant Pathology, Banda University of Agriculture and Technology, Banda – 210 001, Uttar Pradesh
[3]Department of Plant Pathology, Acharya Narendra Deva University of Agriculture and Technology, Kumarganj, Ayodhya – 224 229, Uttar Pradesh
****e-mail:jaisvalpathocsa805@gmail.com***

Introduction

The 100 species of the genus Linum, which includes linseed (*Linum usitassimum* L.), are members of the family *Linaceae*. It is one of the most significant oilseed crops, produced in many nations for both its oil and fiber. Their plant growth characteristics differ significantly. The tall, slender stem fiber plants provide a lot of high quality fiber but bear seeds with low quality oil content. It is India's second-most significant oilseed crop for *Rabi*, or winter. Linseed oil, often known as flaxseed oil or flax oil (when it's edible), is an oil that may vary in colour from colourless to yellowish and is made from the dried, fully developed the flax plant's seeds. Its oil is an edible oil that is popular as a nutritional supplement because it contains the omega-3 fatty acid, alfa-linolenic acid. It can be used for many industrial, culinary, and medical purposes. However, because of its poor production, the area used for linseed cultivation is declining. Diseases are the main abiotic and biotic stressful event impacting its productivity among others. Sometimes they result in complete losses. In India, the most significant diseases affecting the linseed crop are mostly fungal diseases and can result in crop losses of up to 80–100 per cent in an epidemic. Linseed Fusarium Wilt as a result of *F. oxysporum* f. sp. *lini*, Rust (*M. lini*), Alternaria blight (*Alternaria linicola*), Powdery mildew (*Oidium lini*) are the most important ones. The less common diseases are anthracnose, *Sclerotium* stem rot, *Pythium* damping off, Botrytis rot and *Macrophomina/Rhizoctonica* induced wilt/collar rot. Cuscuta, a phanerogamic parasite, can potentially be a major issue in some places. The study discusses diseases that affect the economically significant oilseed crop in India, crop losses as well as crop management. The diseases significantly reduce the production of the linseed crop. It may be managed by using methods and procedures such agronomic practices, sowing dates, growing of resistant varieties, biological control, biopesticides and chemical fungicides as the final option.

Table 20.1: List of Diseases of Linseed and their Causative Agent

Sl.No.	*Name of Disease*	*Causal agent*
A.	**Fungal diseases**	
1.	Rust	*Melampsora lini*
2.	Fusarium wilt	*F. oxysporum* f. sp. *Lini*
3.	Alternaria blight	*Alternaria lini*
4.	Powdery mildew	*Oidium lini*
5.	Anthracnose	*Colletotrichum lini*
6.	Damping off	*Sclerotinia fuckeliana*
7.	Macrophomina stem rot	*Macrophomina phaseolina*
8.	Cercospora leaf spot	*Cercospora linicola*
9.	Root rot, foot rot and seedling blight	*Sclerotium rolfsii*
10.	Pasmo or septoria leaf spot	*Septorialinicola*
B.	**Phytoplasma diseases**	
1.	Aster yellows	Phytoplasma
C.	**Viral diseases**	
1.	Crinkle	Oat blue dwarf virus
D.	**Phanerogamic parasites**	
1.	Doddar/Cuscuta	*Cuscuta epilinum* and *C. Hylina*

Fungal Diseases

1. Linseed Rust

Linseed rust is a most important and regular disease of linseed growing regions of the world. Linseed is a major oilseed crop cultivated in the region of each the states of India. The crop is typically grown from October to April.

In general, the disease begins to appear in February or later. but Butler (1918) reported that the disease appear on first week of November in Central India. As a result of telial sori, the damaged plants appear burned when the crop is harvested. The majority of the local region's linseed fields are quickly affected once the disease established itself in one field. The disease often appears as an epidemic.

Occurrence and Economic Importance

Melampsora lini (Pers.) Lev. rust is a common occurrence that has been reported in practically all nations where linseed is grown. Persoon (1801) first gave the identity of *M. Lini* under the name *Uredo minuta.* Butler (1914) provided the first report of its occurrence from the India. It was common occurrence in MP and UP. In 1961, the disease was present in the Indo-gangetic plains of Northern India in an epiphytotic form Jammu and Kashmir, the northern mountain regions of Himachal Pradesh and UP, the neighbouring regions of PB, BR, WB, Odisha, and some portions of UP are currently suffering from a significant rust problem. In moderately to highly susceptible forms, the disease was observed to produce 15 -100 and 75 - 100 per cent yield losses, but no loss in resistant varieties. The disease affects the oil content and fiber quality in addition to reducing production. Oil content does not decrease with minor infection, but it does with moderate infection, where it is decreased by 10–34 per cent.

Causal Organism

The disease caused by *Melampsora lini* (Pers.) Lev.

Systematic Position

Kingdom- Fungi,

Division-Basidiomycota,

Class- Pucciniomycetes,

Order- Pucciniales,

Family- Melampsoraceae,

Genus- *Melampsora*

Species- lini

It is autoecious in nature and macrocyclic type rust which produces all the five spore forms (pycniospores, aeciospores, uredospores, teliospores and basidiospores) in a single host only. The mycelium occurs in the parenchymatous tissue of the host and is intercellular, dikaryotic, branching, and septate. Uredia of a bright orange colour are formed on leaves and stems, either singly or in bunch. The majority of the telia develops on the stem and is irregularly elongated, fused, and brown to black in colour. Pycnia are flask shaped and pale yellow in colour. They are often formed on leaves, although they can also grow on stems. Only uredospores and teliospores are found in India.

Symptoms

The linseed rust parasite is an autoecious and long cycle fungus that grows on the same plant and produces pycnia, aecia, and telia. Pycnia and aecia frequently produce on leaves and stems in the early stages of the growth season as light yellow to orange yellow sori. On both sides of the leaves, it appears in bright yellow or orange uredopustules, which may also affect the capsules and stem. The seriously damaged leaves resemble scorched emergence. The leaves prematurely die away and becoming chlorotic.

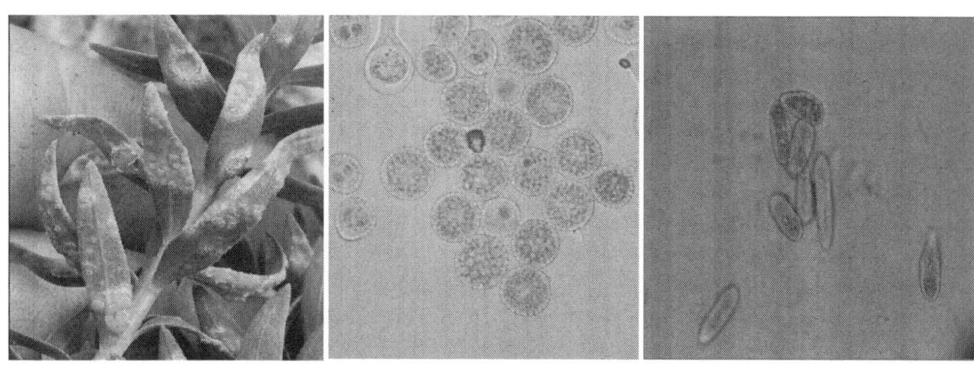

Rust Symptom   Uredospore   Teliospore

Figure 20.1

Teleutopustules, which become visible as encrustations which are brownish-black on the stem, take the place of uredopustules in the latter stages of infection. Although they shed by the time the teleutosori form, the teleutosori can be seen on any leaves that remain. They are not usually found on leaves. The entire physiology of the plants has been disturbed, which leads to severe yield losses, withered seeds, and low-quality fiber.

Favorable Condition

When temperatures are cool at night and warm during the day, dikaryotic urediniospores will germinate on a leaf's surface. According to reports, the disease occurs most frequently in cool, humid conditions, where it can result in losses of up to 16–100 per cent during an epidemic condition. The amount of yield loss depends on how soon the crop gets infected. Oil content and fiber quality of the seeds are also decreased in addition to the loss of production.

Disease Cycle

M. lini has a long-cycle or macrocyclic life cycle that includes all five spore stages with the sexual and asexual phase of the cycle occurring on the same host species (autoecious). When H. H. Flor published the findings of an inheritance study in 1946, showing that single allelic gene pairs determine the avirulence/ virulence phenotype on host lines with specific resistance genes, he put forth his now famous "gene-for-gene" hypothesis, which helped *M. lini* obtain popularity.In India the uredospores and teleutospores do not survive in plains due to unfavorable weather situation. Nevertheless, they may survive in hills throughout the year. Clean seeds, particularly following seed dressing, have been suggested since the disease was spread on by linseed straw fragments that carry viable teleutospores in stored seeds, which is widespread in the hills. Temperature, leaf moisture, and light all affect the uredospore's ability to germinate and survive. Secondary infection is carried on by uredospores that are created as a result of the primary infection. Since the uredospores take very little time to germinate the secondary infection is caused within a short time. Along with the primary infection, secondary infections cause serious damage the seed harvest. The teleutospores that are formed at the end of the growth season do not appear to help to the rust's potential to continue its life cycle. At temperatures between 13° and 21°C, the disease appears in seven days, whereas it takes 18 days at 0° to 10°C. They are killed on exposure for 24h at 38-43°C and in 9h at 43-50°C. Teleutospores develop and germinate best at 18-24°C. The leaf wetness and light intensity also take part in a vital role in infections.

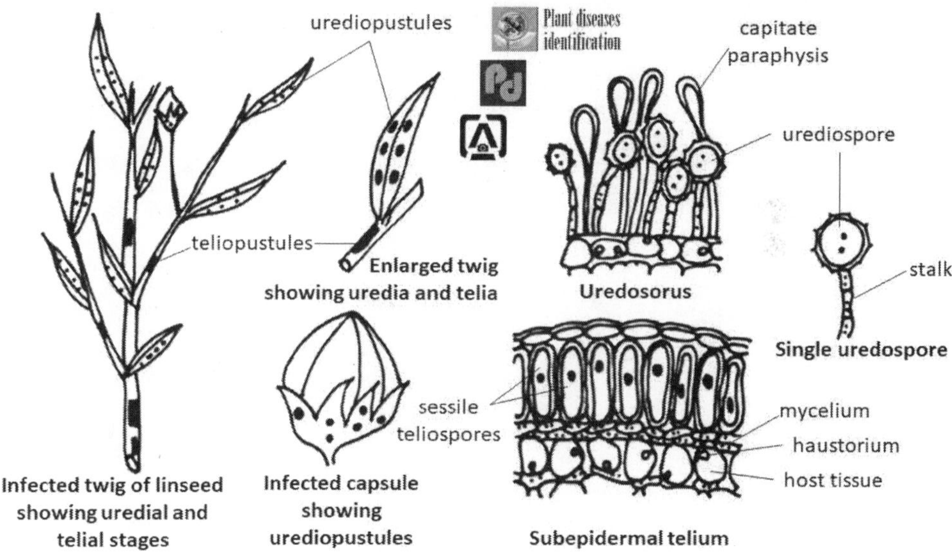

Figure 20.2: Rust of Linseed (*Melampsora lini*) Showing Symptoms, Uredospores and Teliospores

Management

1. Early on sowing, burning and destroying of contaminated plants.
2. Rotating crops and controlling the spread of the disease involves removing or burying the flax waste.
3. Avoidance of excessive nitrogenous manures is application.

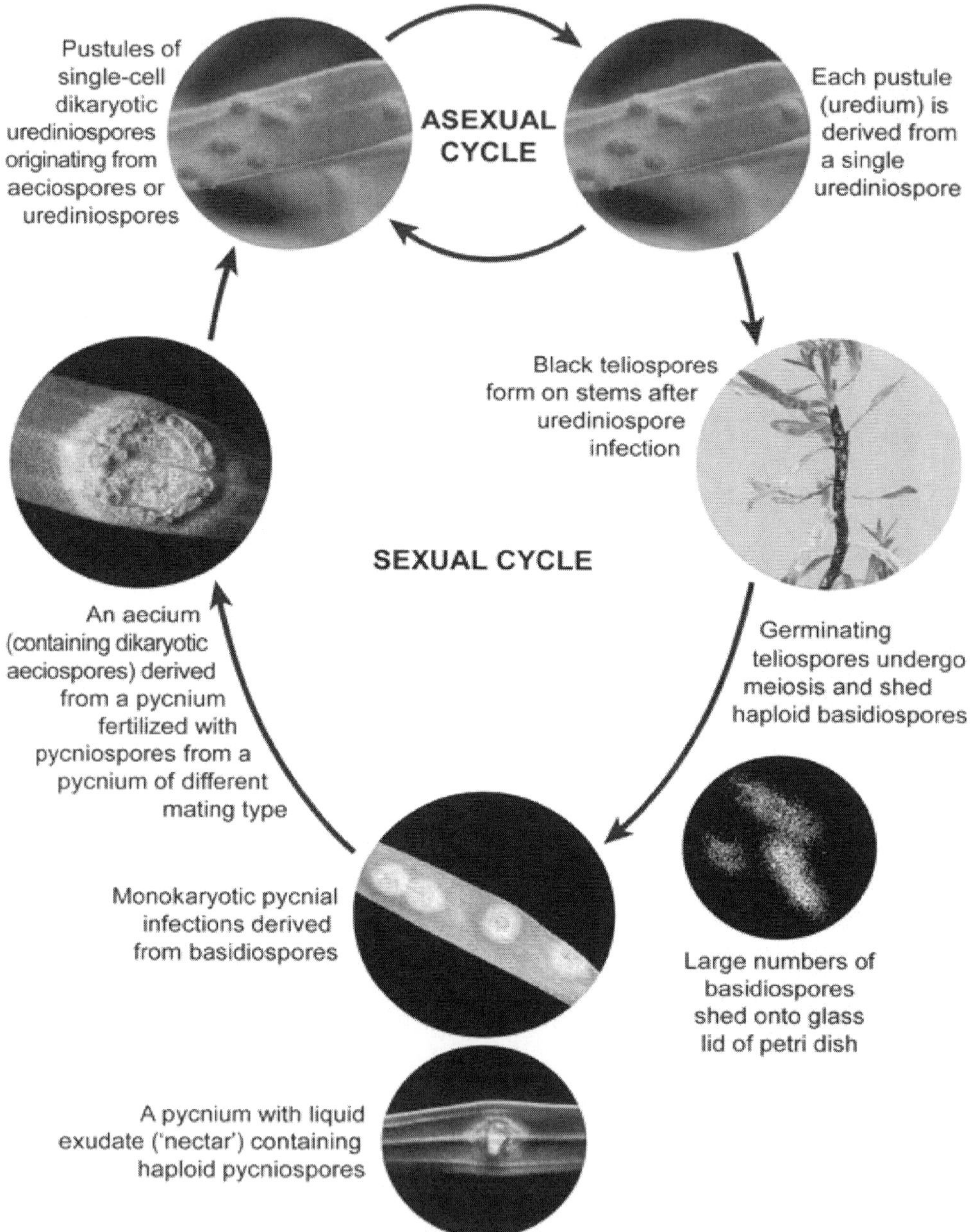

Figure 20.3: Life Cycle of Flax Rust (*Melampsora lini*).

4. The following varieties have been suggested as rust-resistant *viz.*, Jawahar-7, J-23, R-552, Jawahar-17, LC-185, LC-54, Garima, K-2, Kiran, Himalini, Janki, Surabhi (KL-1), Meera, Nagarkot, Rashmi, Sheela, Padmini and Sweta.

5. Seed treated with Oxycarboxin @ 0.1 per cent.
6. Application of Thiram 75 per cent WS or Carbendazim 50 per cent WP at a rate of 2.5 grams per kilogram of seeds.
7. Spraying Hexaconazole 5 per cent EC or Mancozeb 75 per cent WP at 2.5 gm/lit of water on the affected area.

2. Fusarium Wilt of Linseed

Economic Importance and Occurrence

The wilt disease of linseed caused by *F. oxysporum* f. sp. *lini* (Bolley) Snyder and Hansen is found throughout the world's where linseed crop producing. It is one of the most severe linseed diseases were first reported by Luggar (1890) from Minnesota, USA. He establishes that the disease was spread from infected to uninfected fields via water or old straw. In India, Madhya Pradesh was the first state to report it, then Maharashtra and Rajasthan. Every part of the country where linseed is grown has now reported cases of the disease. The timing of the attack has an important effect on the amount of damage. Under ideal circumstances, yield loss ranges from 80 per cent and 87 per cent was recorded.

Causal Organism

Linseed wilt disease is caused by *Fusarium oxysporum* f. sp. *lini* which belongs to Kingdom- *Fungi,* Division- *Ascomycotina,* Class-*Sordariomycetes,* Order- *Hypocreales,* Family-*Nectriaceae,* Genus- *Fusarium,* Species-*oxysporum,* and Forma specialis- *lini.* The fungus has similarities with other Fusarium wilts in terms of morphology. The septate, branching, intracellular mycelium develops into stroma with various shades. Conidiophores have branches on them. They originate in the sporodochia. Hyaline conidia are typically 1-2 celled. Pathogen are produced both micro as well as macro conidia. The dimensions of macroconidia are 21.0-53.0×2.4-5.6µm, while micro conidia measure 4.8-14.4×2.2-4.8µm. Microconidia are rare, and conidia are fusiform to falcate, 3-septate, hyaline to light pink in mass, with short, branching conidiophores that usually form in erumpent sporodochia (Saharan *et al.,* 2005).

Symptoms

Plants may be attacked by the pathogen at any stage, from seedling through maturity, when the surrounding temperature is typically high, the three weeks after seeding is usually when immature seedlings are affected. After this, the immature seedlings wither and eventually die off. The disease shows up on the leaves of older plants as tiny, poorly defined dark green regions. If the plants are still standing, they eventually die because their leaves wither, their branches droop, and so on. Partial wilting, where one side of the plant becomes brown while the other side stays green, is also occasionally seen. Splitting of the stained basal stem and roots reveals the vascular tissues brownish discoloration. The mycelium of pathogens is widely distributed in the discolored, diseased vascular tissues. Wilted plants top frequently curve downward to create a "shepherds crook".

Figure 20.4: A. Wilt Symptom; B. Macroconidia.

Epidemiology and Disease Cycle

Fusarium oxysporum f. sp. *lini* is both soil and seed borne disease. Due to its high competitive saprophytic capacity, these soil-borne organisms have been found to survive for a number of years even without a host (Arora, 1980). The fungus attacks immature plants via the roots and mostly grows in xylem vessels. High temperatures as well as low humidity have been main contributors to the disease development and the emergence of resistance in the majority of flax varieties (Tisdale, 1916, 1917). The disease is seed-borne by nature as well, it spreads to another season via seed. Temperature and soil moisture are key factors when the disease first appears. The best temperature range for infection is between 25 and 28 °C, with an ideal of 24 °C. Even in susceptible cultivars in India, the damage is limited at adverse soil temperatures of 12 °C and 38 °C (Gupta, 2002).

Management

1. Deep ploughing in the summer and rotation of crop follow minimum three years.
2. The inoculum of fungi is reduced by soil solarization and soil amendments with neem or mustard cake.
3. Plants that have been inoculated with arbuscular mycorrhizal (AM) fungus exhibit wilt resistance.

4. Seed treated with biological agents like *Trichoderma viride* and *T. harzianum* @ 4 g/kg of seed.
5. Cultivating resistant varieties such LC-185, K2, Himalini, LC-54, Jeevan, Kiran, Jawahar -23, Meera, R-552, Padmini, Sheela, Sheetal, Rashmi, Surabhi, *etc.*
6. Soil treatment with carbendazim at 0.1 per cent or COC at 0.25 per cent, or seed treated with thiram or carbendazim at 2.5g/kg seed.

3. Alternaria Blight

Economic Importance and Occurrence

Three *Alternaria* spp. including *Alternaria lini* (Dey), *Alternaria linicola* Groves and Skolko and *Alternaria infectoria*, are responsible for the disease known as Alternaria blight or black bud in linseed. In Indian environments, *A. Lini* predominates. The first occurrence of the disease was documented in Kanpur, followed by Gorakhpur, both in UP (Dey, 1933). Afterwards, reports from IARI-Delhi, Punjab and Jabalpur. Now, reports of it have come from every area where linseed is grown. In regions with persistently high humidity, particularly in Indo-gangetic tracts, it is a grave illness. Losses could range from 27.9 to 59.6 per cent as a result of the illness. Buds and capsules are most severely impacted. Up to 90 per cent of the crop might be lost due to bud infection (Chauhan and Srivastava, 1975). In Uttar Pradesh, it is a devastating disease that can reduce crop yields by up to 40.6 per cent.

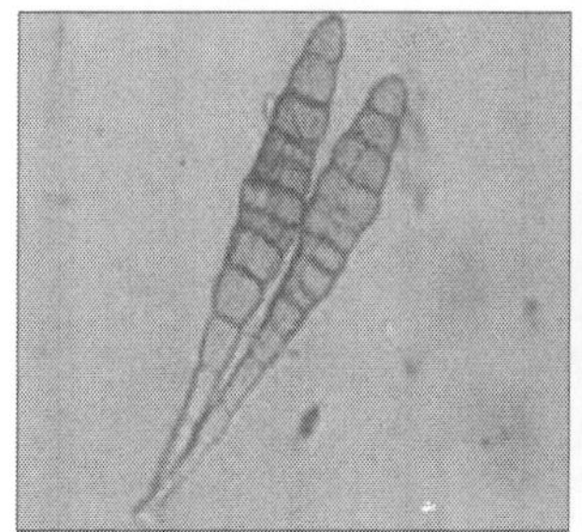

Conidia of *A. Linicola*

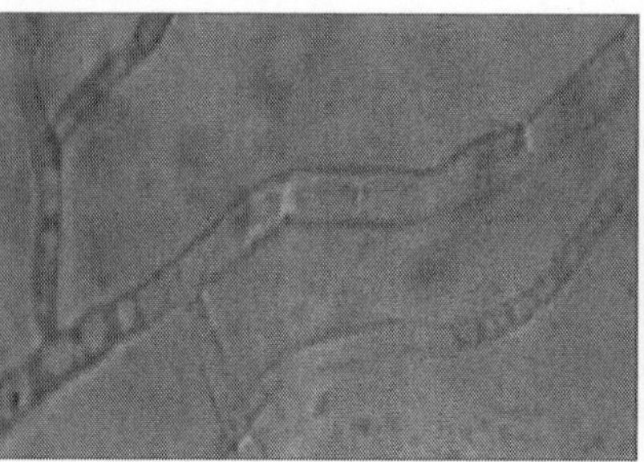

Mycelium of *Alternaria* spp.

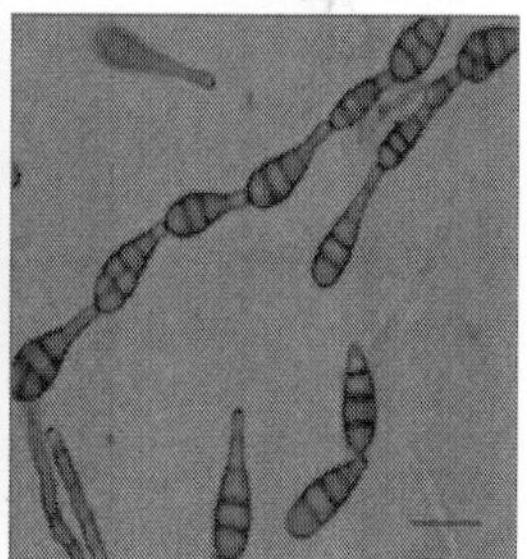

Conidia of *A. lini*

Figure 20.5

Causal Organism

This disease is caused by fungi *Alternaria lini* and *A. linicola*. The fungus belongs from Kingdom-*Fungi*, Division- *Ascomycota*, Class- *Dothideomycetes*, Order- *Pleosporales* and Family- *Pleosporaceae*. When examined under a microscope, infected leaves and capsules revealed septate hyphae, conidiophore, conidia, and sporadically chlamydospores. High magnification microscopic analysis of conidia showed the presence of longitudinal and transverse septa.

It was noticed that some *Alternaria* spp. conidia had long beaks, while other species had short or no beaks. Additionally, it was noted that conidia with a long beak were more common than those with a short or absent beak. Based on the morphology of the conidia, conidiophores, hyphae and patterns of conidia production, fungal species associated with linseed leaves and capsules were identified as *Alternaria linicola* Groves and Skolko and *Alternaria lini* Dey.

Symptom

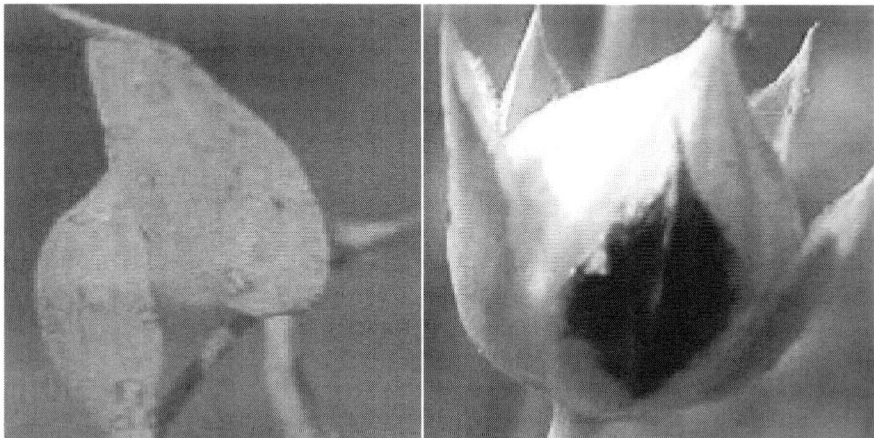

Figure 20.6: Symptoms of Alternaria Blight Produced on Leaves and Pods of the Linseed.

Approximately two to three weeks after sowing on cotyledons, the first symptom of disease develops. On the bottom leaves of linseed, Alternaria blight symptoms first showed as a black spot that subsequently grew in size to take on a round, oval, or irregular form. In extreme circumstances, the leaves actually get blighted and dry out. Near the base of the calyx, which enlarges as it crosses over to the buds and pedicel, dark brown blotches start to emerge on floral parts. When the buds don't open, all the flower parts, like the petals, sepals, and other floral parts, completely die and become a powdery black mass. Young buds that have been attacked are unable to grow and dry out. Sometimes, diseased plants with only half of their capsules have been seen. On the stem, the symptoms first emerged as a light brown, linear area with a darker edge that later grew more linear and deeper in colour.

Epidemiology and Disease Cycle

A. linicola survived as thick-walled chlamydospores in hyphal or conidial cells on infected linseed stem detritus that was either above ground or buried under it. When conidia grew on this substrate under favorable conditions, they were not only viable but damaging to early linseed seedlings. Infected stem debris increased the number of diseased seedlings that merge from infected seed, especially if the debris was visible on the soil's surface rather than underground. The most favorable climatic conditions for the disease's severity and the pathogen's growth were determined to be a temperature range of 26 to 33 °C and high relative humidity (85 to 90 per cent). Disease development is restricted if relative humidity (RH) is less than 75 per cent. The amount of seed containing inoculum had a favorable relationship with the disease incidence on seedlings at the time of emergence from infected seed, but the proportion of seedlings that emerged had an inverse relationship with the disease occurrence of *A. linicola* on the seed.

Management

1. Early sowing of crop.
2. Adopting crop rotation with non-host crop.
3. Recommended dose of fertilizer application and maintain crop density.
4. Intercropping linseed with wheat (4: 2 row ratio).
5. Seed treatment with *Trichoderm aviride* @ 4 g kg^{-1} seed.
6. Foliar spray propiconazole @ 0.2 per cent or mancozeb @ 0.25 per cent considered the most cost-effective and efficient way to manage this disease.
7. Sprays containing Neem leaf extract (5 per cent w/v) were also shown to be highly efficient at reducing the severity of the disease and improving the seed yield.
8. To effectively manage this disease, seed treatment with Bavistin @ 2 g/kg or Iprodione @ 1.5 g/kg removes the seed-borne inoculum, and then foliar sprays Dithane M-45 @ 0.2 per cent during blossoming, with additional sprays applied every 15 days.

4. Powdery Mildew

Economic Importance and Occurrence

Powdery mildew of linseed is the most devastating disease after the rust. The majority of linseed types that are resistant to rust are extremely susceptible to powdery mildew. Early development stages where the disease initially presents itself result in a very large yield decrease. A plant that has been seriously diseased produces extremely poor seed and fiber. The incidence of linseed powdery mildew (*O. lini* Skoric) was first reported in Yugoslavia (Skoric, 1926). The leaves, which have been covered in dense, powdery masses, show indications of drooping and twisting before drying off. If the ailment develops during an early development stage, substantial decreases in yield result. Grain shriveling might be caused by severe infection. In tolerant and susceptible varieties, yield losses were observed ranging from 13.81 to 31.49 percent, respectively.

Causal Organism

The powdery mildew of flax or linseed is caused by *Oidium lini* fungus which belongs to Kingdom- *Fungi,* Phylum- *Ascomycota,* Class- *Leotiomycetes,* Order-*Erysiphales,* Family- *Erysiphaceae,* Genus- *Oidium,* and Species- *lini.*

Symptoms

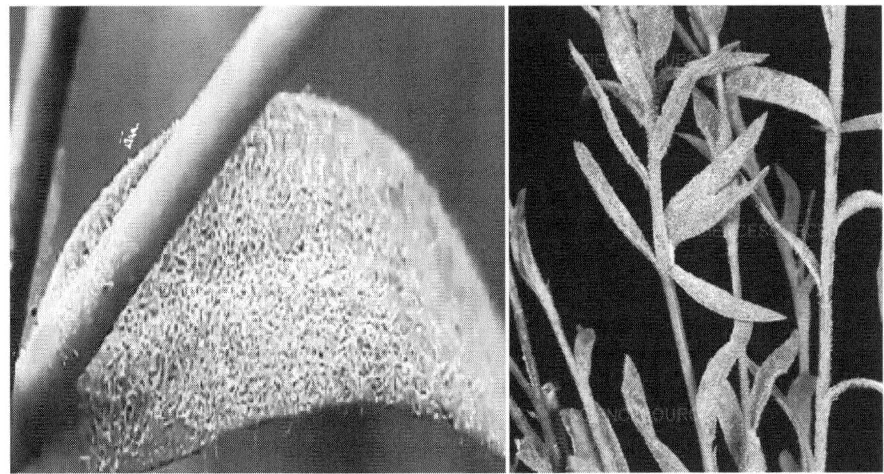

Figure 20.7: A. Powdery Appearance on Leave Surface; B. Diseased Plant.

The first indication of the disease is little, white, floury patches that appear on the upper surface of the leaves. Then, these patches grew to cover the stem, leaves, and capsules on the entire surface of the plant. Leaves with severe infection wither and die. Early disease infections can cause the linseed plant to become defoliated, which lowers the quantity and quality of seeds that grow to cover the entire surface of the plant, including the stem, leaves, and capsules. It results into decreased photosynthesis and amplified respiration. The leaves enclosed in thick powdery masses appear symptoms of twisting and drooping before drying up. Although, severely impacted plants do not die, the yield is significantly decreased.

Disease Cycle

The pathogen survives by developing cleistothecia in soil-borne sick plant debris. Climatic circumstances are favorable during the next crop season, the asci and ascospores are discharged, starting the main disease. The disease begins to appear in the country's center and southern regions, when temperatures rise early, before spreading to the north. The predisposing parameters are a temperature range between 20 and 25 °C, humidity below 65 per cent, and less rainfall. Additionally, conidial germination of *Oidium lini* started after two hours at 30°C and 40 or 80 percent relative humidity (RH) and appresoria production peaked at 17.00 per cent after 14 hours at 70 per cent relative humidity.

Management

1. Use of resistant varieties for cultivation of crop.
2. Early sowing of crop could prevent infection and disease development.
3. Weekly foliar applications of Wettable Sulphur @ 0.25 per cent, Karathane (0.2 per cent), Bavistin (0.2 per cent), Sulfex (0.3 per cent), or Sulfex (0.2 per cent), depending on the intensity of the disease.

5. Anthracnose

Across the world, this disease affects both the seed and flax varieties in humid cool climates.

Causal Organism

Colletotrichum linicolum

Symptoms

The symptoms are distinguishing of in general of the anthracnose disease. Round, zonated, sunken brown patches called cankers first appear on the cotyledons and then, in chilly, damp conditions, spread to the cotyledons and stem apex. Pre-emergence or post-emergence seedling blight may occur, with the latter one situation usually causing a stem canker near the soil surface level. During the growth season, stem cankers and leaf spots are common in high-moisture environments. In typically moist conditions, the center of the spots shows a pinkish mass of spores. On the capsules, brown spots emerge, and the seed develops with fewer apparent spots. On the developed lesions of the stem, the acervuli development develops.

Disease Cycle

The main causes of infection are contaminated seed and crop residue. When infected seeds are sowed, lesions form on the cotyledons and release spore masses. These cause the crop to obtain the disease. Plant debris decomposes slowly in cold temperature zones, and if the trash is not buried fully in the soil, fungi can survive on it for up to two years. The primary inoculum is provided by conidia formed on the cankers on seedlings. Whenever during a good weather condition, secondary infection happens. In determining the severity of the disease during the growing season, the percentage of infected seed plays a crucial role, as the amount of secondary inoculum created will correspond with the amount of primary inoculum. The disease damages fiber and reduces stands.

Management

1. Sowing of disease free seeds.
2. Removed the plant debris.
3. Use the crop rotation.

4. Utilize resistant varieties for cultivation.
5. Applying @4g/kg of *Trichoderma viride* for seed treatment.
6. Thiram 75 WP, Carbendazim 50 WP, Carbendazim + Thiram, and Mancozeb 75 WP at 2.5 g/kg seed were used for seed treatment.

6. Damping Off

Causal Organism

Sclerotinia fuckeliana

Symptoms

When the crop is in the seedling stage, the disease of the damping off pathogen begins to appear as basal leaf and stem rot. On the stem, Subsequent lesions show as a brown or tan spot soaked in water, which may turn gray as it dries. A distinguishing feature of the fungus is the widespread gray-brown sporulation that develops on grew older, diseased tissue. Rotting of plant products during harvest or storage results in major losses. Blight of the stem, leaves, and blossoms can also happen and cause dieback. On woody plant components, a canker development can occasionally be seen.

Disease Cycle

The fungus over winter sclerotia on or within infected tissues that have fallen to the ground, as well as mycelium in live or dead plants. Sclerotia germinate in the spring or early summer, producing cylinder stalks that end in a tiny cup or disk shaped apothecium. Over a period of two to three weeks, the apothecia release an enormous number of ascospores into the atmosphere.

7. Macrophomina Stem Rot

This disease caused by fungus *Macrophomina phaseolina*. The sclerotial stage *i.e., Rhizoctonia bataticola* cause seed and seedling rot. *R. Bataticola* (*M. phaseolina*) was found associated with linseedroots. Under the right moisture and temperature circumstances, the disease can develop at any stage of development. In the Himachal Pradesh district of Kangra, it has been observed to induce death in the linseed crop of up to 80 per cent.

Symptoms

Symptoms of the early growth stage include partial or total turning brown on of the root system, stunting, and yellowing and withering of the leaves. Later on, a gloomy brown discoloration of the roots and stem's base is seen. The shout of the roots peels off, and the bark, wood, and basal portion of the stem all have tiny sclerotia.

Management

1. Use of resistant cultivars like NP-5.

2. *Pseudomonas fluorescens* @ 10g/kg seed was used for seed treatment.
3. Seedling or sprout mortality was reduced by 0.25 per cent of Captan.

8. Cercospora Leaf Spot

The disease was found to be severe for the first time in India in the Varanasi area of Uttar Pradesh. *Cercospora linicola* is the pathogen that causes the disease. The disease originates with small circular leaf spots on the leaf surface. The pathogen persists in crop debris through mycelium.

9. Pasmo Disease

Pasmo disease is caused by *Septorialinicola*. It is a harmful disease of flax. Depending on the severity of disease, it losses 50 per cent to 70 per cent yield. Delayed infections reduce the seed production. At the peak of infection it cause weak and break fiber; infected bolls either do not produce any seeds or produce tiny, useless ones.

10. Root Rot, Foot Rot and Seedling Blight

Linseed root rot caused by *Sclerotium rolfsii*.

References

Ajithkumar, K., Biradar, S. A., Rajanna, B., Singh, P. K., and Goud, I. S. (2017). Management of linseed powdery mildew through soil and foliar application of sulphur. *The Indian Society of Oilseeds Research*, pp. 116-120.

Anonymous (2022). https://oilseeds.dac.gov.in/Linseed.aspx accessed on 11 December 2022.

Arya, H. C., and Prasada, R. (1948). Alternaria blight of linseed (Doctoral dissertation, IARI, Division of Mycology and Plant Pathology, New Delhi).

Chauhan, L. S. and Srivastava, K. N. (1975). Estimation of loss of yield caused by blight disease of linseed. *Indian Journal of FarmScience*, **3**: 107-109.

Dey, P. K. (1933). An Alternaria blight of linseed plant. *Indian Journals of Agricultural Science*, **111**: 881-896.

Gill, K. S. (1987). Diseases of linseed. In: Linseed, *Indian Council of Agricultural Research*, New Delhi, pp. 317-341.

Gupta, V. K. and Paul, Y. S. (2002). Disease of field crops. Indus Publishing Co., New Delhi.

Hirua, M. (1924). On the flax anthracnose and its causal fungus, *Colletotrichumlini*(West.) Toch. *Jour. Bot. Japan*. **2:** 113–132.

Hora, T. S., Chenuleu, V. V. and Munjal, R. L. (1962). Studies on assessment of losses due to *Melampsoralini*. *Indian OilseedsJournal*, **6**: 196.

Lawrence, G. J., Dodds, P. N., and Ellis, J. G. (2007). Rust of flax and linseed caused by *Melampsoralini*. *Molecular Plant Pathology*, **8**(4): 349-364.

Pethybridge, G. H. and Lafferty, N. A. (1918). A disease of flax seedling caused by a species of *Colletotrichum*and transmitted by infected seed. *Sci. Proc. Roy. Dublin. Soc.*, **15:** 359–384.

Prasada, R. (1940). Aecidial stage of the rust of linseed. *Current Science*, **9**: 328-329.

Prasada, R. (1948). Studies in linseed rust, Melampsoralini (Pers.) Lév. in India. *Indian Phytopathol.*, **1**: 1-18.

Singh, S. K., Singh, D. and Kumar, A. (2015).*Plant disease and their management*. Kalyani Publishers, Ludhiana (Punjab)-141008.

Singh, J., Singh, P. K. and Srivastava, R. L. (2017). Diseases of linseed (*Linum usitatissium* L.) in India and their management a review. *The Indian Society of Oilseeds Research*, 52.

Singh, J. and Singh, J. (2002). Effect of fungicidal sprays against foliar diseases of linseed. *Ann. Plant Protec. Sci.*, **10:** 169–170.

Tisdale, W. H. (1916). Relation of soil temperature to infection of flax by *Fusariumlini*. *Phytopath.*, **6:** 412–413.

Chapter 21

Diseases of Sunflower Crop

Tanya Rathore[1], Kirti Vardhan Pandey[2], Manvendra Sharma[3] and Puskar Shukla[1]*

[1]*Department of Plant Pathology, Chandra Shekhar Azad University of Agriculture and Technology, Kanpur – 208 002, Uttar Pradesh*
[2]*Department of Agricultural Meteorology,* [3]*Department of Seed Science and Technology, Acharya Narendra Deva University of Agriculture and Technology, Kumarganj, Ayodhya – 224 229, Uttar Pradesh*
**e-mail: tanyarathore959@gmail.com*

Introduction

In India, the sunflower has become increasingly popular as a source of oil. It is particularly well-suited for regions with low rainfall due to its short growing season. Its appeal lies in its high oil content and superior oil quality. In areas where other oilseed crops like groundnuts cannot thrive, sunflowers are grown either with irrigation or relying on rain. However, in some regions of India, this crop faces challenges from fungal and viral infections, which hinder its expansion. Cultivated sunflowers are annual plants that vary in size but are typically tall, upright, mostly unbranched, and have a coarse texture. They possess deep root systems with thin lateral roots. While the taproot can extend up to 3 meters into the ground, its diameter decreases as it goes deeper. Sunflowers thrive in various climate conditions and are often considered insensitive to day length, as they can flower under a wide range of daylight hours. Compared to maize, which doesn't use light efficiently at high light levels, sunflowers are believed to be more efficient in utilizing light.

Diseases like leaf spot, rust, head rot, downy mildew, powdery mildew, root rot, and basal rot are significant threats to sunflower crops in India, causing substantial economic losses globally. Sunflowers serve as an excellent backup crop in case of adverse weather conditions, making them an important choice for diversification. They can be sown when the primary crop fails. The main goal is to provide comprehensive information on these diseases and facilitate effective disease management strategies.

Table 21.1: List of Sunflower Diseases their Causal Agent

Sl.No.	*Name of Disease*	*Causal Agent*
A. Fungal diseases		
1	Leaf spot	*Alternaria helianthi*
2	Rust	*Puccinia helianthi*
3	Powdery mildew	*Erysiphe cichoracearum*
4	Head rot	*Rhizopus spp.*
5	Downy mildew	*Plasmopara halstidii*
6	Verticillum wilt	*Verticillum albo-atrum*
7	Base rot/Collar rot	*Sclerotium rolfsii*
8	Septoria leaf spot	*Septoria helianthi*
9	White blister blight	*Albugo tragopogonis*
10	Charcoal rot	*Rhizoctonia bataticola*
11	Basal rot	*Sclerotium rolfsii*
B. Bacterial diseases		
1	Bacterial leaf spot	*Pseudomonas syringae* pv. helianthi
2	Bacterial wilt	*Pseudomonas solanacearum*
C. Nematode diseases		
1	Root knot	*Meloidogyne* spp.
2	Stunt	*Tylenchorhynchus nudus*
D. Viral diseases		
1	Sunflower mosaic	Cucumber mosaic virus
2	Chlorotic mottle	Sunflower chlorotic mottle virus

Fungal Diseases of Sunflower

1. Rust

Economic Importance and Occurrence

This disease is predominantly found in India and is caused by a fungus. It has been increasingly observed in both the Rabi and Kharif seasons. When combined with Alternaria blight of leaves, it can lead to yield losses of up to 40 per cent. The severity of the disease is typically higher when the daytime temperature ranges from 25.5°C to 30.5°C, with relative humidity (RH) between 86 per cent to 92 per cent, which intensifies the rust attack.

Causal Organism

This disease is attributed to *Puccinia helianthi* Schw., a member of the Fungi kingdom, Basidiomycota phylum, Pucciniomycetes class, Pucciniales order, Pucciniaceae family, Puccinia genus, and helianthi species. It's an autoecious fungus. The uredospores are round or obovoid, with a dark cinnamon-brown wall. Teliospores are elliptical or oblong, two-celled, chestnut brown, and have a smooth, long, and colorless pedicle.

Symptoms

The initial symptoms involve small reddish-brown pustules (uredia) covered with a rusty-colored dust appearing on the undersides of leaves. Later, the infection spreads to all leaves and even extends to the green parts of the head, causing them to turn yellow and dry up. The pycnial and aecial stages are typically found on volunteer crops and remnants of the previous crop.

Figure 21.1: Rust of Sunflower.

Epidemiology

The initial infection arises from various sources, including sporidia, teliospores, aeciospores, or surviving uredospores on volunteer plants, and it is transmitted through wind. Subsequent infections occur through the continuous production of uredospores during the crop season.

Disease Cycle

The pathogen's survival occurs in volunteer sunflower plants and in plant debris in the soil, primarily in the form of teliospores. The disease is then disseminated through wind-carried uredospores originating from infected crops.

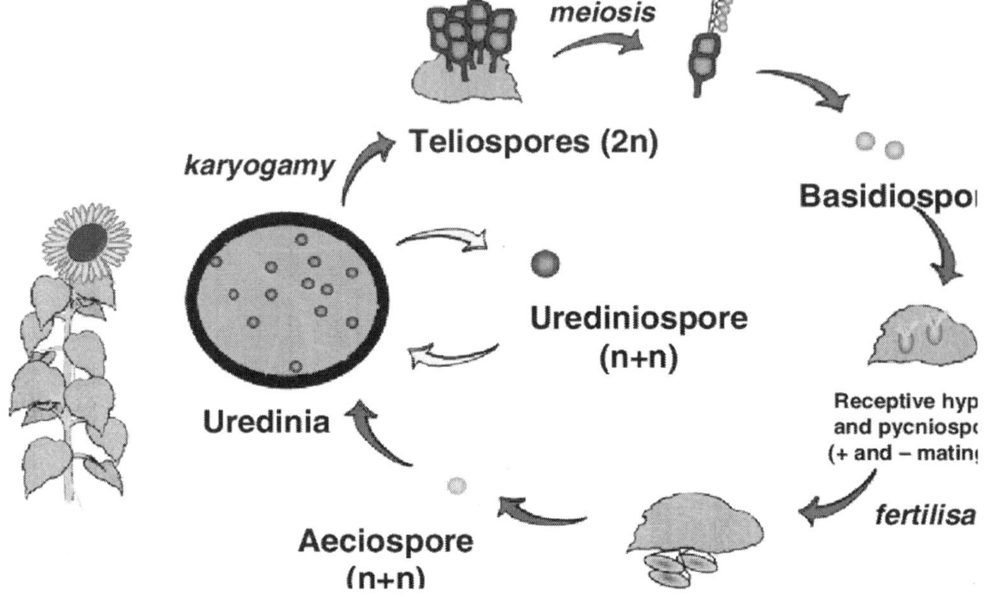

Figure 21.2: Rust of Sunflower (*Source*: Sendal *et al.*, 2006).

Management

1. Eliminate volunteer seedlings.
2. Dispose of the previous crop residue and apply chemical treatments like 0.1 per cent Dithane M-45 or Orthiovit.
3. Always use disease-free and healthy seeds.
4. Avoid sowing during periods of high air temperature, short-term rains, and dews.
5. Opt for resistant varieties such as Admiral and Advent.
6. Continuously implement crop rotation measures.
7. Apply two or three foliar sprays of Dithiocarbamate fungicides like maneb, mancozeb, or zineb at a concentration of 0.2 per cent to 0.3 per cent to effectively control the disease.
8. Use sulfur-based fungicides like lime-sulfur, wettable sulfur (0.2 per cent), or sulfur dust (15 kg/ha).
9. Employ systemic fungicides like benodanil and oxycarboxin for effective control.

2. Leaf Spot

Economic Importance

Leaf spot diseases are widespread in various varieties, particularly during the cool and rainy seasons. These diseases are prevalent in almost all varieties and tend to rapidly spread during the rainy season. Cases of this ailment have been reported worldwide, with a notable presence in India. Under certain favorable conditions, Alternaria leaf blight is responsible for more than 80 per cent of yieldloss.

Causal Organism

The disease is caused by *Alternaria helianthi* Hansf., classified under the Kingdom Fungi, Division Ascomycota, Class Dothideomycetes, Order Pleosporales, Family Pleosporaceae, Genus Alternaria, and Species helianthi. It is also referred to as Drechslera (Helminthosporium) helianthin. The fungus produces numerous conidia on standard nutrient media. The conidiophores are 25-80 × 8-11, typically up to 5 septate, cylindrical, pale grey-yellow, and may be straight or curved, simple or branching. Conidia range from 40-110 x 13-28, generally rounded at both ends, and can be cylindrical to long-ellipsoid. They are often 1 to 11 septate, frequently displaying transverse or longitudinal septa, and occasionally both. Conidia do not develop in chains.

Symptoms

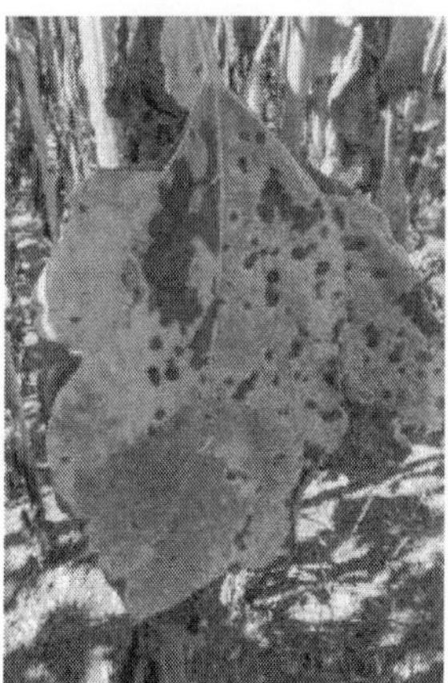

Figure 21.3: Leaf Spot of Sunflower (*Source*: extensionpublications.unl.edu).

The symptoms manifest as brown pustules on the leaves and can also affect the stem, sepals, and petals. Lesions on the leaves are dark brown with pale margins surrounded by a yellow halo. Over time, the pustules enlarge in size with concentric rings, often irregular in shape. Several pustules may merge to form larger, irregular lesions that result in drying and deformation.

Epidemiology

Teliospores of Puccinia helianthi predominantly inhabit fallen leaves or the upper layer of the soil. While there is a possibility that the fungus could persist as teliospores or uredospores on the surface of seeds, there is currently no definitive proof of seed transmission. In the remnants of the previous crop, namely, uredia, sporidia, pycnia, and aecia may manifest on volunteer seedlings, and it's highly probable that the pathogen survives through such volunteer plants. Sporidia, teliospores, aeciospores, or uredospores may potentially germinate on plants at higher elevations and get carried by air currents, leading to primary infections. The recurring production of uredospores during a crop season results in secondary infections. Epidemiological research conducted in Coimbatore suggests that the intensity of rust attacks increases when daytime temperatures range between 25.5°C and 30.5°C, and relative humidity levels are between 86 per cent and 92 per cent

Disease Cycle

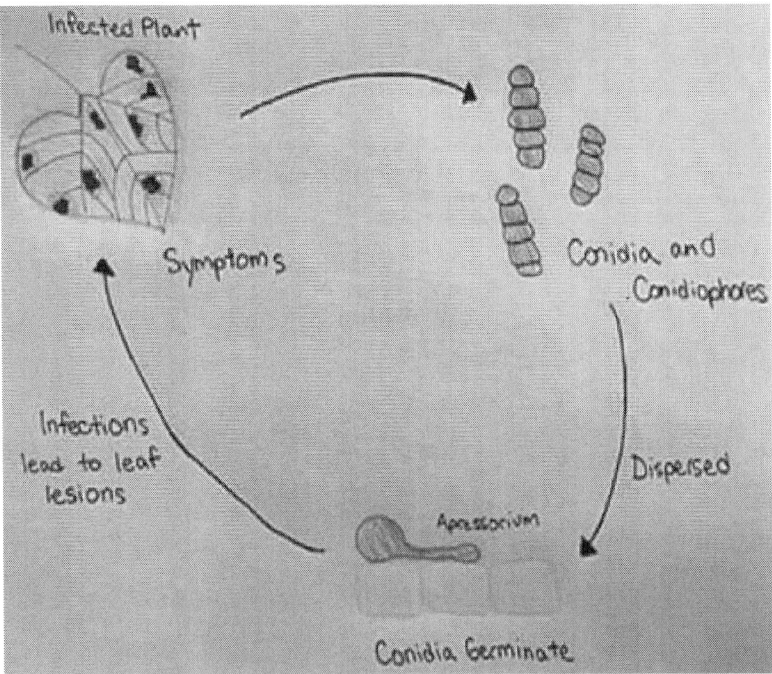

Figure 21.4: Disease Cycle of Leaf Spot of Sunflower.

Management

1. Remove and dispose of the infected plants.
2. Conduct thorough ploughing during the summer season.
3. Employ clean farming practices and ensure field cleanliness.

4. Implement crop rotation strategies.
5. Maintain appropriate plant spacing to prevent nutrient and space competition.
6. Apply 0.2 per cent Dithane M-45, thiovit, or copper-based fungicides on the 30th, 40th, and 50th days.

3. Head Rot of Sunflower

Economic Importance and Occurrence

In Maharashtra, this disease becomes a notable issue during rainy conditions, leading to decreased crop yields. To minimize losses due to biodeterioration, it is advisable to store sunflower kernels with a moisture content not exceeding 6.5 per cent and a relative humidity level below 70 per cent.

Disease Symptoms

Figure 21.5: Head Rot of Sunflower (*Source*: sunflowernsa.com).

Causal Organism

The primary culprit responsible for this disease is *Rhizopus* spp. This fungus features a coenocytic, brown mycelium that produces both sporangia and sporangiophores. These sporangia generate spherical, flattened spores with colors varying from brown to black. The wind then disperses these spores, leading to the infection of new host plants.

Etiology

To initiate an infection, it is essential for the flower head to be damaged. Even when the disease-causing agent is present, injury occurring before flowering or during the early stages of head growth is less likely to promote infection. As the flower head matures, it becomes increasingly susceptible to infection. Notably, the most evident rotting occurs at the soft dough stage. Depending on the maturity of the head at the time of Rhizopus infection and the development of rot, seed development is significantly impeded. It has been observed that *Heliothis armigera* larvae make flower heads more prone to infection. Additionally, the presence of birds visiting the heads in search of seeds is positively linked to the spread of the disease.

Management

1. To reduce the disease incidence, it is recommended to spray Fenthion (0.1 per cent) and Thiovit (0.2 per cent) when the head begins to develop.
2. Starting from the end of the flowering stage, it is advisable to use a combination of a compatible pesticide and fungicide to efficiently control the disease.
3. Whenever feasible, take precautions to prevent injury to the flower head to protect it from infection.

4. Downy Mildew

Economics Importance and Occurrence

Up until 1984, the illness was unknown in India. It was reportedly present in Maharashtra in a significant form in 1985. *Plasmopara helianthi*, the causative fungus, is thought to have originated in North America. The seed trade has quickly spread it. Initial findings in Maharashtra point to the potential for a significant drop in yield brought on by the disease.

Symptoms

Figure 21.6: Downy Mildew of Sunflower (*Source*: ipmimages.org).

The disease exhibits several distinctive signs, including seedling damping off, systemic infection, localized foliar lesions, and basal root or stem galls. In susceptible sunflower seedlings, damping off occurs when the pathogen in the soil attacks, typically in cool and moist conditions. Early indications include yellowing of the initial pair of true leaves. Young seedlings with high humidity can develop a whitish downy mildew on their cotyledons. Sunflower plants with systemic infections are severely stunted, with entirely chlorotic upper leaves. Extensive downy growth of the fungus coincides with epiphyllous yellow patches. The stem becomes fragile, and secondary rootlets decrease. Infected plants typically produce sterile flower heads with rare seed development. Symptoms may not appear until head formation in older plants. Infected leaves may show small, greenish-yellow angular lesions, and microbial growth on the lower leaf surface, although these localized infections are generally of minor importance.

Etiology

All of the damaged tissues have fungus oospores, but the infected roots have the most oospores of any organ. The pathogen lives on oospores on seeds in systemically infected plants or in the leftovers of the previous sunflower harvest. Transmission of seeds has been amply proved. The majority of the time, when oospores are wet, zoospores are produced. Infection is caused by zoospores. Rarely do plants become systemically infected as a result of seed transmitted illness. To a considerable part, soil-borne inoculum is to blame for the disease's emergence in its severe form. Important parameters in this context include the amount of rain during the seedlings' early growth stages and their age at the time of infection. Heavy soils with limited drainage make infections more likely.

Disease Cycle

The pathogen resides in the soil as asexually produced oospores. The round to oblong thick walled oospores are formed in root tissue shortly after infection and serve as survival structures. They are highly resistant to adverse environmental conditions and remaining dormant in soil for up to 10 years. When soil as saturated and cool (54-57 °F) oospore can germinate and form zoosporangia that release motile zoospores that swim through water and facilitate infection to the healthy plants.

Management

1. The best control is provided by the use of resistant cultivars.
2. The disease can also be somewhat controlled by choosing planting locations and properly removing agricultural wastes that have been infected
3. Seed treatment with metalaxyl at a rate of 3 g per kg of seed provides efficient control.

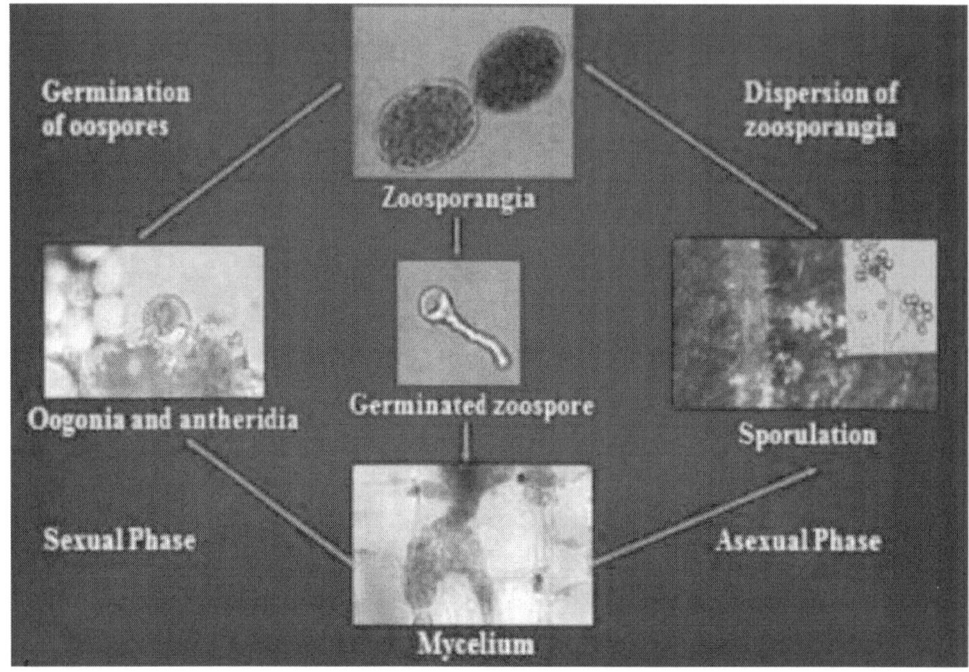

Figure 21.7: Disease Cycle of Downy Mildew of Sunflower.

5. Root Rot or Charcoal Rot

Economic Importance

Particulary in the arid regions, where there has been recorded seed weight loss of 30 to 46 per cent, charcoal rot is important economically.

Causal Organism

Rhizoctonia bataticola (Pycnidial stage: *Macrophomina phaseolina*) is the causal agent of this disease. The fungus produces a large number of black, round to irregular shaped sclerotia. The pycnidia are dark brown to black with an ostiole and contain numerous single celled, thin walled, hyaline and elliptical pycnidiospores.

Symptoms

The pathogen is seed-borne nature and primarily causes seedling blight and collar rot in the initial stages. After flowering, mature plants begin to exhibit symptoms as well. The affected plants have drooping leaves, and spots of death appear. The lower stem and root bark is torn, and several sclerotia are present along with it. The lowest part of the stem also develops tiny, dark-colored pycnidia.

Figure 21.8: Root Rot of Sunflower (*Source*: extension.sdstate.edu).

Favourable Conditions

Temperature increases and moisture stress favor the development of the disease.

Disease Cycle

Sclerotia and pycnidia allow the pathogen to persist in soil and agricultural residue that have been infected. The pathogen is transmitted by seeds and is the main cause of infection is Conidia carried by the wind spread. Secondary spread through rain splash, irrigation water, and tools can all disseminate the soil-borne sclerotia.

Management

1. Closer planting of the seedling should be avoided.
2. Optimum nutrition should be provided to maintain the plant growth.
3. Whenever the soil becomes dry and the soil temperature rises then irrigation should be provided.
4. Seed treatment with *Trichoderma viride* formulation at 4 g/kg seed.
5. In endemic areas long crop rotation should be followed.
6. Spot drench with Carbendazim at 500 mg/lit

6. Powdery Mildew

Economic Importance and Occurrence

Erysiphe cichoracearum caused powdery mildew has been regarded as a disease with significant economic impact. It has been documented from several regions of the world and affects the majority of commercial cultivars being grown. The fungus survives the winter on diseased plant components and in debris like crumpled leaves. It might create cleistothecia, which are resting structures resistant to the hard winter weather. Small black spots will show up among the white, powdery patches.

Causal Organism

Erysiphe cichoracearum

Symptoms

White powdery development on the leaves is a symptom of thedisease. On the top surface of older leaves, there is a white to grey mildew. Black pinhead-sized growths can be seen in regions with white mildew as the plant ages. Affected leaves develop more luster, curl, get chlorotic, and eventually die.

Figure 21.9: Powdery Mildew of Sunflower (*Source*: enviroselects.com).

Favorable Conditions

The disease is more under dry condition to the end of the winter months.

Management

1. Removal of infected plant debris.
2. Complete field and crop sanitation.
3. Chose early varieties of sunflower.
4. Application of Karathane or calixin 1lit/ha or wettable sulphur 2 kg/ha is found effective in reducing the disease incidences.

7. Basal Rot

Economic Importance and Occurrence

Due to its wide host range, rapid growth rate, and sclerotia as extremely useful survival structures, the soil-borne pathogen Athelia rolfsil causes significant economic losses throughout the world. Around the world, hosts include more than 500 species from more than 100 plant groups, including sunflower, legumes, maize, sorghum, cucurbits, wheat, numerous weeds, and ornamental flowers. A. rolfsit, though frequently a minor disease of sunflower, is particularly harmful in tropical and subtropical areas, where it may also affect rotational weeds and seasonal crops. Hot weather favors A. rolfsil greatly, and as temperatures rise, A. rolfsil distribution and severity have been rising in places like Australia.

Causal Organism

Sclerotium rolfsii.

Symptoms

40 days after planting, the disease symptoms start to show. The sickly look of the diseased plants makes them easy to spot. Due to the disease infection, plants dry out. White or brownish white fungal colonies have colonized on stem lowest part. In the worst scenarios, the plants wilt and die. Near the ground, dark brown lesions develop on the stem base and cause it to wither. There are several sclerotia present.

Favourable Conditions

1. Infection occurs in the crop in the month of July and August.
2. The fungus survives through sclerotina in soil and plant debris.

Figure 21.10: Basal Rot of Sunflower (*Source*: extension.sdstate.edu).

Disease Cycle

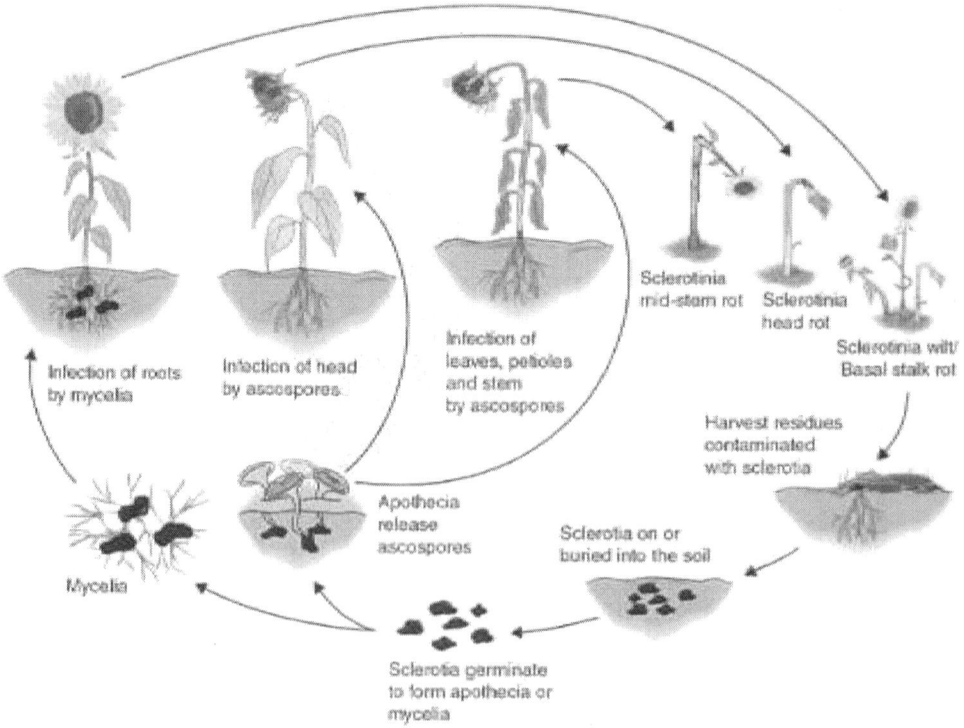

Infection of roots by mycelia

Infection of head by ascospores

Infection of leaves, petioles and stem by ascospores

Sclerotinia mid-stem rot

Sclerotinia head rot

Sclerotinia wilt/ Basal stalk rot

Harvest residues contaminated with sclerotia

Apothecia release ascospores

Mycelia

Sclerotia on or buried into the soil

Sclerotia germinate to form apothecia or mycelia

Figure 21.11: Disease Cycle of Basal Rot of Sunflower.

The base of the stem may have sclerotia, which are the overwintering structures of the fungus. Sunflower stem with severe rotting and white fungus growing at its base. A close up of Sclerotinia basal rot showing the sclerotia, the overwintering structures of the basal rot fungus. The *sclerotinia* pathogen survives in the soil and in infected plant debris as sclerotia. The sclerotia germinates after stimulation from sunflower roots and decays the roots. The fungus then progresses through the stem causing above ground symptoms. Contact with infected roots will spread the disease among the neighboring plants. Although the sclerotia can remain viable in the soil for a few years, rotating into non-hosts 3-5 years will eventually reduce the inoculum in the soil.

Management

1. Collect and remove plant debris.
2. Total crop and field sanitation.
3. Deep summer tillage.
4. The use of resistant breeds.
5. Use Trichoderma to lessen wilt on seeds and soil.
6. Before planting, apply and include the fungus *Coniothyrium minitans* since it invades the soil and eliminates the pathogen.
7. *Pseudomonas fluorescens* or *P. putida* strains applied to the seeds protect sunflower seedlings from Sclerotinia infection.
8. Applying captan or thiram to seeds at a rate of 3 g/kg of seed.
9. Spray 3 g of chestnut compound per litre of water on the plant's base.
10. *Trichoderma harzianum* 10 g/kg soil addition, 0.2 per cent carbendazim seed treatment, and 0.2 per cent carbendazim spraying on seedlings that are at least 15 days old.

References

Anonymous (2023). Wikipedia.https://en.mwikipedia.org

Bhattacharya, U.P. (2020). Plant pathology at a glance. *Kalyani Publishers*, Pp 203.

Rangaswami, G. and Mahadevan, A. (2022). Diseases of crop plants in India, Fourth Edition. *PHILearningprivateLimited, Delhi*, pp. 339-340.

Satyagraha Maiti, M. R., Hegde and S. B. Chattopadhyay (1988). Handbook of annual oilseed crops. *Oxford and IBHP PublishingCo. Pvt. Ltd.*, pp 176-183.

Chapter 22

Diseases of Cotton Crop

Dipake Shital*, S.K. Biswas and Tanya Rathore

Department of Plant Pathology, Chandra Shekhar Azad University of Agriculture and Technology, Kanpur – 208 002, Uttar Pradesh
****e-mail: shitaldipake14@gmail.com***

Introduction

Cotton (*Gossyipum* spp.) is a tropical shrub and belongs to family Malvaceae. Cotton is the most important fiber crop in India. Cultivation of cotton is mostly in central and western part of India. Cotton is also known as "White Gold". Cotton fiber supply a raw material for textile industries. India has first rank in cotton production but now-a-days yield of cotton is reduced due to attack of many diseases on cotton causes economic losses. The cotton plant has upright, branching branches, alternately arranged leaves and spectacular, big blooms with five petals that are primarily white or creamy-purple in colour. With enough moisture, cotton may grow to a depth of 60 cm and has a tap root system. The fruit resembles a capsule and has 3-5 leathery valves. Cotton seeds have an ovoid shape and are covered in long threads or fibers. Cotton seeds can be as large as 10 X 6 mm and as heavy as 80 mg. It possesses a tough seed coat that is cuticle covered. There are 50 recognized cotton species, only four of which are grown commercially worldwide such as *Gossypium hirsutum, Gossypium herbaceum, Gossypium barbadense* and *Gossypium arboretum*; the rest are found growing wild in tropical and subtropical region (Gotmare *et al.*, 2000). The length, maturity, strength and micro-naire of the fiber vary among the species of cotton. In India, major devastating diseases such as Wilt, Bacterial blight, Anthracnose and Root rot affect cotton fiber quality. The cash crop, cotton affected by fungal, bacterial, nematode, viral diseases and physiological disorder as mentioned below:

A. Fungus Diseases

1. Wilt: *Fusarium oxysporum* f. sp. *vasinfectum*
2. Root rot: *Rhizoctonia bataticola* or *Rhizoctonia solani*
3. Verticillium wilt: *Verticillium dahliae*
4. Anthracnose: *Colletotrichum capsici*
5. Grey or Aerolate mildew of cotton: *Ramularia areola*
6. Ascochyta blight: *Ascochyta gossypii*
7. Cercospora leaf spot: *Cercospora gossypina*
8. Helminthosporium leaf spot: *Drechslera gossypii*
9. Alternaria leaf spot: *Alternaria macrospora*
10. Rust: *Phakospora desmium*
11. Sooty mould of cotton: *Capnodium* spp.
12. Powdery mildew: *Leveillula Taurica*
13. Stem rot: *Chaetomium amphitrichum*
14. Leaf spot: *Pestalotia gossypii*
15. Damping off: *Pythium debaryanum* and *Phytophthora parasitica*
16. Boll rot: *Nematospora nagpuri*
17. Sclerotium wilt: *Sclerotium rolfsii*

B. Bacterial Disease

1. Angular leaf spot or Black arm or Blight: *Xanthomonas axonopodis* p.v. *malvacearum*

C. Viral Disease

1. Cotton leaf curl – Geminivirus

D. Phytoplasma Disease

1. Cotton virescence

E. Nematode Disease

1. Reniform nematode disease – *Rotylenchulus reniformis*

Physiological Disorder

1. Red leaf : Nitrogen deficiency
2. Crinkle leaf: Excess of Manganese
3. Dwarfing and yellowing disease of cotton: Sulphur deficiency

Major Diseases of Cotton in India

1. Wilt of Cotton

Economic Importance and Occurrence

This disease originated from Mexico or Central America. In India, this disease first recorded by Evans at Nagpur, India in 1908 (Rangaswamy *et al.*, 2022). During the agricultural season, the disease manifests in severe forms in heavy soils with soil temperatures between 20 °C and 30°C.

Causal Organism

Fusarium oxysporum f.sp. *vasinfectum* belong to Kingdom-*Fungi*, Division-*Ascomyvotina*, Class- *Sordariomycetes*, Order- *Hypocreales*, Family- *Nectriaceae*, Genus- *Fusarium*, and Species- *oxysporum* (Kirck *et al.*, 2008).

Symptoms

At seedling stage, yellowing and browning of the cotyledons, base of petiole shows brown central ring after that death and falling off of the affected part. In cotton plants, leaves lose their turgidity colour, start wilting and finally drop off. Discoloration of leaves start from margins to mid-ribs. Wilting may complete or partial. The most remarkable and diagnostic symptoms of the disease is the browning and blacking of the vascular tissues. The transverse section of the affected branch reveals a discolored ring in the woody tissue of the plant.

Figure 22.1: Wilt of Cotton (*Source*: APS, cottoninc.com, Solverchem.com).

Favourable Condition

Soil temperature (20-30 °C), soil moisture, soil texture greatly influences the incidence of the disease. Hot and dry periods for long duration followed by rains provide favourable condition for maximum wilt. Black soil is more susceptible.

Disease Cycle

The fungus enters through wounds and insect injuries in host roots then fungus multiplies and move towards xylem vessels where abundance of mycelium formed which interrupting the transport of nutrients towards above ground plant parts. Therefore, Plant growth stunted. Continuous break of nutrients flow to plants leads to wilting. Fungus induces tyloses production in xylem vessels which prevents the movement of nutrients. The toxin Fusaric acid responsible for wilt symptoms.

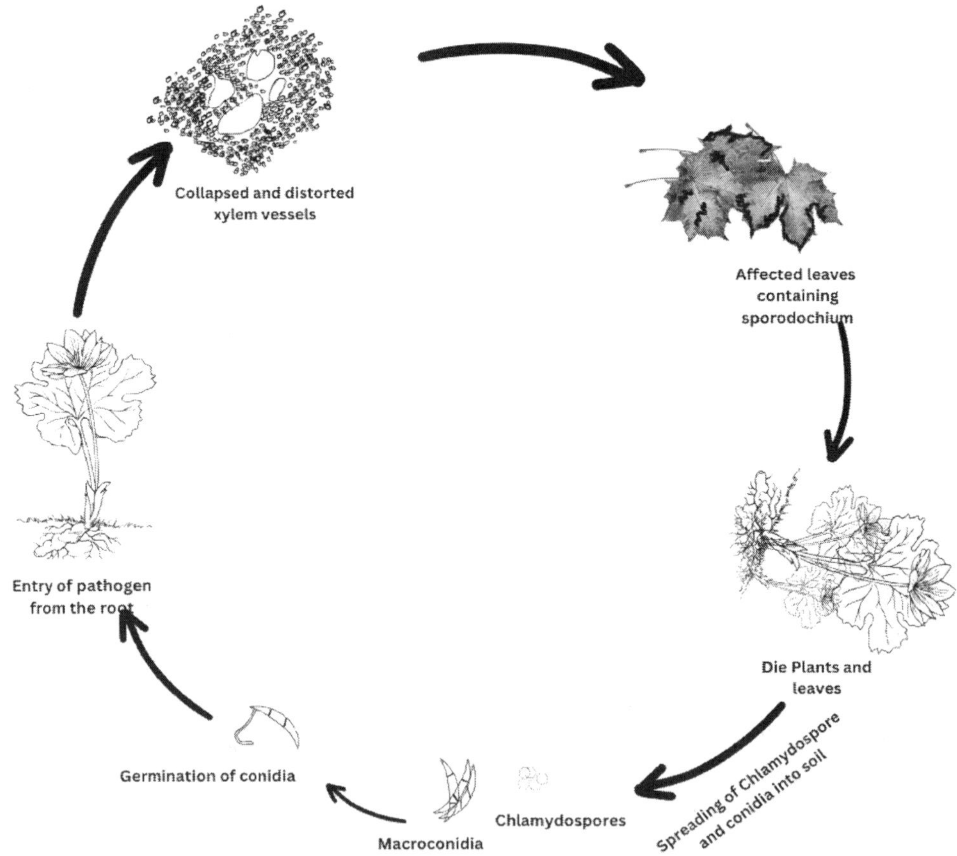

Figure 22.2: Disease Cycle of Cotton Wilt.

Management

1. Follow soil solarization to reduce soil born pathogen.
2. Grow resistant variety is the best method to control cotton wilt.
3. Apply FYM to growth of antagonistic organisms which suppress pathogen in soil.

4. *G. arboretum* and *G. herbaceum* varieties of cotton are susceptible to wilt hence avoid these varieties in cultivation.
5. Application of high dose of potash with balanced nitrogenous and phosphatic fertilizers control the disease.
6. Seed treatment with *Bacillus subtilis* + *Trichoderma harzianum* at 4 g/kg seed.
7. Seed treatment with chemical Raxil MD Extra –W (Tebuconazole + Metalaxyl + Imidacloprid insecticide) 2 g/Kg of seed effective against wilt causing pathogen.

2. Root Rot of Cotton

Economic Importance and Occurrence

One of the worst diseases affecting cotton, widespread throughout India, is root rot. It also happens in the United States, certain regions of Africa, Venezuela, and other nations. In India, it mostly affects Karnataka, Tamil Nadu, Madhya Pradesh, and Uttar Pradesh also occurs in Punjab and Gujarat in more severe forms.

Causal Organisms

Rhizoctonia bataticola or *Rhizoctonia solani* belong to Kingdom- *Fungi*, Division- *Basidiomvcotina*, Class- *Agaricomycetes*, Order- *Cantharellales*, Family- *Ceratobasidiaceae*, Genus- *Rhizoctonia*, and Species- *solani/bataticola* (Kirk *et al.*, 2008).

Symptoms

Figure 22.3: Root Rot of Cotton
(*Source*: Jayant H Meshram, 2021).

The main sign is a sudden, total wilting of the plants. In the field, the disease appears on leaves in concentric circles. The entire root system of the affected plants may be discovered to be decaying if they are removed and investigated, with the decay being more complete on the lateral and thinner roots. Only the tap root remains attached to plants that have been severely damaged; the rest of the root system has decomposed and been retained in the soil. The tap root is typically sticky and damp, with shredded bark and yellowed woody fibers. The woody parts of severely harmed plants may turn dark and brittle. On the surface of the woody tissue and on the rotting bark, the fungal sclerotia frequently appear as tiny black dots. Shredding of the bark and the development of sclerotial bodies may occur when the rotting spreads to the basal stem in some circumstances.

Favourable Condition

1. **Soil moisture:** 15-20 per cent
2. **Soil temperature:** >35°C
3. **Soil type:** Alluvial and clays

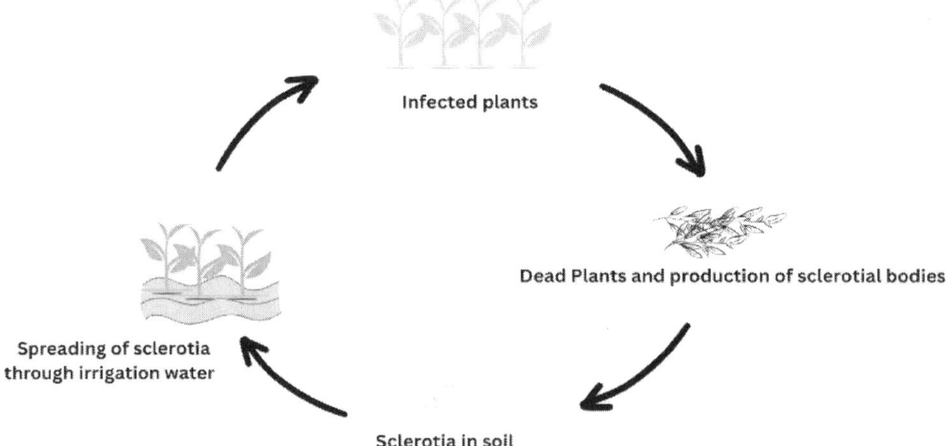

Figure 22.4: Disease Cycle of Root Rot of Cotton.

Disease Cycle

The two fungus can live saprophytically for many years on soil organic materials and are facultative pathogens. A soil moisture percentage between 15-20, a soil temperature above 35 °C during plant growth, and specific soil types like alluvial and clay are predisposing factors for the beginning and spread of the illness. The fungi take control inside the roots as soon as the plants are susceptible to infection. The hyphae multiply quickly in the cortical tissues, extend up to the pith, and occasionally enter the xylem arteries. There is an increased buildup of calcium and iron in damaged plant roots, and several harmful compounds are thought to contribute to the symptoms of wilt. On the host root in the soil, the fungus persist saprophytically. They also proliferate on numerous more hosts, including pathogenic vegetable and oil seed crops. If the predisposing parameters are favorable and the cotton crop is replanted in the soil, the viruses will reawaken and infect the plants. It is uncertain what function the pycnidial and perfect stages of the corresponding fungi play in the persistence of the illness.

Management

1. Resistant varieties not succeeded because disease found on Indian and American cottons.
2. Adjusting the sowing dates.
3. Mixed cropping with sorghum and *Phaseolus aconitifolius* L. are commonly used.
4. Drenching early symptomatic plants and surrounding plants with *Trichoderma* spp. 1 per cent WP @ 50 g/10 litres of water.
5. Seed treatment with Carboxin 37.5 per cent + 37.5 per cent DS@ 3.5 g/ kg of seeds for root rot.

3. Anthracnose of Cotton

Economic Importance and Occurrence

Globally, this cotton disease is present. Butler first mentioned it in Bihar, India, in 1918; since then, it has been mentioned in practically all cotton-growing regions. The illness might be so severe in some years that the yield is significantly reduced.

Causal Organisms

Colletotrichum capsici belong to Kingdom- *Fungi*, Division- *Ascomyvotina*, Class- *Sordariomycetes*, Order- *Glomerealles*, Family- *Glomerellaceae*, Genus- *Colletotrichum*, and Species- *capsici* (Kirk *et al.*, 2008).

Symptoms

Small reddish circular patches on the cotyledons and main leaves are the first signs of the illness when it affects seedlings. The collar region of the stem is girdled by the lesions, which results in wilting and death in the seedlings. The fungus infects the stem of mature plants, causing it to split and lose its bark. Boll spotting is the most prevalent symptom and initially manifests as round, slightly depressed, water-soaked reddish-brown lesions on the bolls. These lesions enlarge, darken, and have red borders. A number of spots may combine to completely cover the bracts and other parts of the boll. The boll infections are pervasive, affecting the seeds and lint, turning the lint from yellow to brown and frequently causing it to decay. Such bolls are tiny, malformed, and dry too soon. The seeds are discolored brown and the lint changes into a pile of brittle fiber when it is cut apart and studied.

Figure 22.5: Anthracnose of Cotton (*Source*: Plantix, agritech.tnau.ac.in).

Disease Cycle

The disease typically infects seeds and becomes active when the seed is sown, and conidia that are air- and soil-borne are used for secondary dispersion. The fungus spreads quickly in the field during wet conditions, resulting in boll rot, stem lesions, and seedling wilt. When the bolls are harvested, the seeds from healthy bolls become contaminated by the infected seeds from diseased bolls. Investigation needs to be done into the role that the fungus' ideal stage plays in the persistence of the disease. *Aristolochia bractiata* Retz., and *Hibiscus diversifolius* Jacq., two widespread weeds in cotton fields, are also infected by *C. capsici*, and they may help the virus survive during the off-season.

Management

1. The disease can also be controlled by removing the field's collateral weed hosts.
2. Sowing seeds that have been viable for two years is also beneficial because the fungus can survive on the seed for roughly a year.
3. The reduction of humidity surrounding the boll caused by defoliation with calcium cyanamide spray at a rate of 30 kg per hectare aids in preventing boll infection.
4. Spray 1 per cent Bordeaux mixture after boll formation to control disease.
5. Seed treatment with conc. Sulphuric acid, sulphur or other dressing agent to remove of inoculum on seeds.
6. Spray the crop at boll formation stage with Mancozeb 2kg or Copper oxychloride 2.5 kg or or Carbendazim 500g/ha

4. Angular Leaf Spot/Black Arm of Cotton

Economic Importance and Occurrence

Blackarm of cotton is one of the most serious bacterial plant diseases. Atkinson from the United States first characterized it in 1891, but it wasn't 1901 that American E. F. Smith that the disease's bacterial origins were established. Since that time, reports of the illness have come from South America, Egypt, Sudan and other African nations as well as the USSR, China, Australia, Sri Lanka, Pakistan, and India. In Tamil Nadu, India, in 1918, the illness was first noted. Since then, it has been thoroughly investigated by numerous researchers in numerous locations throughout India as well as in other nations. Maharashtra, Karnataka, Andhra Pradesh, Tamil Nadu, and Madhya Pradesh are states that are extremely concerned about this disease.

Causal Organisms

Xanthomonas axonopodis pv. *malvacearum* belong to Domain- *Bacteria*, Phylum- *Pseudomonadota*, Class- *Gammaproteo bacteria*, Order- *Xanthomonadales*, Family- *Xanthomonadaceae*, Genus-*Xanthomonas*, Species- *axonopodis* and Pathovar - *malvacearum* (Young *et al.*, 1992).

Symptoms

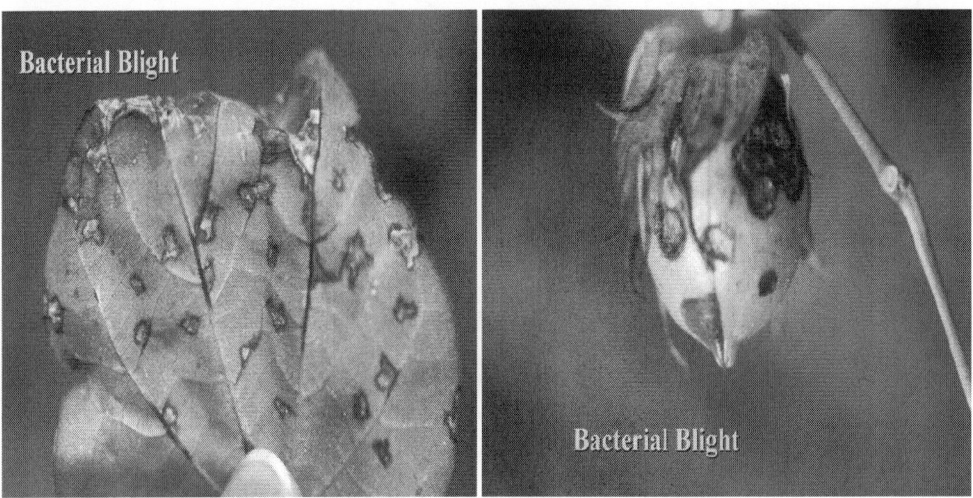

Figure 22.6: Bacterial Leaf Blight Symptoms on Leaves and Ball.

The first sign of the disease appears on the seedling as water-soaked circular lesions that subsequently get larger and turn irregular and brown, distorting and withering the cotyledons. Seedlings that are severely harmed wilt and wither. Lesions appear on the blade and petiole of newly created leaves as a result of the illness spreading there as well. Water-soaked lesions first emerge on the leaves of mature plants, where they develop into angular growths that are bound by veinlets and turn from dark brown to black. The size of the spot might range from 1-5 mm, and it's common for numerous spots to combine to produce erratic, black patches that wither and dry. Dark brown to black, linear, deep lesions are produced on the stem and fruiting branches, girdling the stem and branches to promote early leaf drop, giving the characteristic blackarm or dry black twigs standing in the place of healthy plants. The bacterium also affects the bolls, generating lesions that are water- soaked and later develop into irregular, dark-brown-black patches. These stains are deeply embedded, and the bacteria spread throughout the lint fiber and seed inside the boll. The degree of harm varies according to the plant's infection stage and the weather. If the bacterial illness affects young plants in addition to the leaves and stems, there may be an 80 percent yield loss in some places.

Favourable Condition

High humidity, Moderate temperature at 28°C important factor for disease development. Moisture availability is most important for secondary infection for first 48hours. Disease development occurs mostly when frequent rains followed by cloudy weather. Dry and hot weather decreases disease development.

Disease Cycle

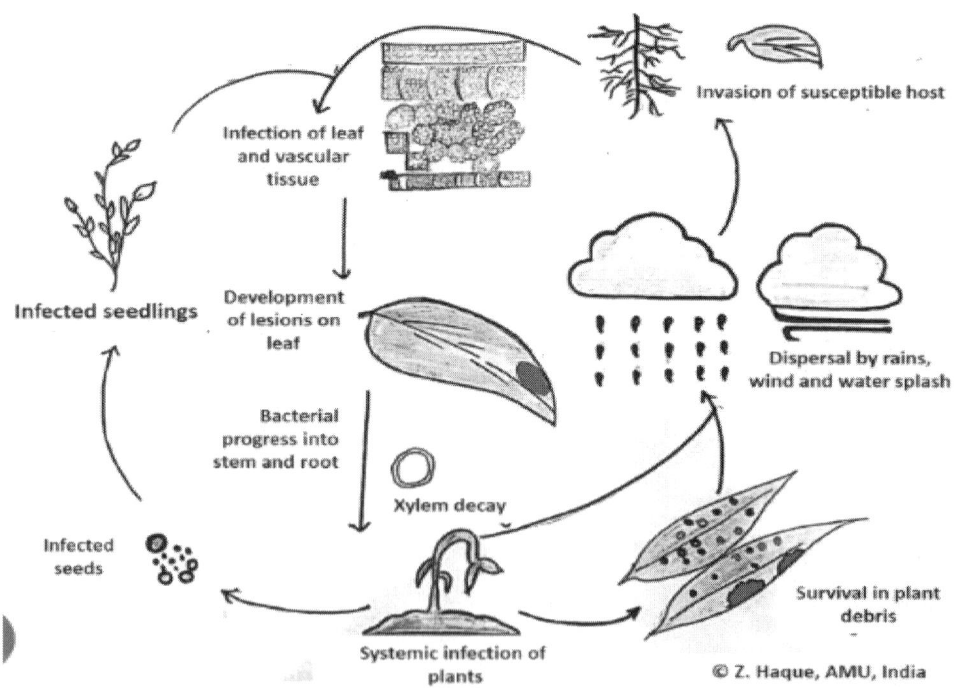

Figure 22.7: Disease Cycle of Angular Leaf Spot/ Black Arm of Cotton (*Source*: Ziaul haque, 2022).

Seed-borne bacteria are the principal source of the initial infection. On the seed coat's fuzz, the bacteria continue to exist as a slimy mass. After the seed is sown, it quickly multiplies and infects the seedling. When seeds are involved, the bacteria may enter the seed coat via the micropyle and infect the seedlings. Primary infection can also come from stray plants that sprout up as a result of bolls that ripen too early. However, the bacteria that the host tissue introduced to the soil only survive for a brief time. Wind, water, and other physical and biological forces that enter the host through natural openings or accidental or insect-caused wounds are the primary means by which the bacteria are transmitted out in the field. This bacteria infect *Thurbaria thespesioides* A. Gray, *Eriosendron anfructuosum* DC. and *Jatropha curcas* L. these collateral hosts for perpetuation.

Management

1. Use resistant varieties.
2. Seed treatment with conc. Hydrochloric acid or benlate.
3. Soaked seeds in 100 ppm of streptomycin sulphate or Agrimycin overnight to remove internal infection.
4. Secondary spread suppress by spraying 500 ppm of streptomycin sulphate at intervals of 3 weeks in fields.
5. Spray fungicides as Dithane or Copper oxychloride @ 2-2.5 kg/ha.

5. Cotton Leaf Curl

Economic Importance and Occurrence

Diseases are becoming more common thus has almost universally become a serious issue for both Bt and non-Bt cotton hybrids in the north zone. Indian subcontinent, have seen significant losses to their cotton harvests due to CLCuD, which has grown to be a serious danger to cotton output. Only the northern cotton-growing region of the nation is subject to the disease, which has expanded across the whole north zone over an area of roughly 1.9 million hectares.

Causal Organisms

The cotton leaf curl viral disease (CLCuD) is caused by complexes of begomoviruses (family Geminiviridae) that infect cotton and have a monopartite genome with circular ssDNA that is closely linked to satellite molecules, specifically the alpha satellite and beta satellite. There are around 26 genera of weed, ornamental, and vegetable hosts for CLCuD, which aid in the virus' ability to spread from one plant to another plant and from one season to the next. CLCuD is spread by the whitefly, I (Gennadius) (Hemiptera: Aleyrodidae). It is a highly polyphagous pest that lowers the yield by direct feeding and the growth of black sooty mold as a result of honeydew excretion on leaves. Crop plants' ability to photosynthesize are greatly diminished by sooty mold, which also shades leaves and produces a financial loss.

Symptoms

On young leaves at the top of the plant and on the branches, the disease symptoms appear. Small and massive vein thickening and darkening are the first signs of the disease, which are visible from underneath the side of the leaves. The virus is transmitted by the whitefly, Bemisia tabaci, which can show symptoms as early as the seedling stage (about 30 days after planting) and at any stage thereafter. The CLCuD affected cotton plants exhibit distinctive upward/downward curling along with small vein thickening type (SVT), or the thickening and darkening of veins when viewed against sunlight. In addition to having larger dark green veins on their leaves, damaged plants also have translucent veins that are thicker than those on healthy plants. Some of the leaves form secondary, leaf-like structures known as enations that come from the nectaries. The infected plants continue to develop in a stunted manner, giving the appearance of dwarf plants with decreased inter nodal length.

Figure 22.8: A. Cotton Leaf Curl Infected Le; B. Downward Curling.

Management

1. For successful disease and insect-pest control, weeds and other alternative hosts are regularly monitored for the prevalence of illnesses, pests, and CLCuD vectors.
2. Keep the field, field bunds, and irrigation channel clear of weeds and weed-free since weeds deprive the cotton crop of nutrients, harbor sucking pests like whiteflies, and serve as a source of CLCuD inoculum.
3. Eliminate volunteer, ratoon, and sick cotton plants as well as weed hosts from fallow lands and areas near irrigation channels, canals, and bunds.
4. Use crop rotation to disrupt disease life cycles, lessen disease inoculum from soil-borne pathogens, and improve the fertility and health of the soil.

5. Avoid planting whitefly and CLCuD host crops next to cotton crops: To prevent the accumulation and spread of pests and diseases in cotton, avoid cultivating crops like cucurbits, bhindi, moong, arhar, castor, dhaincha, and kinnow orchard near cotton.
6. Avoid early and late sowing because they attract insect pests, especially late sowing, which is more susceptible to illnesses and insect pests like CLCuD.

References

Ahmad, S. and Mirza, H. (2020). Cotton production and uses. *Springer Nature Singapore Pte Ltd.* Pp 1-271.

Anonymous (2023). Wikipedia. https://en.mwikipedia.org

Bhattacharya, U. P. (2020). Plant pathology at a glance. *Kalyani Publishers,* Pp. 212-213.

Dastur, R. H., Asana R. D., Sawhney, K., Sikka, S. M., Vasudeva, R. S., Khan, Q. and Rao, V. P. (1960). Cotton in India: A monograph. *Indian Central cotton committee, Bombay,* Pp 164-209.

Gotmare, V., Singh, P. and Tule, B. (2000). Wild and cultivated species of cotton. In: Technical bulletin, vol **5**. Central Institute for Cotton Research, Nagpur.

Monga, D. and Sain, S. K. (2020). Integrated disease management in Cotton. AICRP on Cotton-Technical Bulletin 1/2020. ICAR-CICR, Regional Station, Coimbatore-641003 Tamil Nadu. Pp. 38.

Rangaswami, G. and Mahadevan, A. (2022). Diseases of crop plants in India, Fourth Edition. *PHI Learning private Limited, Delhi -10092. Pp 425-436.*

Sain, S. K., Banu, G., Valarmathi, P., Gawande, P., Nagrale, D. T., Hiremani, N. S., Monga, D., Khan, M. R. and Sharma, R. K. (2021). Important Disease and Nematode Problems In: M.R. Khan, Z Haque and F. Ahamad (Eds.) Diseases of Nationally Important Field Crops, Pp. 501-537.

Chapter 23

Diseases of Jute and Mestha Crop

Shivam Kumar[1], *Ravi Kumar*[1], *Gajendra Pratap*[1*], *Ram Kumar*[2], *and Puja Sharm*[3]

[1]*Department of Plant Pathology, 2Department of Entomology Chandra Shekhar Azad University of Agriculture and Technology, Kanpur – 208 002 Uttar Pradesh*
[2]*Department of Mycology and Plant Pathology, Institute of Agricultural Sciences, Varanasi, 221005 (U.P.)*
[3]*Department of Forestry, College of Agriculture, Jawaharlal Nehru Vishwa Vidhyalaya, Jabalpur – 482 002, Madhya Pradesh*
**e-mail: rkm46327@gmail.com*

Introduction

Jute, scientifically known as *Corchorus olitorius* L. and *C. capsularis* L. is commonly referred to as the "Golden fiber" crop. It is primarily cultivated in the eastern region of India, covering an area of approximately 0.77 million hectares. The annual jute production amounts to 10 million bales, with each bale weighing 180 kilograms and a dry fiber yield of 2,329 kilograms per hectare. Jute is predominantly grown as a crop before kharif in some states such as West Bengal, Bihar and Assam, contributing 81.29 per cent, 11.63 per cent, and 6.25 per cent, respectively, to the national jute fiber production (De, 2019). As the second most affordable and readily available fiber crop, jute is a valuable source of biodegradable and renewable lignocellulose fiber (Niu *et al.,* 2015). Bast fibers, which are primarily found in the stems or barks of plants, consist of interconnected sclerenchyma cells forming either small or large strands or tissue bundles, typically arranged in layers. In India, significant bast fiber crops include jute (*C. olitorius* and *C. capsularis*), Mesta (which includes *Roselle,*

Hibiscus sabdariffa and Kenaf, *H. cannabinus*), *Crotalaria juncea* (Sunnhemp) and *Boehmeria nivea* (Ramie) (sarkar and Gawande, 2016).

The production and productivity of these bast fiber crops are hindered by various abiotic and biotic stresses, with one particularly destructive factor being *Macrophomina phaseolina*. This pathogen poses a significant threat as it can attack on the entire plant at any stage of growth from germination of seed to harvesting of crop. Consequently, it leads to detrimental effects such as damping off, seedling blight, collar rot, leaf blight, stem rot and root rot (Roy *et al.*, 2008). Approximately 10 per cent yield loss has been reported, but when disease happen in its peak it may increase to 35-40 per cent loss. It becomes severe in hot (34 ±1°C) and humid weather (Mandal, 1990).

Major Diseases of Jute and Mesta

Table 23.1: Diseases of Jute Crop (*Corchorus olitorius* L. and *C. capsularis* L.)

	Fungal diseases	
1	Stem rot	*Macrophomina phaseolina*
2	Anthracnose	*Colletotrichum corchorum; C. gloeosporioides*
3	Black band	*Botryodiplodia theobromae*
4	Soft rot	*Sclerotium rolfsii*
5	Tip blight	*Curvularia subulata*
6	Stem gall	*Physoderma corchori*
7	Mildew	*Oidium* spp.
8	Sooty mould	*Cercospora corchori, Corynespora cassicola, Alternatia* spp.
	Bacterial Disease	
1	Hooghly wilt	*Ralstonia* (*Pseudomonas*) *solanacearum*
	Viral disease	
1.	Jute mosaic	A Begomo virus, family Gemini viridae, vector: Whitefly (Bemisia *tabaci*)
2.	Jute Chlorosis	A member of Tobra virus genus
3.	Yellow vein	A bipartite *Begomo virus*, vector: whitefly (*Bemisia tabaci*)
	Nematode Disease	
1.	Root Knot	*Meloidogyne incognita, M. javanica*

1. Stem Rot and Root Rot

Economic Importance and Occurrence

Stem rot is the most important disease of jute equally affecting both the species *viz.*, tossa jute) (*Corchorus olitorius*) and white jute (*C. capsularis*). It is common throughout the world's jute-growing regions. This illness is very serious in India, especially in the states of Assam, West Bengal, Bihar, and

Orissa. It drastically decreases the fiber's yield and quality. The disease causes an average yield loss of 10 per cent, but in severe cases, it can reach 35–40 per cent (Roy *et al.,* 2008).

Causal Organism

Macrophomina phaseolina (Tassi) is responsible for the severe occurrence of stem rot and root rot in jute, posing significant threats to both cultivated varieties of *C. olitorius* and *C. capsularis* (Roy *et al.*, 2008; De, 2013; Suryawanshi *et al.*, 2015). The pathogen belongs to Kingdom – *Fungi*, Phylum – *Ascomycotina*, Class- *Dothideomycetes*, Order- *Botryosphaeriales*, Family- *Botryosphaeriaceae*, Genus- *Macrophomina*, Species- *phaseolina* (Kirk *et al.,* 2008). M. phaseolina contains a variety of hyphae, including light- or dark-colored hyphae with septa (divisions) and hyaline (transparent) hyphae with thin walls. Hyphae branch at an angle to their mother hyphae and frequently exhibit constriction at the point of genesis. The fungus produces microsclerotia, which are compact masses of hardened mycelium. These microsclerotia are light brown at first, turning dark brown (to black) as they deteriorate. They can be round, oval, or oblong in shape. Compared to microsclerotia, pycnidia are larger and are infrequently seen in the wild. They have an uneven or globose shape, a rough texture, and a dark brown to black color. Beaked and ostiolated, or having a tiny aperture or pore, are further characteristics of pycnidia (Lakhran *et al.,* 2018). On artificial conditions, the fungus varies throughout isolates in terms of mycelium color, microsclerotia distribution, pycnidia production, and chlorate phenotypes. It has been suggested that the morphological changes in fungus are caused by their reactions to environmental stimuli or variations in the host species they come into contact with (Tok, 2019; Pandey *et al.,* 2020).

Symptoms

The pathogen has the capacity to infect various parts of the jute plant, leading to a variety of symptoms, which include Damping-off, Seedling blight, Stem rot, Collar rot, Leaf blight, Root rot, and Pod spot, particularly in seed crops. Damping-off results in the decay of newly emerged seedlings, affecting both the parts above and below the soil surface. Seedling blight is characterized by the browning and death of cotyledons, which are the seedling's first leaves. Stem rot infections typically occur in the mid of June and August, primarily due to high temperatures and humidity, and are identifiable by the development of brown spots on leaves. These infections can lead to the formation of dark brown lesions on the green stems, which can reach heights of 10-15 cm or more. Plants affected by this condition exhibit wilting, with drooping leaves that are susceptible to breakage in strong winds. Infected stems tend to dry up and change in color from green to dark brown or black. Furthermore, the collar region of the plants becomes discolored and rots, resulting in collar rot symptoms. In the root rot phase, plants experience wilting, defoliation, a transformation from brown to black, and eventually death, leaving behind bare stems. In jute seed crops, the pathogen also induces the formation of dark brown to black color of spots on the pods, which can subsequently lead to seed infection.

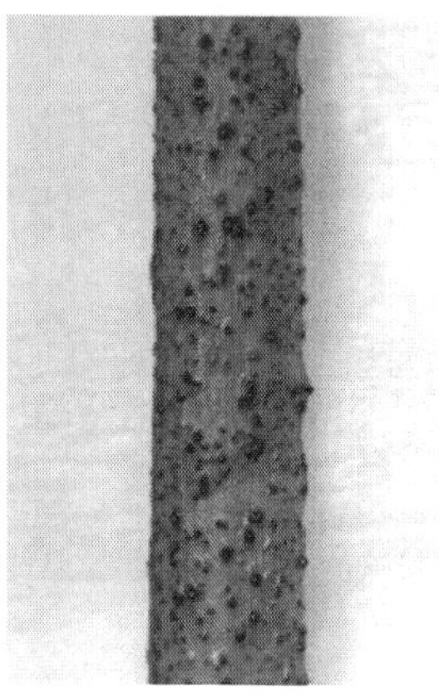
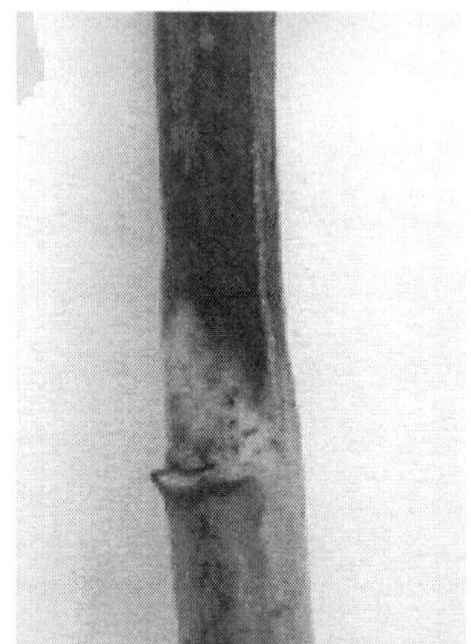

Figure 23.1: Stem rot and Rot rot
(*Source*: Sarkar *et al.*, 2014) (*Source*: Kritchai7752/Shutterstock.com)

Favourable Condition

M. phaseolina is more likely to infect alluvial and lateritic soils with low pH levels (5.6–6.5), high nitrogen contents, heavy rainfall, and high humidity. Older plants are more vulnerable to higher soil temperatures and lower soil moisture levels. More crops sown in March suffer than those sown later (De, 2012).

Disease Cycle

In jute crops, the primary cause of secondary disease spread has been attributed to the production of pycnidiospores on early infected stem and leaf tissues. The initial source of infection is microsclerotia, which are found in the soil, infected seeds, or host tissues. Throughout the growing season, microsclerotia sprout; this is especially true when the temperature is between 28 and 35 degrees Celsius. On the surface of the plant's roots, where germ tubes form specialized structures called appressoria, sclerotia germinate. These appressoria function to break through the epidermal cell walls of the host plant, either by means of enzymatic digestion and mechanical pressure or by taking use of natural openings. By releasing root exudates, this process causes the host plants to become infected. Fungal hyphae proliferate intracellularly through the xylem as the jute plant enters its flowering stage. These growths could impede the plant's vascular vessels by producing microsclerotia (Short *et al.*, 1978; Mayek –Perez *et al.*, 2002). The fungus produces necrotic lesions on the stems, branches, and peduncles of the infected plants by aggressively disrupting the host cells. The fungus enters the grains through the pod peduncles, travels to the pods, and then grows there. The pathogen is propagated via the mycelium and microsclerotia generated within the infected plant material, including plant leftovers. These microsclerotia are mostly situated at a depth of 0–20 cm and are frequently observed in clusters on the soil's surface. For two to fifteen years, the microsclerotia can live, depending on the surroundings. Variations in soil moisture content, low carbon-to-nitrogen ratios, and recurrent cycles of freezing and thawing the soil all have a detrimental effect on the propagules' ability to survive.

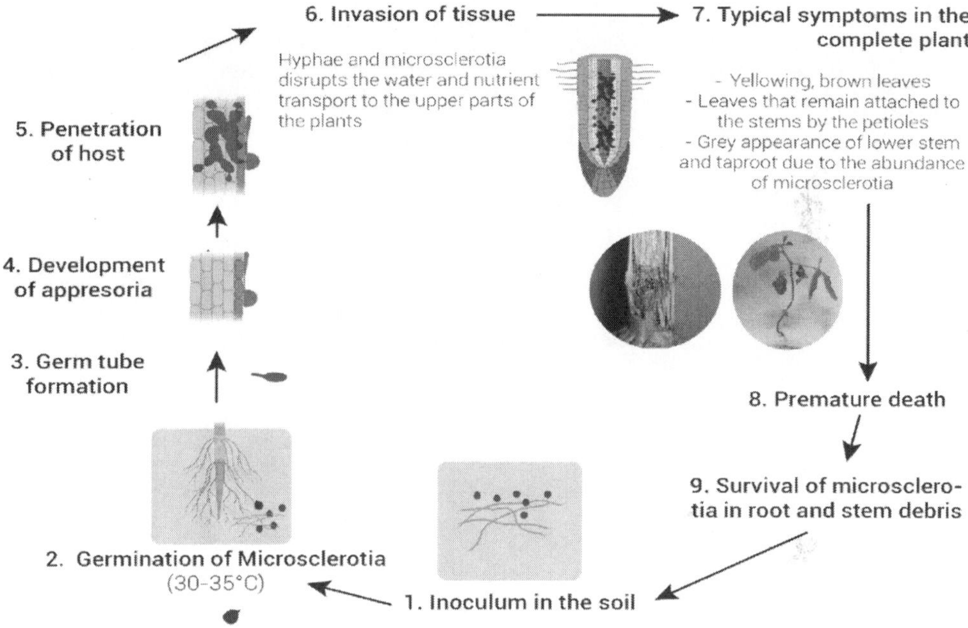

Figure 23.2: Disease cycle of Macrophomina phaseolina (*Source*: Maquez *et al.*, 2021).

Management

1. Applying soil treatment (mixing 1 kg of Trichoderma powder with 100 kg FYM) and treating seeds with 10g/kg of *Trichoderma viride* When it's time for the final plowing, spread the mixture around the field after covering it with polythene for seven days, watering it down and turning it every three to four days.
2. Seed treatment with tebuconazole @ 1.5 ml/kg seed or azoxystrobin + difenoconazole @ 1.0 ml/kg seed.
3. Spraying of tebuconazole @ 0.15 per cent at 40-45 or spraying of Azoxystrobin + Difenoconazole @ 0.075 per cent at 40-45.
4. 5-7 g/l Copper oxichloride spray after appearance of symptoms at interval of 15 days.

2. Anthracnose

Economic Importance and Occurrence

In India's capsularis jute-producing regions (Assam, North Bengal, Bihar, Uttar Pradesh), this disease is a persistent problem. It's common in Bangladesh as well. Inadvertently, the disease was brought to India in the 1930s by Southeast Asian jute germplasm, specifically from Taiwan. In 1945, 'Jap-Red,' a capsularis species imported from Formosa (Taiwan), was the first known victim of the disease in Dacca, Bangladesh. The illness then proceeded to other regions of Bangladesh and eventually made its way into India via Assam. The disease is more likely to spread in areas with constant rainfall, high humidity, and temperatures close to 35°C. The exotic variety 'Japanese Red' in Chunchura, West Bengal, was reported to have an anthracnose epidemic in 1950–1951, while detailed details regarding the amount of fiber yield loss are not available. Assam's olitorius jute is sensitive, but Uttar Pradesh's capsularis jute is the most afflicted. When the crop gets closer to harvest, olitorius jute usually suffers a higher prevalence of the disease, which causes very little loss. On the other hand, several spots form on the lower part of the stem when the disease first manifests, which causes the development of specky or knotty fiber (Sarkar and gawande, 2016).

Causal Organism

The disease is caused by *C. gloeosporioides* and *C. corchorum* brlongs to Kingdom *Fungi*, Division - *Ascomycota*, Class - *Sordariomycetes*, Order - *Glomerellales*, Family- *Glomerellaceae*, Genus – *Colletotrichum*, Species - *corborum* and *gloeosporioides*. The mycelium of fungus are immersed hyaline at young satage but pale brown to dark at maturity, septate and branched. Acervuli are epidermal or sub epidermal septate or confluent, composed of hyaline to dark brown, thick or thin walled cells, bearing conidiophores, setae and conidia. A needle like structure, black and smooth septate acutely pointed at the apex found in acervuli. Hyaline and aseptatae conidida are produced on conidiophores. Conididophore are short aseptate, developed on stroma, brown in colour and cylindrical in shape.

Symptoms

At the seedling stage, the disease manifests as brownish spots and streaks on the leaves and stems, followed by desiccation. On mature plants, the initial signs are light yellowish patches on the stem, which gradually darken into brown or black sunken spots. These spots are irregular in shape and size, and sometimes they merge, causing deep necrotic areas that result in cracks on the stem. In severe cases, the merging spots can encircle the stem, leading to plant breakage and eventual death. Surviving affected plants exhibit necrotic wounds across the stem. The fibers harvested from these impacted plants are classified as very low grade (cross bottom) due to their poor quality, speckly and knotty appearance. The disease also causes depressed patches on the pods of afflicted plants, and it contaminates the seeds that are extracted from these pods. These contaminated seeds have a paler hue, have shrunk, and have poor germination. According to Sarkar and Satpathy (2014), the illness caused by *C. olitorius* commonly affects crops in their later stages of growth, beginning in mid-July and lasting until September.

Favourable Conditions

Rain and cloudy weather encourage the disease's rapid spread in densely populated crops. The seedling stage occurred when the infection was at its most severe. Splashes of rain aid the fungus's spores on the field. According to Kumar *et al.* (2015), the pathogen needs high relative humidity, temperatures between 30 and 32 °C, and significant rainfall to be conducive to infection.

Disease Cycle

Infected seed and crop residue serving as primary source of infection the infected stem show acervulus setae and conidia. Secondary spread of disease happen due to air born conidia.

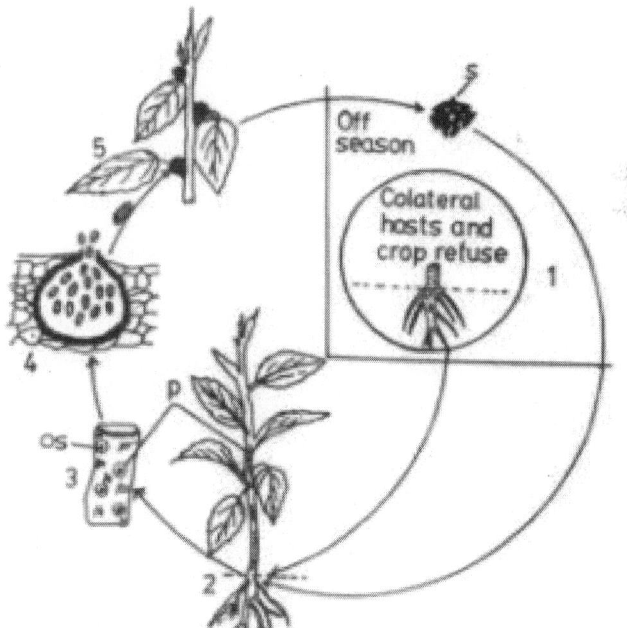

Figure 23.3: Disease cycle of Jute and Mesta (*Source:* Hasan, 2019).

Management

1. Remove all the weeds, sanitate to field.
2. Avoid overhead irrigation.

3. Treatment of seeds with Carbendazim 50 WP @ 2g/kg powder.
4. Foliar application of Copper oxychloride @ 5-7 g/l water or Carbendazim 50 WP @ 2 g/l of water or Mancozeb @ 4-5 g/l water can be used for management of disease.

3. Soft Rot

Economic Importance and Occurrence

Although stem rot of mesta and jute was a minor disease, at CRIJAF Farm in Barrackpore, the disease's incidence in seed crops reached 50 per cent in 2012–2014. The afflicted plant produces no seeds or fiber.

Causal Organism

Disease caused by *Agroathelia rolfsii* (*Sclerotia rolfsii*), belongs to the Kingdom - *Fungi*, Division- *Basidiomycota*, Class - *Agaricomycetes*, Order - *Amylocorticiales*, Genus –*Agroathelia*, Species- *rolfsii*

Symptoms

Figure 23.4: Soft rot of Jute and Mesta (*Source*: ag.purdue.edu).

Water-soaked spots on different sections of the stem are the disease's initial manifestation. These gradually turn into brown patches that are visible from a distance. The stem portion above or below these areas seems healthy at first. But the infection spreads, and before long it covers a foot or more of the stem. The affected tissues become flexible and easily break apart as a result of this decaying process. In the end, the stem segments above the impacted area could break off, dieback, and wither. White mycelial strands, which produce a cushion-like structure in the branch axils, cover the surface of the affected areas. Furthermore, on this mycelial mat, black-colored sclerotia are visible. These tough sclerotia can accumulate in the pith area of the stem. It is noteworthy that the bolls contained these sclerotia as well. Usually starting in December, the disease steadily develops over time until it eventually causes boll production.

Favorable Condition

In the colder months of December and January, the disease manifests itself in its most severe form. December 2012 and January 2013 had relative humidity levels of 76 per cent and 72 per cent, respectively, and mean monthly temperatures of 18 °C and 17 °C. From the area of contamination, the fungus *Sclerotria sclerotiorum* (Lib.) de Bary was isolated, and a pathogenicity test revealed a positive result. The sclerotia are able to endure in the soil, and when the temperature drops in December and January, they germinate and grow apothecia, which are the structures that produce ascospores and asci. Ascospores release particles that lead to infection (Tripathi *et al.*, 2013).

Disease Cycle

Host and crop residue serving as initial source of infection. The infected plant with pycnidia, pycnidia on stem, a portion of stem showing pycnidia and ostiole, stem showing pycnidia and pycnidiospore, capsule and seed infection by air born pycnidiospores caused secondary infection.

Management

1. Use healthy seeds for cultivation.
2. Clean and weed free cultivation, summer ploughing should be done.
3. Foliar spraying of Copper-oxychloride (COC) @ 5-7 g/l water at base region reduces the disease incidence.

4. Foot and Stem Rot

Economic Importance and Occurrence

All of India's mesta-growing regions, including West Bengal, Andhra Pradesh, Odisha, and Bihar, are prone to food and stem rot. It is the most significant mesta disease in India, accounting for a loss of fibre output of roughly 10–5 per cent. More than 40 per cent crop loss in roselle was noted in extreme circumstances (De and Mandal, 2007). The disease begins attacking the plant very early on and keeps going until it reaches maturity, altering the amount and quality of fibre. In comparison to kenaf (*H. cannabinus*), it is more severe in roselle (*H. sabdariffa*).

Causal Organism

The disease foot and stem rot is caused by soil borne pathogen, *Phytophthora parasitica* var. *sabdariffae*.

Symptoms

The symptoms typically manifest on the stem, usually just above the ground, although these spots can also be found at different heights on the stem. These spots exhibit a deep brown to blackish coloration and come in varying sizes. In cases where the spots are larger, they frequently encircle the stem, leading to the breaking of the plant at the point of infection. Such plants do not yield any fiber.

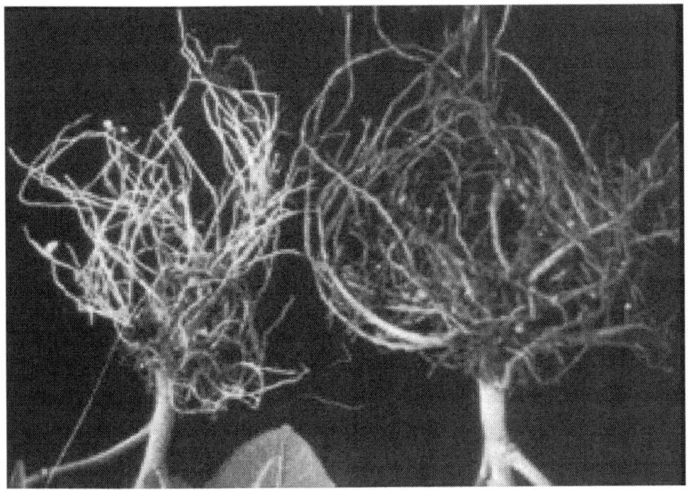

Figure 23.5: Foot and Stem rot of Jute and Mesta (*Source*: apnikheti.com).

Favourable Condition

High temperatures (30–35 °C) and constant drizzle are favourable for it. In West Bengal, mesta growth season is characterised by frequent water stagnation, which makes plants more susceptible to infection. The epidemic may be caused by persistent rain, heavy precipitation, and overcast weather from May to September (De and Mandal, 2007).

Disease Cycle

The pathogen is transmitted through the soil, and the soil contains its spores. Phytophthora infestans has an asexual life cycle that consists of a succession of alternating phases, such as hyphal growth, sporulation, sporangia germination (which can happen by emergence of the germ tube from the sporangium or by zoospore release), and then resuming hyphal growth. There additionally exists a sexual cycle that starts when isolates of the opposing mating kinds (A1 and A2) come into touch with one another. Hormonal communication initiates this sexual cycle, which produces oospores–sexual spores–in the process. The survival and spread of *P. infestans* are significantly aided by these diverse spore forms. *P. infestans* can travel between different host plants more easily thanks to the dispersal of sporangia caused by wind or water. When zoospores are released from sporangia, they can travel farther on water films found in the soil or on leaves because they are chemotactic and bifurcated. It is important to remember that oospores can live for many years, while sporangia and zoospores have comparatively brief lifespans.

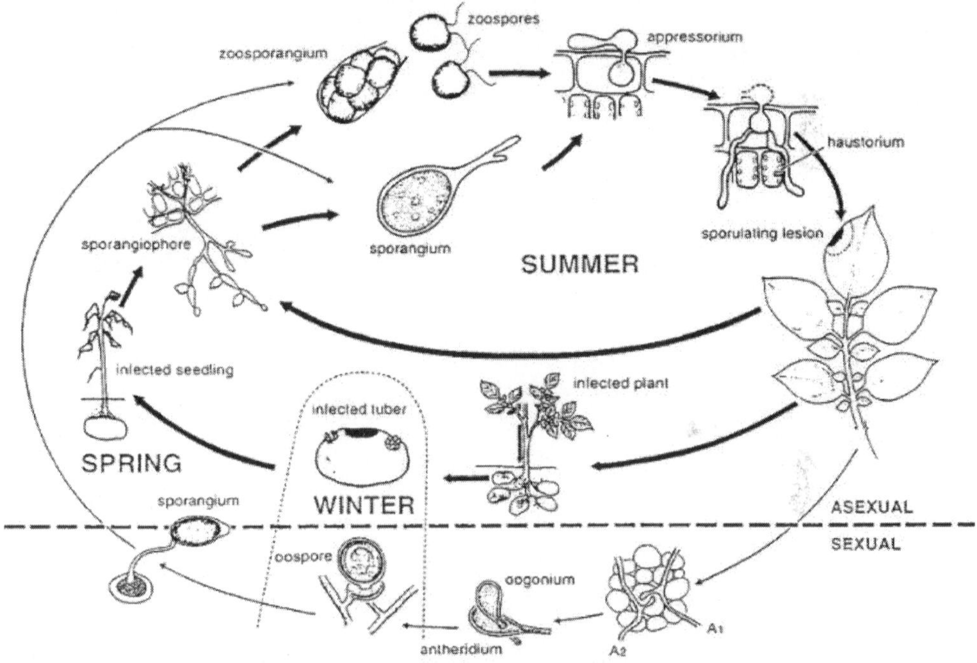

Figure 23.6: Disease Cycle of Foot and Stem rot of Jute and Mesta (*Source*: Drenth and Guest, 2004)

Management

1. None of the cultivated varieties are resistant to the disease, but there are some like AMV 1, Roselle Type 1 and AP 481 were seems to be moderately resistant while Red bristled (*H. sabdariffa*) lines are more resistant.
2. Seed treatment with Mancozeb @ 5.0 g kg^{-1} of seed followed by soil drenching (0.2 per cent) and spraying of Copper oxychloride (COC) 50 WP @ 5.0–7.0 g/l or Carbendazim @ 2.0 g/l of water (Mukherjee and Basak, 1973).

3. Copper oxychloride (COC) pre-sowing seed treatment proved to be more successful than the use of carbendazim, mancozeb, metalxyl, ediphos, carbosim, hexaconazole, and thiophanate methyl. At 30 days after sowing and at maturity, copper oxychloride decreases the disease incidence by 50.3 per cent and 45.5 per cent, respectively, and improves the yield of fibre by 20.6 per cent.
4. Before sowing, do seed treatment with Mancozeb@ 3gm/kg of seeds. If infestation is observed take spray of Ridomil@ 2g/l of water.

5. Hooghly Wilt

Economic Importance and Occurrence

The districts of Hooghly, some sections of Howrah, North 24 Parganas, and Nadia were the main places where this disease was found; these areas were especially those where tossa jute was alternated with potatoes or other solanaceous crops. There have also been reports of cases of this disease in Bangladesh. Infection rates in India's Hooghly district exceeded 40 per cent between the late 1970s and the late 1980s. A new Pest and Disease Control Centre was founded in Tarakeswar, Hooghly, in reaction to the disease's severity in the late 1940s and early 1950s. It is important to note, though, that this disease is no longer a serious worry as of right now. Between 1950 and 1954, jute crop losses were reported to have ranged from 30 per cent to 34 per cent on average. Infection rates in the Hooghly district's Kamarkundu area varied from 5 to 37 per cent in the late 1980s and early 1990s, while in some parts of Nadia and North 24 Parganas, they ranged from 2 to 20 per cent.

Causal Organism

The principal pathogen is *Ralstonia solanacearum* (= *Pseudomonas solanacearum*), with companion pathogens including *R. bataticola* and *Meloidogyne incognita*. Because these pathogens cause wounds in the roots, making it simpler for the primary bacterial pathogen, *R. solanacearum*, to enter, their presence enhances the disease. These pathogens are root rot (*R. bataticola/M. phaseolina*) and root knot nematode (*M. incognita*).

Symptoms

This disease is marked by the emergence of small yellow spots on the leaf surface during its early stages. These spots progressively expand and intermingle with green areas, resulting in a mosaic-like yellow appearance on the leaves. In certain instances, leaves may even develop small outgrowths along their mid-veins. In severe cases, the affected plants become stunted, causing a reduction in plant height of up to 20 per cent (Das *et al.*, 2008). According to Ghosh *et al.* (2007), the incidence was almost 50 per cent on some popular cultivars, including JRC 214 and 7447.

Management

1. Common hosts such as potatoes and other solanaceous crops grown in jute rotation should be avoided. The most efficient rotation is Jute: Paddy: Paddy or Jute: Paddy: Wheat.
2. Burning dead plants and rotting potato tubers, as well as removing wilt-affected plants from the field, are crucial cultural practises to contain the disease.
3. Application of potash @ 30 kg K_2O/ha reduce disease incidence.
4. Good drainage should be ensured. Capsularis jute is resistant to this disease and may be grown instead of olitorius.
5. Carbendazim 50WP @ 2 g/kg seed treatment and the same fungicide @ 2 g/l water spraying both lower the prevalence of root rot, which reduce the occurrence of wilt.

References

ACIAR Monograph No. 114, 238p

Almomani, F., Alhawatema, M. and Hameed, K. (2013). Detection, identification and morphological characteristic of Macrophomina phaseolina: the charcoal rot disease pathogens isolated from infected plants in Northern Jordan. Arch. Phytopathol. Plant Prot., 46: 1005–1014.

Bhattacharya, D., Dhar, T. K., Siddiqui, K.A.I. and Ali, E. (1994). Inhibition of seed germination by *Macrophomina phaseolina* is related to phaseolinone production. *J. Appl. Bacteriol.,* **77**:129–133.

Das, S., Ghosh, R., Paul S., Roy, A. and Ghosh, S.K. (2008). Complete nucleotide sequence of a monopartite begomovirus associated with yellow vein mosaic disease of mesta from north India. *Arch. Virol.* **153:** 1791-1796.

De, R.K. (2019). Manage disease of Jute with cheap and easy input. *Indian Farming* **69**(02): 43–47.

De, R.K. and Mandal, R.K. (2012). Identification of resistant sources of jute (*Corchorus olitorius* L.) against stem rot caused by *Macrophomina phaseolina* (Tassi) Gold. *J. Mycopathol. Res.,* **50**: 217-222.

Drenth, A. and Guest, D.I. (2004). Diversity and management of *Phytophthora* in South east Asia. ACIAR Monograph No.114. pp. 238

Ghosh, R., Paul, S., Das, S., Palit, P., Acharyya, S., Mir, J.I., Roy, A. and Ghosh, S.K. (2007). *ssociation of begomovirus complexes with yellow mosaic disease of* *jute in India*. In: Proceedings of Xth Plant Virus Epidemiology Symposium, October 15-19, ICRISAT, India. pp 115.

Kumar, S., Singh, V.K., Tripathi, A.N. and Kumar, P. (2015). Diseases of jute and sunhemp crops and their management. *Crop Diseases and Their Management*, pp. 345.

Lakhran, L., Ahir, R. R., Choudhary, M. and Choudhary, S. (2018). Isolation, purification, identification and pathogenicity of *Macrophomina phaseolina* (Tassi) goid caused dry root rot of chickpea. *J. Pharmacogn. Phytochem.*, **7**: 3314–3317.

Mandal, R.K. (1990). Jute diseases and their control. In: Proceedings of National Workshop cum Training on Jute, Mesta, Sunn hemp and Ramie. Central Research Institute for Jute and Allied Fibres, Barrackpore, West Bengal, India.

Marquez N., Giachero, M.L., Declerck, S. and Ducasse, D.A. (2021). Macrophomina phaseolina: General Characteristics of Pathogenicity and Methods of Control. Front. Plant Sci., 12:634397.

Mayek-Pérez, N., Garcia-Espinosa, R., LópezCastañeda, C., Acosta-Gallegos, J.A. and Simpson, J. (2002). Water relations, histopathology, and growth of commonbean (*Phaseolus vulgaris* L.) during pathogenesis of *Macrophomina phaseolina* under drought stress. *Physiol. Pl. Pathol.*, **60**:185–195.

Niu, X. (2015). Selection of reliable reference genes for quantitative real-time PCR gene expression analysis in jute (*Corchorus capsularis*) under stress treatments. *Front. Plant Sci.*, **6:** 848.

Pandey, A.K., Burlakoti, R.R., Rathore, A. and Nair, R.M. (2020). Morphological and molecular characterization of *Macrophomina phaseolina* isolated from three legume crops and evaluation of mungbean genotypes for resistance to dry root rot. *Crop Prot.*, **127**:104962.

Rout, M. K., Dash, G. B., Jena, S.and Das, S.K. (2018). *Trends in Biosciences*. **11** | (19): 2863-2866.

Roy, A., De, R.K. and Ghosh, S.K. (2008). Diseases of bast fibre crops and their management in jute and allied fibres, pp. 327. In: Karmakar, P. G., Hazara, S. K., Subramanian T. R., Mandal, R. K., Sinha, M. K. and Sen, H. S (Eds.). *Updates Production Technology*, Central Research Institute for Jute and Allied Fibres, Barrckpore, West Bengal, India.

Sarkar, S.K. and Gawande, S.P. (2016). Diseases of Jute and allied fibre crops and their management. *J. Mycopathol. Res.*, **54**(3): 321-337.

Sarkar, S.K., Satpathy, S. and Mitra, S. (2016). Present status of jute anthracnose (*Colletotrichum corchorum* and *C. gloeosporioides*) in India *Indian Phytopath.*, **69**(4): 252-254.

Sarkar, T.S., Biswas, P., Subrata, Ghosh, S.K. and Ghoshm, S. (2014). Nitric Oxide Production by Necrotrophic Pathogen *Macrophomina phaseolina* and the Host Plant in Charcoal Rot Disease of Jute: Complexity of the Interplay between Necrotroph-Host Plant Interactions. *PLOS ONE*, **9**(9): 107348.

Short, G.E. and Wyllie, T.D. (1978). Inoculum potential of *Macrophomina phaseolina*. *Phytopathol.*, **68**:742–746.

Tok, F.M. (2019). Relationship between morphologic, phenotypic and pathogenic characteristics in Macrophomina phaselina Isolates from cucumber plants. Int. J. Innov. Approaches Agric. Res., 3: 651–660.

Chapter 24

Diseases of Sugarcane Crop

Saurabh Kumar*, Shivam Kumar, Ravi Kumar and Sandeep Kumar

Department of Plant Pathology, Chandra Shekhar Azad University of Agriculture and Technology, Kanpur – 208 002, Uttar Pradesh
****e-mail: shivamkumar224229@gmail.com***

Introduction

An ancient crop of the Austronesian and Papuan tribes was sugarcane. The warm temperate and tropical regions of India, Southeast Asia, and New Guinea are the native habitats of sugarcane. Prehistoric Austronesian sailors brought it to Madagascar, Polynesia, and Island Melanesia. Sometime between 1200 and 1000 BC, Austronesian traders brought it to southern China and India as well. Between the sixth and fourth centuries BC, the Greeks and Persians came upon the well-known "reeds that produce honey without bees" in India. They took up sugarcane farming and spread it. Sugar, seen as an opulent and costly spice, was first traded by merchants from India. Sugarcane plantations first appeared in the Caribbean, South America, Indian Ocean, and Pacific island nations in the eighteenth century. Among the economically significant plant families that comprise sugarcane are the Poaceae, which also includes a variety of fodder crops as maize, wheat, rice, and sorghum. Large-scale migrations were largely driven by the need for laborers for the sugar harvest; some individuals chose to accept indentured service voluntarily, while others were forcibly transported as slaves. Specialized mill facilities are used to extract sucrose, also known as table sugar, from sugarcane. In addition to being eaten straight in confections, it is also used as a preservative in jams and preserves, a sweetener for drinks, a finishing touch for cakes and pastries, and a raw material in the food industry. It can undergo fermentation to yield ethanol, which is utilized to create alcoholic beverages. The immature, unexpanded flower head of Saccharumedule (duruka) is consumed raw, steamed, toasted, or prepared in a variety of ways throughout Southeast Asia, including some Indonesian island communities and Oceanic nations like Fiji. The majority of research was focused on three main diseases that were affecting the production of sugarcane at the time: red rot, grassy shoot, and smut. The tall, perennial grass that is utilized to produce sugar is a member of the Andropogoneae tribe and genus Saccharum. The plants have robust, jointed, fibrous stalks that are rich in sucrose, which builds up in the stalk internodes. The plants range in height from 2 to 6 meters (6 to 20 feet). Tropical and subtropical climates are where it is grown worldwide. India is the world's second-biggest producer of sugarcane, after Brazil, making it the greatest crop in terms of output volume. In India, sugarcane production occupies 21 per cent of the country's agricultural land. Approximately 22 per cent of the world's sugarcane is produced in this nation. The state of Uttar Pradesh is India's biggest producer of sugarcane. *Saccharum officinarum* and its hybrids provide around 70 per cent of the sugar produced. There is There is interbreeding among all sugarcane species, and the majority of commercial cultivars are intricate hybrids. The majority of the most prevalent diseases that impact the sugarcane crop in India are fungal infections, which can cause epidemic crop losses of up to 80–100 per cent. Even though some minor diseases of sugarcane include damping off, downy mildew, ringspot, lead scald, and some bacterial diseases like red stripe (*Pseudomonas rubrilines*), the major diseases of sugarcane are Red Rot (*Colletotrichum falcatum*), Smut (*Ustilago scitaminea*), Sett rot or Pineapple disease (*Ceratocyctis paradoxa*), and some phytoplasmal diseases like grassy shoot. Agronomic techniques, planting dates, cultivating resistant cultivars, biological control, bio-pesticides, and chemical fungicides as a last resort are some of the techniques and procedures that can be used to manage it. On the other hand, the following provides a thorough overview of sugarcane diseases, including their causes, the disease cycle, and integrated management strategies:

Table 24.1: List of Sugarcane Diseases and their Causal agent

Sl.No.	*Name of Disease*	*Causal Dgent*
A.	**Fungal disease**	
1.	Red Rot	*Colletotrichum falcatum*
2.	Smut	*Ustilago scitaminea*
3.	Pineapple disease	*Ceratocyctis paradoxa*
4.	Wilt	*Cephalosporium sacchari*
5.	Rust	*Pucciniaerianthi*
B.	**Viral disease**	
1.	Sugarcane mosaic virus	Potyvirus

Sl.No.	Name of Disease	Causal Dgent
C.	**Phytoplasmal disease**	
1.	Grassy shoot	Phytoplasma
D.	Bacterial disease	
1.	Red stripe	*Pseudomonas rubrilines*
2.	Gummosis	*Xanthomaonas axonopodis* pv. *vasculorum*

A. Fungal Disease

1. Red Rot of Sugarcane

Economic Importance and Occurrence

One serious sugarcane disease is red rot. Dr. C.A. Barber conducted pioneering research on the condition in India under the Madras Presidency, when it first appeared. At the Imperial Agricultural Research Institute in Pusa, Bihar, India, Butler, the Imperial mycologist, conducted in-depth research on the causative organism and its portals of entry into the cane stalk. He gave the disease the name "Red rot" because of its most defining characteristic, which is the rotting of the internal stalk tissues with a reddish discoloration. In order to lessen crop damage in India, Barber and Butler both understood the significance of the illness and developed management techniques that included using healthy seed and avoiding waterlogging. Sugarcane was severely damaged by severe red rot epiphytotics in the Godavari Delta and **North** Indian plains, but this situation led to Dr. Barber founding the Sugarcane Breeding Institute (SBI) in Coimbatore, India, in 1912 to create red rot-resistant varieties through interspecific hybridization. Asia, Africa, North America, the Caribbean, South and Central America, Europe, and Oceania. It has been reported from the following countries: Guam, Federated State of Micronesia, America Samoa, Cook Islands, Fiji, French Polynesia, Australia, Papua New Guinea, Samoa, Solomon Islands, Tonga, and Vanuatu. During the 1938–1939 growing season, a red rot disease was observed in the subtropical regions of Uttar Pradesh (UP) and Bihar, which is India's primary sugarcane growing region. The significant commercial variety Co 213 collapsed as a result of this devastation, which destroyed thousands of hectares.

Causal Organism

Colletotrichum is a member of the Melanconiaceae family, Melanconiales order, class Coclomycetes, Eumycota division, and Deuteromycotina subdivision. Sutton (1973) listed eleven species of Colletotrichum. Colletotrichum has been reported in 1000 species by Alexopoulos and Mims (1979). As per the most recent report by Baxter *et al.* (1985), this genus has 21 species.

Symptoms

The initial outward sign typically manifests on the third or fourth leaf, which eventually withers away at the margins' tips. When a stalk is broken lengthwise, the internodes can reveal typical red rot symptoms. Among these is the reddening of the interior tissues, which are often stretched perpendicular to the stalk's long axis. The key feature for determining the disease is the existence of across white spots. Additionally, the infected cane smells foul and acidic. The stalk becomes hollow and covered in white mycelial growth as the disease progresses. Subsequently, the rind becomes thinner throughout its length, revealing tiny, smooth black fruiting bodies that emerge from it. Additionally, the infection causes little crimson lesions with dark spots on the upper surface of leaves. The lesions have dark borders and a blood red color at first, but they eventually turn straw in the center. The diseased leaves frequently break at the lesions and dangle, covered in a vast number of tiny black spots.

Figure 24.1: A. Red Rot of Sugarcane (*Source*: toppr.com).

Favourable Conditions

Sugarcane monoculture, consecutive ratoon cropping, conditions of standing water, and insect-related injuries.

Disease Cycle

In addition to being spread via setts, the fungus lives on infected clumps and stubbles in the soil as latent mycelium and chlamydospores. The majority of the initial infection comes from infected setts. Tools used in cultivation and irrigation provide secondary dissemination in the field. Conidia from sick plants spread to healthy plants in the field with the aid of raindrops, air currents, and dew. The fungus also thrives on *Sorghum vulgare, S. halepense*, and *Saccharum spontaneum*, which serve as collateral hosts. Conidia may germinate and penetrate the leaves through multiple types of wound forms if they land on the leaves. Both root primordia and insect bores can cause stem infection. The fungus that is carried by the soil can potentially infect the shoots early by entering the healthy setts through the cut ends. Even though the fungus's perfect form has been seen in nature, the role of ascospores in the disease cycle is not understood.

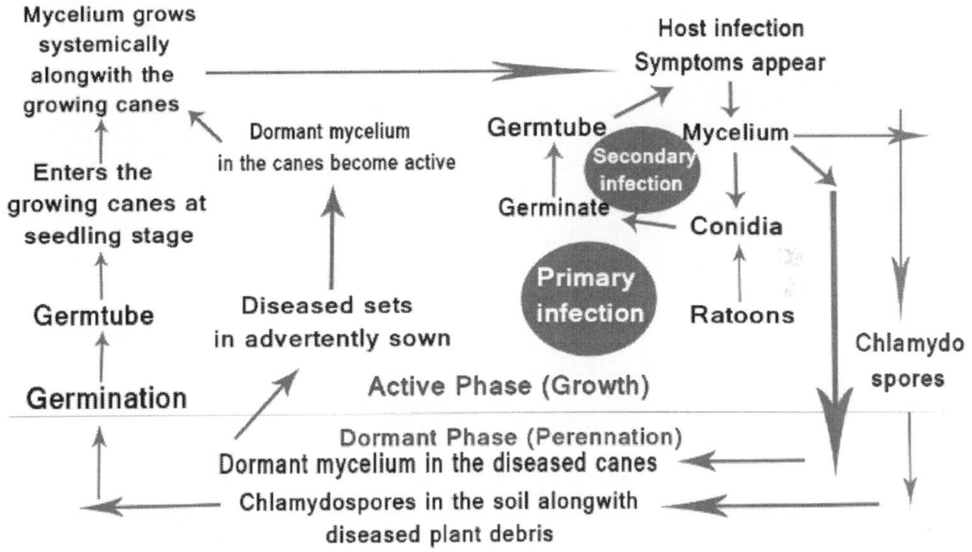

Figure 24.2: Disease Cycle of Red Rot of Sugarcane (*Source*: botanyyworld.com).

Management

1 Aviod ratooning of the diseased crop.
2. Implement agricultural rotation by planting crops such as rice and green manure.
3. Choose the setts from the disease-free regions or fields.
4. Before planting, soak the setts in a 0.1 per cent carbendazim or 0.05 per cent triademefon solution for 15 minutes.
5. Grow the moderately resistant CO 8001 and CO 8201 kinds and the resistant CO 62198 and CO 7704 varieties.
6. Aerated steam at 52 °C can be used to treat sets for four to five hours, and moist hot air at 54 °C can be used for two hours.

2. Smut of Sugarcane

Economic Importance and Occurrence

The fungus *Sporisorium scitamineum* is the source of sugarcane smut, a fungal disease of the crop. In regions of the world where sugarcane is grown, sugarcane smut is a terrible illness. One of the most major diseases affecting sugarcane globally, sugarcane smut is caused by *Sporisorium scitamineum* (Syd.) M. Piepenbr., M. Stoll and Oberw. In vulnerable types, it can result in large yield losses (Comstock 2000). Since its discovery in 1877 in Natal, South Africa, this illness has spread to other nations that grow cane, including Indonesia, Central and South America, Brazil, Australia, and other cane-growing regions of Africa (Braithwaite *et al.*, 2004; Hoy *et al.*, 1986). Anecdotal evidence indicates that I and *S. spontaneum*, two wild cousins of contemporary sugarcane (Saccharum hybrid), were the principal hosts of smut during its origins in India. A number of nations simultaneously reported smut for the first time in a number of decades; for example, reports of smut were made in the United States (including Texas, Hawaii, Louisiana, Florida, and Florida), Central America, and the Caribbean between 1970 and 1980. Smut was not found in Papua New Guinea until 2016 (Tom *et al.*, 2017) or Australia until 2006 (Croft *et al.*, 2008a; Magarey *et al.*, 2006). Although close neighbors like Indonesia, China, Malaysia, Taiwan, and the Philippines have a history of smut infestation dating back to the early to mid-20th century, the incursion in the latter two nations has been delayed (Ferreira and Comstock, 1989). The original sugarcane cropping species, S. officinarum, or Noble cane, is thought to have originated in Papua New Guinea (Daniels and Roach, 1987). This region was free of smut until 2016 (Tom *et al.*, 2017). The only nation that grows sugarcane and is yet devoid of smut is Fiji.

Causal Organism

The disease that is produced by *Sporisorium scitamineum* is classified as belonging to to Domain – Eukaryota, Kingdom –*Fungi*, Division – *Basidiomycota*, Class – *Ustilaginomycetes*, Order – *Ustilaginales*, Family – *Ustilaginaceae*, Genus – *Sporisorium*,and Species - *scitamineum.* The majority of the fungal hyphae are intercellular; they gather as a dense mass in between the host cell's vascular bundles and release microscopic black spores. The host epidermis is represented by the thin membrane covering the Smut whip. The light brown, spherical, echinulate smut spores have a diameter of 6.5 to 8.5 um. When smut spores germinate, they form a hyaline, oval-shaped promycelium with four cells and four pointed ends called sporidia.

Symptoms

It is a culmiculous smut that causes stunting in the affected plants and transforms the central stalk into a long, dusty, black structure like a whip. The whip can range in length from a few inches to several feet. This structure has a thin, papery, white membrane covering it in the early stages. The Whip could have a small curvature or be straight. When it reaches maturity, it bursts, releasing and allowing the wind to disperse millions of tiny black smut spores (Teliospores). Typically, afflicted plants are slender, rigid, and maintain an acute angle. Every shoot or tiller that emerges from the cluster has the potential to form the whip-like structure that symbolizes the central shoot with its varied leaves may be produced by each one of the shoots/tillers arising from the clump. The smutted clumps also yield mummified arrows, with the upper section of the rachis transformed into a typical smut whip and the lower portion consisting of a typical inflorescence with typical flowers. Smut sori can occasionally appear on the stem and leaves as well.

Figure 24.3: Smut of Sugarcane (*Source*: Plantix.net).

Favourable Conditions

Monoculturing of sugarcane, continuous ratooning and dry weather during tillering stage.

Disease Cycle

Teliospores have a 10-year maximum lifespan in the soil. The soil-borne plant materials that are infected also contain spores and sporidia. Additionally present in or on the diseased setts are the latent mycelium and smut spores. Infected seed bits are the main way that the disease spreads (Setts). Furthermore, soil-borne infections are caused by sporidia and spores that are present in the soil and spread by irrigation and rainfall. With the help of air currents, the smut spores that form in the whips are the primary means of secondary dissemination in the field. With the help of air currents, smut spores that have formed in the whips are the primary source of the fungus field. Additionally, the fungus thrives on secondary hosts such *as Cyperus dilatatus, Saccharum pontaneum, S. robustum, Sorghum vulgare,* and *Imperata arundinacea.*

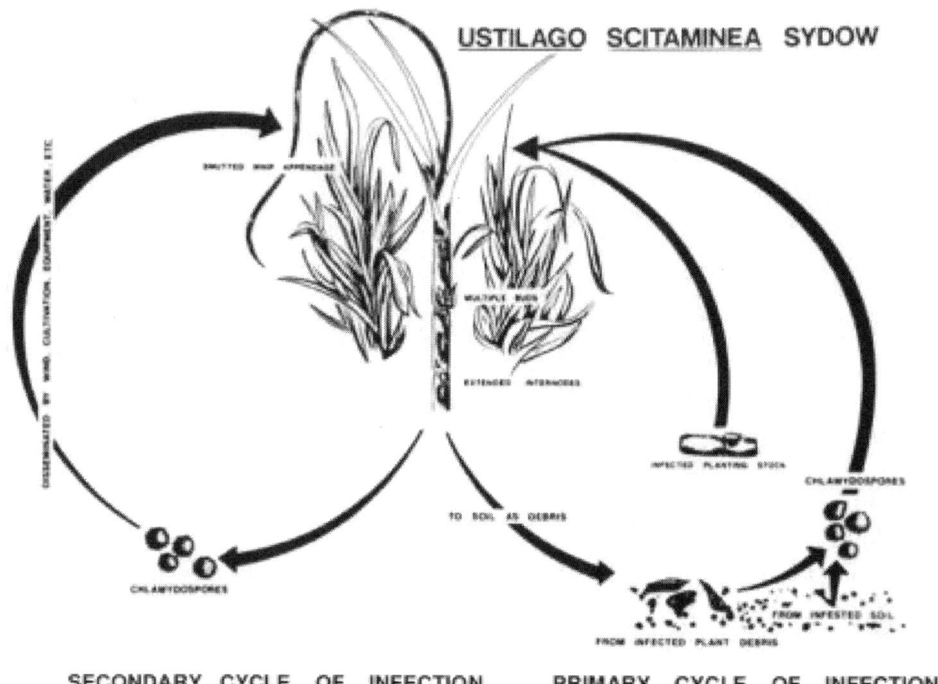

Figure 24.4: Life Cycle of *Ustilaginaceaes citamineum* (*Source*: ccmedia.fdacs.gov).

Management

1. Use healthy plant setts.
2. Remove and destory the smutted clump.
3. Prevent the ratooning of crops that are diseased and have an infection rate of more than 10 per cent.
4. Crop rotation with using green manure crops.
5. Grow resistant varieties like Co 7704 and moderately resistant varieties COC 85061 and COC 8also survives on collateral hosts like *Saccharum spontaneum, S. robustum, Sorghumvulgare, Imperata arundinacea* and *Cyperus dilatatus*.

3 Sett Rot

Economic Importance and Occurrence

According to Talukder *et al.* (2007), pineapple disease is one of the most destructive diseases that affect sugarcane among other illnesses. The odor of a mature pineapple connected to sick cane setts is the source of the phrase "pineapple disease" (Chhamaet al., 2014). The disease-causing organism, *Ceratocystis paradoxa*, produces ethyl acetate through metabolic processes, which is the source of this odor. (Chhama and others, 2014). According to Yadahalli (2006), the illness is also occasionally referred to as pineapple sett rot or just sugarcane sett rot. According to Raid (2013), *C. paradoxa* is transmitted by soil and setts, and infections typically originate from wounds caused by insect pests on standing cane or from the cut ends of setts. Asia, Africa, Europe, Oceania, North, South, and Central America. Sugarcane from Australia, Fiji, and Papua New Guinea has been reported to have black rot, also known as pineapple disease. Maize, mango, pineapple (black rot or water blister), potato, sweet potato, taro, cocoa (pod rot), coconut (stem bleeding), banana and plantain (black end or fruit rot), and sugarcane (black rot or pineapple disease).

Causal Organism

Ceratocystis paradoxa or Black Rot of Pineapple is a plant pathogen that is a fungus, part of the phylum Ascomycota. It is characterized as the teleomorph or sexual reproduction stage of infection. This stage contains ascocarps, or sacs/ fruiting bodies, which contain the sexually produced inoculating ascospores. These are the structures which are used primarily to survive long periods of time or overwinter to prepare for the next growing season of its host. Unfortunately, the sexual stage is not often seen in the natural field but instead the anamorph, or asexual stage is more commonly seen. This asexual stage name is *Thielaviopsis paradoxa* and is the common cause of Black rot or stem-end rot of its hosts. Domain- *Eukaryota*, Kingdom – *Fungi*, Division – *Ascomycota*, Class – *Sordariomycetes*, Order – *Microascales*, Family – *Ceratocystidaceae*, Genus – *Ceratocystis*, Species - *paradoxa*. Both macroconidia and microconidia are produced by the fungus. Conidiophores are thin-walled, linear structures with long terminal cells and short base cells. When the microconidia are small, they are hyaline, but as they mature, they nearly turn black. They are cylindrical, have thin walls, and are generated in chains inside the lengthy conidiophore cells before being expelled one after the other. Macroconidia are formed on short, lateral conidiophores, either singly or in chains. Macroconidia are hyaline to olive green or black in color and can be spherical, elliptical, truncate, or pyriform. 16–19 x 10–12 um in size. The fungus also generates oval, thick-walled, brown chlamydospores on short lateral hyphae in chains. The perithecia have a very long neck and resemble flasks. The perithecium's bulbous base, which measures 200–300 um in diameter and is adorned with knobbed, unevenly shaped appendages, is hyaline or pale yellow in color. Numerous pale-brown, upright, tapered hyphae cover the ostiole. The ascospores are single-celled, hyaline, ellipsoid, and more convex on one side. They measure 7–10 x 2.5–4 um. The asci are clavate and measure 25 x 10 um.

Symptoms

The fungus is soil-borne, enters through cut ends, and multiplies quickly in the parenchymatous tissues; the affected tissues initially turn reddish, but later turn brownish black. Severely affected setts exhibit internodal cavities covered in mycelium and abundant spores; the rotting tissues smell strongly of pineapple. The disease primarily affects the setts, usually two to three weeks after planning. The setts may decay before the buds germinate, or the shoots may die after reaching a height of approximately 6 to 12 inches shoots are stunted.

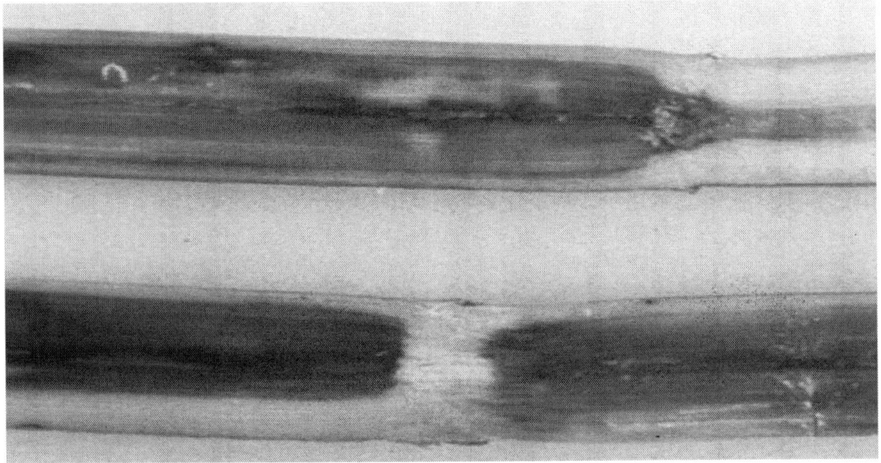

Figure 24.5: Sett Rot (*Source*: pestoscope.com).

Favourable Conditions

Poorly drained fields, heavy clay soils, temperature of 25-30-°C, prolonged rainfall after planting.

Disease Cycle

The fungus lives in the soil and in the tissues of infected, buried cane as conidia and chlamydospores. Field to field transmission of the inoculum occurs via wind-borne conidia, irrigation, or precipitation. It quickly penetrates the parenchymatous tissues inside the sett, resulting in sett rot. Insects that aid in the disease's dissemination include the cane borer (Diatraeadyari). In addition, the pathogen can thrive on arecanut, coconut, cocoa, mango, papaya, coffee, and maize. Insects contribute to the pathogen's spread as well.

Management

1. Soak the setts in 0.05 per cent Carbendazim 15 minutes.
2. Use long setts having 3 or 4 buds.
3. Provide adequate drainage during rainy seasons.

4. Apply carbendazim @ 2gm/lit of water at the root zone area and same as follow at 15 days interval.

4. Red Stripe

Economic Importance and Occurrence

Due to significant changes in climatic circumstances, red stripe/top rot disease, which was previously thought to be a minor sugarcane disease, is becoming more and more significant. Leaf stripes and top rot are the two main types of this disease that can appear separately or in combination in fields with relatively high air humidity. Young rats were found to be more susceptible to infection than plant canes of the same age in Hawaii (Martin, 1938), Java (Bolle, 1929), Taiwan (Okabe, 1933), and Louisiana (Rands and Dopp, 1932), whereas Cottrell-Dormer (1932) reported the reverse case in Australia. Fors (1978) noted thick red lines spreading from red patches on stalks in the vicinity of root primordia to the internodes. After being discovered for the first time in Hawaii by Lyon in 1922, red stripe disease was subsequently reported across much of the cane-growing region (Ricaudet al., 1989; Martin and Wismer 1961). Later on, it became known as bacterial red stripe or red stripe (Lee and Jennings, 1924). Tryon (1923) documented the state of top rot in sugarcane in Queensland, and he subsequently reported from fifty locations worldwide that grow sugarcane. Although the red stripe disease is often regarded as being of little consequence (Edgerton, 1959; Martin and Wismer, 1961), reports have indicated that the pathogen has the ability to generate epidemics (Chona and Rao, 1963; Fors, 1978). Other states where reports of the disease have been made include Louisiana (Christopher and Edgerton, 1930), Florida (Rands and Dopp, 1932), Cuba (Faris, 1927), the Philippines (Lee and Pierce, 1928), Puerto Rico (Cook, 1929), Brazil (Grillo, 1938), India (Subramaniam, 1936), and Taiwan (Matsumoto, 1952). Red stripe disease was first documented in India in 1933 by Desai, and subsequent reports on various sugarcane varieties from several other states (Dange and Payak, 1973; Agnihotri, 1990) were followed. Chona and Rao (1963) reported that the disease was occurring in an epidemic form in the eastern districts of Pun (Ricaud *et al.*, 1989; Martin and Wismer 1961).

Causal Organism

The bacterium *Acidovora xavenae* subsp. *avenae* is a short rod (0.7 X 1.67um), gram negative, non capsulate with a polar flagellum.

Symptoms

The disease initially appears on the underside of the immature leaves. The stripes begin as long, narrow, chlorotic lines soaked in water and turn reddish brown after a few days. These stripes, which run parallel to the midrib, measure 0.5 to 1 mm in width and 5-100 mm in length. The stripes stay limited to the lower portion of the leaf lamina, and white flakes travel to the shoot's growth points where yellowish stripes initially form before eventually turning reddish brown. It is possible for the rotting to start at the tip of the shoot and progress downward. The center hollow forms as the core shrivels and turns a reddish-brown

Figure 24.6: Red Stripe (*Source*: Plantix).

Favourable Conditions

Continuous ratooning and prolonged rainy weather with low temperature (25 °C).

Disease Cycle

Infected plant remnants and the soil both harbor the infection. In addition, the bacterium can thrive on other Saccharum species, sorghum, finger millet, pearl millet, and maize. The main way that the bacteria spreads is through diseased canes. Insects, irrigation water, and raindrop splashing are the principal secondary spreaders. Parenchymatous cells that are infected may collapse, impairing the plant parts' ability to function normally. Numerous grasses, such as bajra and ragi, have been shown to harbor the bacterium, and these hosts might contribute to the pathogen's persistence and dissemination.

Management

1. Whenever the disease is noticed; the affected plants should be removed and burnt.
2. Growing resistant varieties Select setts from the healthy fields*viz.*, Co 86032, Co 86249, CoSi 95071, CoG 93076, CoC 22, CoSi 6, and CoG 5.

3. Avoid growing collateral hosts near the sugarcane fields.
4. Use fungi toxic chemicals like Bavistan, Benomyl, Topsin, and Aretan at 0.1 per cent for 18 min. at 52°C for dipping setts which gave almost complete elimination of rot infection.

Minor Diseases

5. Damping Off

Causal organism - *Pythium aphanidermatum, P. debaryanum, P. graminicola, P.ultimum*

Symptoms

During the pre-emergence phase, immature seedlings and germinating seeds are attacked and killed. Additionally, seedlings have water-soaked sores at the collar region, which cause them to dry up and wither in the post-emergency stage.

6. Downy Mildew

Causal organism - *Peronoscleros porasacchari*

Symptoms

Fungal growth that is downy and has yellow stripes on its upper surface. Affected canes' internodes are rapidly elongating, and older leaves are being shredded.

7. Eye Spot

Causal organism - *Helminthosporium sacchari*

Symptoms

The wet patch appears on leaves, grows longer over time, and eventually takes the shape of a "eye." A reddish brown center encircled by straw yellow tissues is called a spot.

8. Ring Spot

Causal organism - *Leptosphaerias acchari*

Symptoms

The water-soaked patches on the leaves first show as straw-colored patches that are encircled by a dispersed discoloration zone and a narrow band of reddish brown color.

References

Ricaud, C., Egan, B. T., Gillaspie, A. G. and Hughes, C. G. (2012). Diseases of sugarcane: major diseases. *Elsevier.*

Bhuiyan, S. A., Magarey, R. C., McNeil, M. D. and Aitken, K. S. (2021). Sugarcane smut, caused by *Sporisorium scitamineum*, a major disease of sugarcane: A contemporary review. *Phytopathology*, **111**(11): 1905-1917.

Viswanathan, R. and Rao, G. P. (2011). Disease scenario and management of major sugarcane diseases in India. *Sugar Tech*, **13**: 336-353.

Viswanathan, R. (2021). Red rot of sugarcane (*Colletotrichumfalcatum* Went). *CABI* Reviews.

Duttamajumder, S. K. (2008). Red rot of sugarcane. Indian Institute of sugarcane research.

Yonzone, R. and Devi, M. S. (2018). Red stripe/top rot disease of sugarcane: a review. *IJCMAS*, **7**(1): 1469-1478.

Wismer, C. A., Bailey, R. A., Ricaud, C., Egan, B. T., Gillaspie, A. G. J. and Hughes, C. G. (1989). Pineapple disease. Diseases of Sugarcane - Major Diseases, 145-156.

Hughes, C. G. (1978). Diseases of sugarcane - a review. *Pans*, **24**(2): 143-159.

Wismer, C. A. (1951). Controlling pineapple disease of sugarcane. Hawaiian Planters' Record, **54**: 23-53.

Chapter 25

Diseases of Sugarbeat

Ankit Kumar, Shivam[*1], *Harshita Yadav*[2], *Arvind Yadav*[2] *and Shri Kant Rawat*[3]

[1]*Department of Plant Pathology, Chandra Shekhar Azad University of Agriculture and Technology, Kanpur (U.P.) 208002*
[2]*Department of Plant Pathology, Acharya Narendra Dev University of Agriculture and Technology, Ayodhya (U.P.)- 224229*
[3]*Department of Plant Pathology, Banda University of Agriculture and Technology, Banda (U.P.)- 210001*
**e-mail: akkisvbp623@gmail.com*

Introduction

Within the Amaranthaceae family, sugar beetroot, formally known as Betavulgaris, is one of the four cultivated variants of the plant Beta vulgaris. It is mostly farmed for its sucrose content and is the second-largest global source of sugar after sugarcane. It is a well-known practise to produce sugar from sugar beet root juice. Long before its ability to produce sugar was understood, sugar beetroot was historically grown as a vegetable and fodder crop. Thanks to the chemist Andreas Marggraf, the initial successful attempt to make sugar from beets was made in Germany in 1747. But in Silesia (now in Poland), the first beetroot sugar plant was built in 1802. Napoleon had a big impact on the development in beetroot sugar production in 1811 because of the British blockade that prevented France from obtaining raw cane sugar from the Caribbean region. Forty beetroot sugar plants were built in France under his patronage. After Napoleon's defeat, the industry suffered a setback, but the

1840s saw a significant recovery. Throughout Europe, beetroot sugar output grew steadily and by 1880 had surpassed cane sugar production. Nowadays, most sugar produced in the European Union and around one-fifth of all sugar produced worldwide comes from beets. However, if not correctly managed, the sugar beetroot plant is prone to a number of diseases and insect pests, such as Cercospora leaf spot and black root rot, which can severely lower yields and sugar content. Precautions must also be taken to shield the crop from harm from worms as well beetles, and nematodes.

Major Diseases

1.	Aternaria leaf spot	*Alternaria alternata* *A. brassicae*
2.	Anthracnose	*Colletotrichum dematium*
3.	Aphanomyces root rot (black root)	*Aphanomyces cochlioides*
4.	Cercospora leaf spot	*Cercospora beticola*
5.	Charcoal rot	*Macrophomina phaseolina*
6.	Downy mildew	*Peronospora schachtii*
7.	Powdery mildew	*Erysiphe betae*
8.	Southern blight (Sclerotium root rot and stem rot)	*Sclerotium rolfsii*
9.	Rust	*Uromycesbetae*

1. Alternaria Leaf Spot

Economic Importance and Occurrence

Studies conducted in India have shown that growing sugar beetroot as a supplementary sugar crop is a feasible alternative, especially in Northwestern India. In the Rajasthani region of Sriganganagar, sugar beetroot was grown commercially between 1972 and 1998. Every month between April and May, sugar beetroot was processed for sugar manufacturing in this cultivation, producing sugar yields that were favourable. Sugar beets have useful byproducts in addition to sugar; their tops and molasses can be used as cow fodder or fermented to produce vitamin B complex and other pharmaceutical compounds. Diseases are a major obstacle to obtaining viable sugar beetroot yields worldwide, which are necessary for the manufacture of commercial sugar at a reasonable cost with regard to of each crop tonnage and sugar content. Diseases cause 16–20 per cent of the sugar beetroot crop to be lost annually. The existing distribution of the beet sugar business and sugar beet cultivation in many of the nations where sugar beets are grown is largely shaped by the prevalence of these diseases. (Ruppel and Duffus, 1993). These diseases can harm the crop at any point during the growing season, right up until harvest.

Causal Organism

Alternaria leaf spot (ALS) is caused by fungi belonging to the Alternaria genus, mostly *A. alternata* (syn. *A. tenuis*) and *A. brassicae.*

However, it should be noted that *A. tenuis*, which is common on dead or dying plant parts, may easily be distinguished from *A. brassicae* on sugar beetroot. The second fungus produces long chains of spores (section Longicatenatae), each of which is practically beakless, much smaller than a spore of *A. brassicae* (one-fourth to one-tenth the length), and dark in colour. With the use of a hand lens, these various differences can be identified in the field since they are sufficiently distinct.

Symptoms

The disease first appears as dark brown, round to irregular patches that range in size from 2 to 5 mm on the older leaves. These patches enlarge with age and can go as big as 10 mm in diameter. The centres of these spots eventually turn grey and may take on the appearance of a lesion with a modest zonate pattern. The elder spots may or may not have a reddish tint surrounding them; this will depend on the species of sugar beetroot used. The leaf veins frequently serve as boundaries for these areas, however diseases on the veins can spread beyond them on either side. These spots cause affected leaves to become yellow and wither quickly. When the environment is right, many infection centres can form, which will cause the spots to meld together and quickly cause the leaf to necrotize. In extreme situations, this necrosis could worsen to the point where the plant's leaf blades are almost completely destroyed. The petioles are also susceptible to the infection, and if the plant continues to bolt in cool, damp weather, lesions may even develop on the seed stalks. It's feasible that this illness might potentially affect other plant organs.

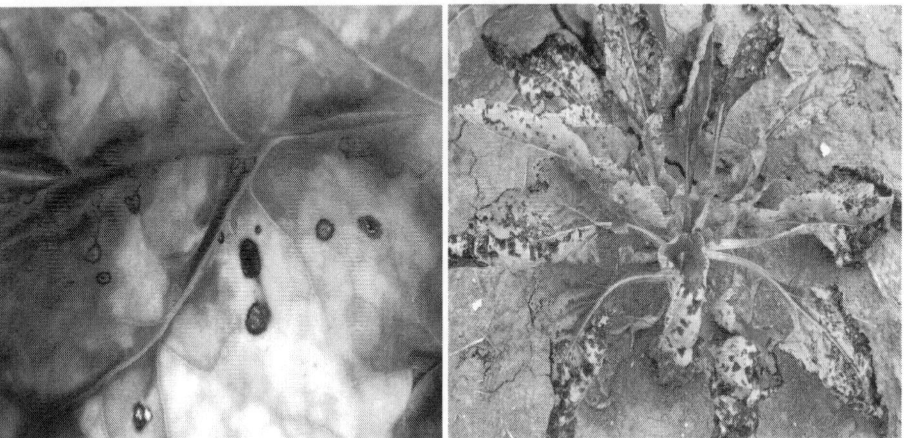

Figure 25.1: Alternaria Leaf Spot of Sugarbeet (*Source*: vegetables.cornell.edu).

Favourable Condition

High humidity and cold to slightly hot temperatures (15.6-25.6 °C or 60-78 °F) are ideal for Alternaria leaf spot. As a result, as the temperature starts to drop towards the end of summer or early autumn, this disease is primarily seen on mature plants.

Disease Cycle

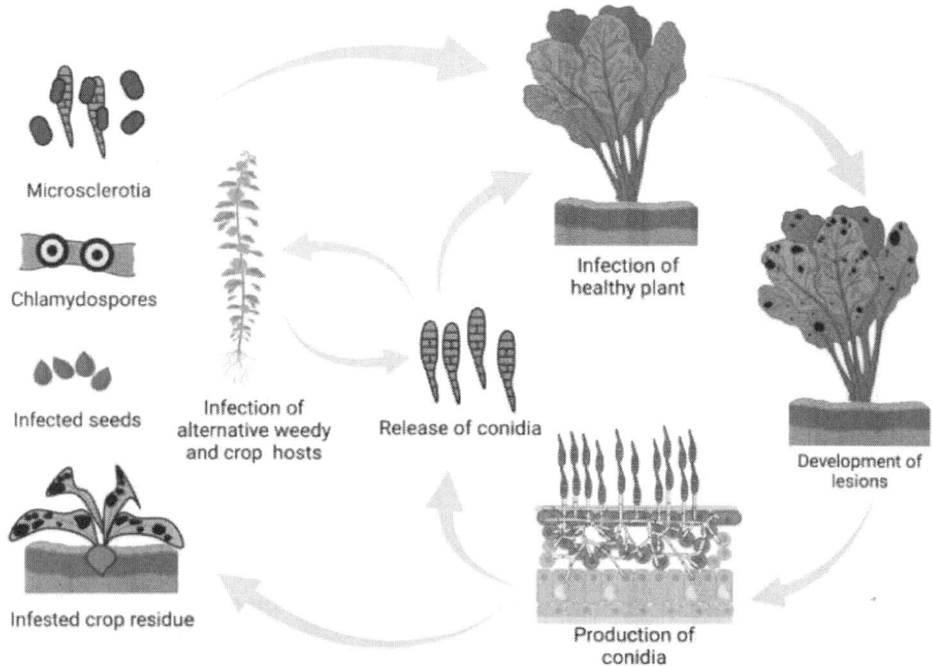

Figure 25.2: Disease Cycle of Alternaria Leaf Spot of Sugarbeet (*Source*: vegetables.cornell.edu).

If the climate is right, the disease may infect the crop at any point during its growth cycle. Because the pathogen is a saprophyte, meaning it lives on dead and decaying organic matter, it can survive in the absence of a host. These specialised structures are responsible for this survival: (1) chlamydospores, which are vegetative spores with thick walls created through the expanding and thickening of hyphal cells; and (2) microsclerotia, which are multicellular, dark, hard structures derived from conidia that have irregular to spherical shapes. The pathogen can withstand extreme circumstances like chilling and dehydration thanks to these structures. Furthermore, the disease can survive as a parasite in weeds and other hosts. It has occasionally been reported to survive in the seeds of particular hosts. As a secondary infection, the fungus usually enters the host through older leaves that have wounds, damaged tissues, or stomata.

It is also possible for healthy leaves to become directly infected. Lesions appear after infection, and it is on these lesions that asexual spores called conidia grow and are grouped in chains. Conidia can be spread by a variety of factors, such as wind, splashes of water, farm machinery, and even worker activity, leading to several infection cycles in one growing season.

Management

1. Using certified and fungicide-treated seeds will help prevent the entry of pathogens.
2. To prevent the fungal spores from being dispersed by splashes, use drip and furrow irrigation rather than overhead sprinklers.
3. Spray propiconazole 2 ml/litre at 15 days of interval give effective management.

2. Cercospora Leaf Spot

Economic Importance and Occurrence

The disease Cercospora leaf spot is widely recognized as the most devastating leaves disease affecting sugar beet crops. It stands as the most important leaves disease affecting sugar beet cultivation, with potential yield reductions of up to 40 per cent or more when environmental conditions are favorable for its development. In the field, it's common to observe lower harvest weights, and the sugar content in affected beets is notably reduced. Less sugar-containing beets don't preserve as well as more sugar-containing ones. Storage issues arise, resulting in decay, increased molasses content, and higher impurity levels. These losses can be attributed to two main factors: first, the reduction in leaf area due to the development of lesions on the leaves, and second, the harmful effects of the toxins cercosporin and beticolin produced by the fungus. As the sugar beet expends more energy on producing additional leaves to compensate for those lost to the disease, it becomes less capable of storing sugars in the root.

Causal Organism

The disease caused by fungi *Cercospora beticola* which belong to Domain-Eukaryota, Kingdom- Fungi, Division- Ascomycota, Class- Dothideomycetes, Order- Capnodiales, Family-Mycosphaerellaceae, Genus- *Cercospora*, Species-*beticola*.

Symptoms

Cercospora leaf spot manifests as small, round lesions, typically measuring 3-5 mm in diameter. The centers of these spots take on a light brown to brown coloration. Upon close examination, it's possible to observe black fungal stroma within the lesions, especially in conditions of high humidity. Under such circumstances, the stroma may also contain white conidia and conidiophores.

Larger affected regions on the leaves may result from the fusion or coalescence of these lesions as the disease progresses. When the leaves are affected by the disease and subsequently die, they often remain attached to the plant. This is due to the rapid leaf death induced by the toxins produced by the fungus. Additionally, these spots can also be found on the petioles, where they may appear more elongated in shape. Complete leaf death may happen in cases of severe diseases, especially in leaves that are important for supplying the energy required for the plant to produce sugar.

Figure 25.3: Cercospora Leaf Spot of Sugarbeet (*Source*: vegetables.cornell.edu).

Favourable Condition

The commencement of a cercospora leaf spot epidemic hinges on two primary factors: the presence of inoculum potential and the prevailing weather conditions. Specifically, the formation of conidia on stromata necessitates specific temperature and humidity conditions. For conidia to develop, a minimum temperature of 15°C is required, along with a relative humidity level exceeding 60 per cent. Conversely, no sporulation takes place when temperatures are below 10°C or above 38°C. These temperature and humidity thresholds are pivotal in initiating and advancing the cercospora leaf spot outbreak, as they create the optimal environment for the disease to develop and propagate (Pool and McKay 1916; Solel and Minz 1972; McKay and Pool 1918; Rees *et al.*, 2007; Jacobsen and Franc 2009; Weiland and Koch 2004;).

Disease Cycle

The process of infection starts when a first conidium, which is formed either directly from the survival structures or through saprophytic mycelial growth, is carried by wind, rain, insects, or splashing irrigation water to the lower surface of sugar beet leaves or petioles (McKay and Pool 1918; Lawrence and Meredith 1970; Khan and Khan 2010). According to recent reports, sugar beetroot roots may also serve as infection sites (Vereijssen *et al.*, 2005). When favourable conditions are met, conidia on a leaf germinate to produce appressoria, from which hyphae emerge, piercing the tissue of the leaf through stomata and developing intercellularly, eventually causing damage and necrotizing the nearby cells through the production of toxins like cercosporin and beticolin as well as the action of enzymes like cellulases and pectinases (Pool and McKay 1916; Pal and Mukhopadhyay 1984; Srobarova and Brillova 1979; Rathaiah 1977; Steinkamp *et al.*, 1979, 1981; Mutto and D'Abra 1981; Shane and Teng 1983; Moser *et al.*, 1990; Mike's *et al.*, 1994; Rossi *et al.*, 2000). The resulting patches, whose haphazard arrangement on leaves is indicative of C. beticola's infection of Beta species (Steinkampet al. 1979; Weiland and Koch 2004), serve as the location for conidiophores, which cause additional conidia to form and spread outward once more. In this way, the spread of Cercospora leaf spot epidemics in fields occurs in a polycyclic fashion, meaning that new infections of sugar beetroot leaves arise from ongoing processes of conidia sprouting, infection, sporulation, and conidia discharge (Rossi *et al.*, 2000; Franc 2010).

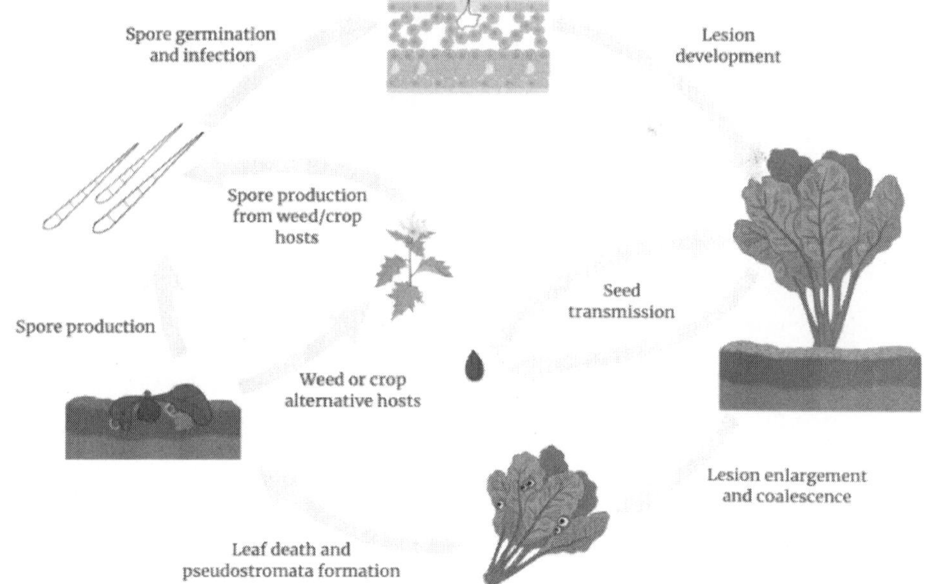

Figure 25.4: Disease cycle of Cercospora Leaf Spot of Sugarbeet (Source: vegetables.cornell.edu).

Management

1. Crop rotation of three years showed effective minimization of inoculum level.

2. Grow resistant varieties is effective strategy to minimize damage.
3. Avoid overhead irrigation.
4. Spray Benzovindiflupyr + difenoconazole500 ml mix in 500 litre water for one hectare at appearance of first symptom.

3. Powdery Mildew

Economic Importance and Occurrence

Powdery mildew is a prevalent disease typically associated with winter season crops, and it can result in significant yield reductions. In severe cases, sugar yields may be diminished by as much as 30 per cent. In India, farmers tend to perceive powdery mildew as less of a destructive disease since the pathogen's spores are readily washed away by overhead irrigation.

Causal Organism

Erysiphe betae is a fungus which responsible for causing powdery mildew in sugarbeet crop.

Symptoms

On the outermost layer of the leaves, a fungal mycelium and spore development that resembles a white to light grey mould normally occurs between August and October. Spores that are at rest can be seen on the afflicted leaves as tiny black spots. The leaves eventually change, becoming yellow and falling victim to the illness. Conidia, or asexual spores, are released by specialised hyphae called conidiophores, which help spread the disease throughout the crop or to neighbouring fields. Later in the season, ascospores (sexual spores) are discharged when the environment is not conducive to the development of conidia.

Figure 25.5: Powdery mildew of Sugarbeet (*Source*: cropscience.bayer.co.uk).

Figure 18.1: A. Early Leaf Spot (Left); B. Late Leaf Spot (Right). (*pg.* 347)

Figure 18.3: Rust Symptoms in Groundnut (*pg.* 350)

Figure 18.5: Collar Rot Symptoms in Groundnut (*pg.* 352)

Figure 18.6: Symptoms of Stem Rot (*pg.* 354)

Figure 18.8: Symptoms of Dry Root Rot. (*pg.* 357)

Figure 18.10: Bud Necrosis Disease: Rings and chlorotic spots, stunting, and necrosis of terminal bud of peanut (Courtesy of B.V. Bhaskara Reddy, Acharya N G Ranga Agricultural University). (*pg.* 359)

Figure 18.11: Symptoms of Groundnut Rosette Disease (*pg*. 360)

Figure 18.12: Symptoms of Root Knot Nematode Healthy (Left); Infected Plant (Right) (*pg*. 361)

Figure 18.14: Iron-chlorosis (*pg*. 364)

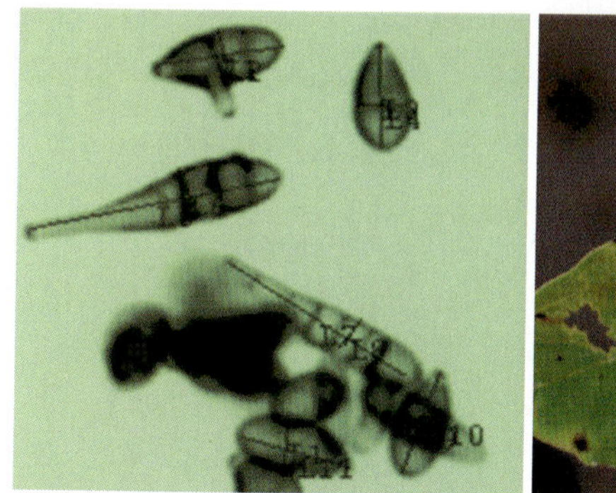

Figure 19.1: Microscopic View and Symptom. (*pg*. 369)

Figure 19.2: Cercospora Leaf Spot of Sesamum. (*pg*. 370)

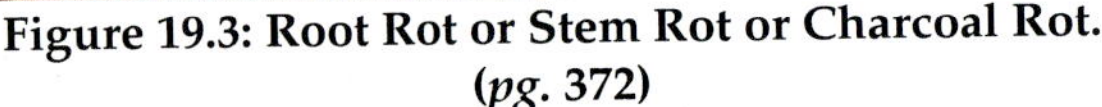

Figure 19.3: Root Rot or Stem Rot or Charcoal Rot.
(*pg.* 372)

Figure 19.4: Sesamum Wilt
(*pg.* 374)

Figure 19.4: Stem Blight
(*pg.* 375)

Figure 19.6: Powdery Mildew of Sesamum.
(*pg.* 377)

Figure 19.7: Phyllody of Sesamum.
(*pg.* 380)

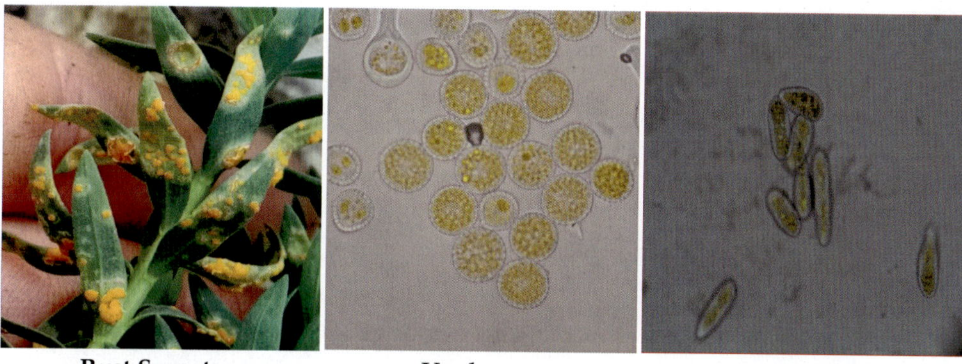

Rust Symptom Uredospore Teliospore

Figure 20.1 (*pg*. 388)

Figure 20.4: A. Wilt Symptom; B. Macroconidia. (*pg*. 392)

Conidia of A. Linicola

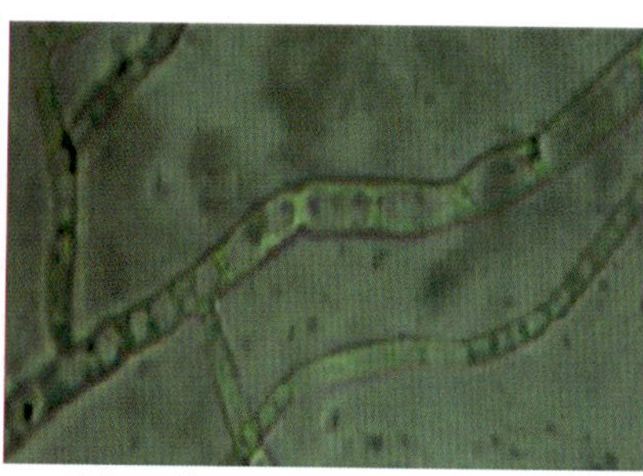

Mycelium of Alternaria spp.

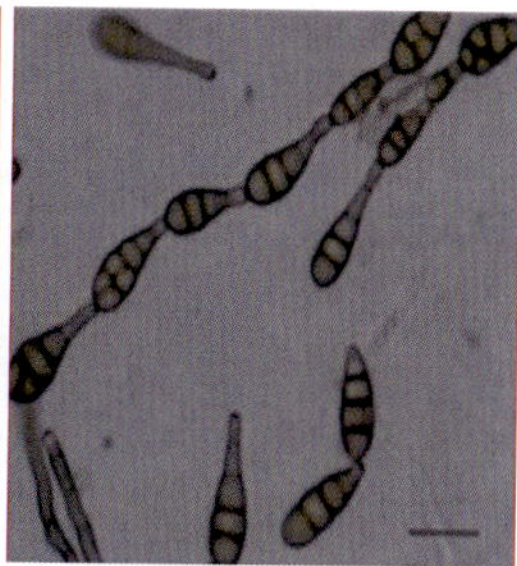

Conidia of A. lini

Figure 20.5 (*pg*. 393)

Figure 20.6: Symptoms of Alternaria Blight Produced on Leaves and Pods of the Linseed. (*pg.* 394)

Figure 20.7: A. Powdery Appearance on Leave Surface; B. Diseased Plant. (*pg.* 396)

Figure 21.1: Rust of Sunflower. (*pg.* 403)

Figure 21.3: Leaf Spot of Sunflower (*pg.* 405)

Figure 21.5: Head Rot of Sunflower (*pg.* 407)

Figure 21.6: Downy Mildew of Sunflower (*pg.* 408)

Figure 21.8: Root Rot of Sunflower (*pg.* 411)

Figure 21.9: Powdery Mildew of Sunflower (*pg.* 412)

Figure 21.10: Basal Rot of Sunflower (*pg.* 414)

Figure 22.1: Wilt of Cotton (*pg.* 419)

Figure 22.3: Root Rot of Cotton (*pg.* 421)

Figure 22.5: Anthracnose of Cotton (*pg.* 424)

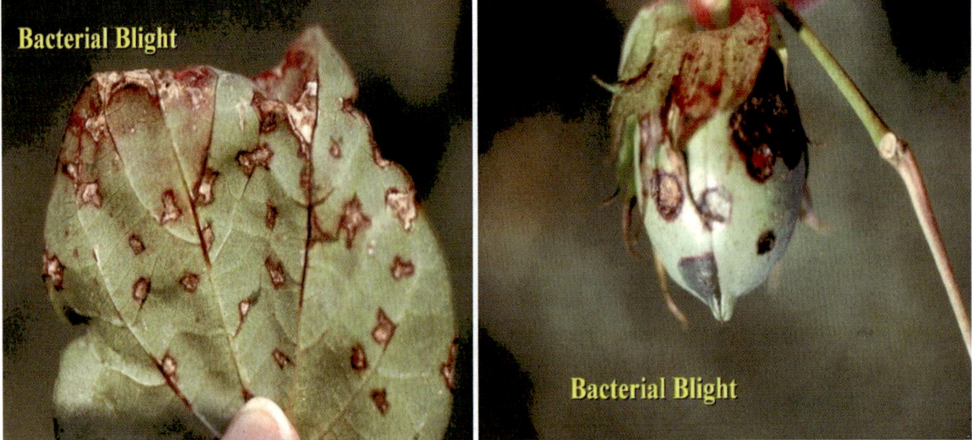

Figure 22.6: Bacterial Leaf Blight Symptoms on Leaves and Ball. (*pg.* 425)

Figure 22.8: A. Cotton Leaf Curl Infected Le; B. Downward Curling. (*pg.* 428)

Figure 23.1: Stem rot and Rot rot (*pg.* 434)

Figure 23.4: Soft rot of Jute and Mesta (*pg.* 438)

Figure 23.5: Foot and Stem rot of Jute and Mesta (*pg.* 440)

Figure 24.1: A. Red Rot of Sugarcane (*pg.* 450)

Figure 24.3: Smut of Sugarcane (*pg.* 453)

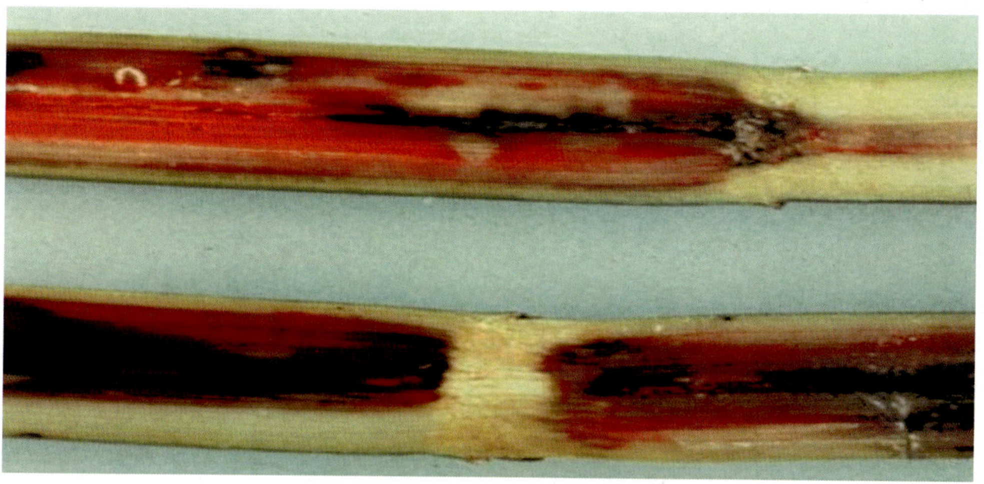

Figure 24.5: Sett Rot (*pg.* 456)

Figure 24.6: Red Stripe (*pg.* 458)

Figure 25.1: Alternaria Leaf Spot of Sugarbeet (*pg.* 463)

Figure 25.3: Cercospora Leaf Spot of Sugarbeet (*pg.* 466)

Figure 25.5: Powdery mildew of Sugarbeet (*pg.* 468)

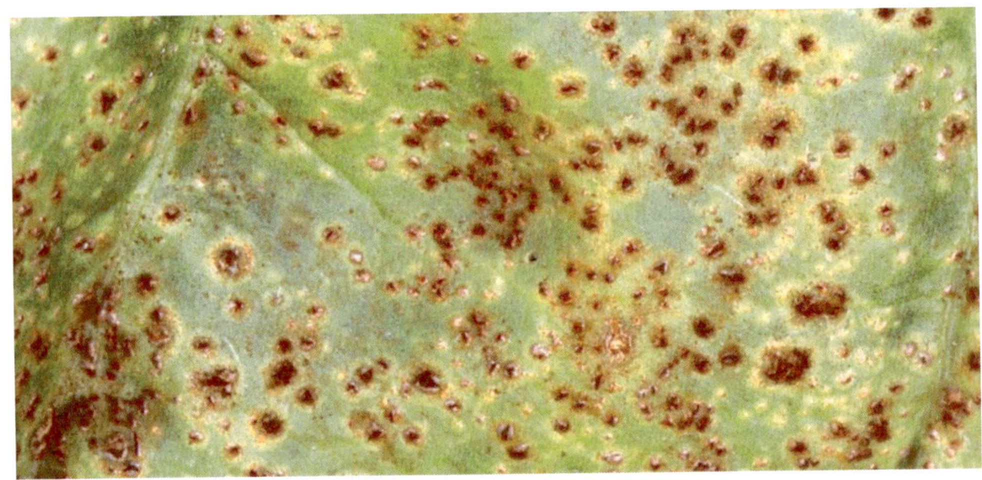

Figure 25.7: Rust of Sugarbeet (*pg.* 471)

Favourable Condition

Conditions that are ideal for the establishment of disease include dry weather, high temperatures (20 °C), and nighttime dew. The frequency of frosts in February and March determines the annual projection of the expected disease pressure for the newly sown crop. Frequent frosts and cold winters are not conducive to the development of illness.

Disease Cycle

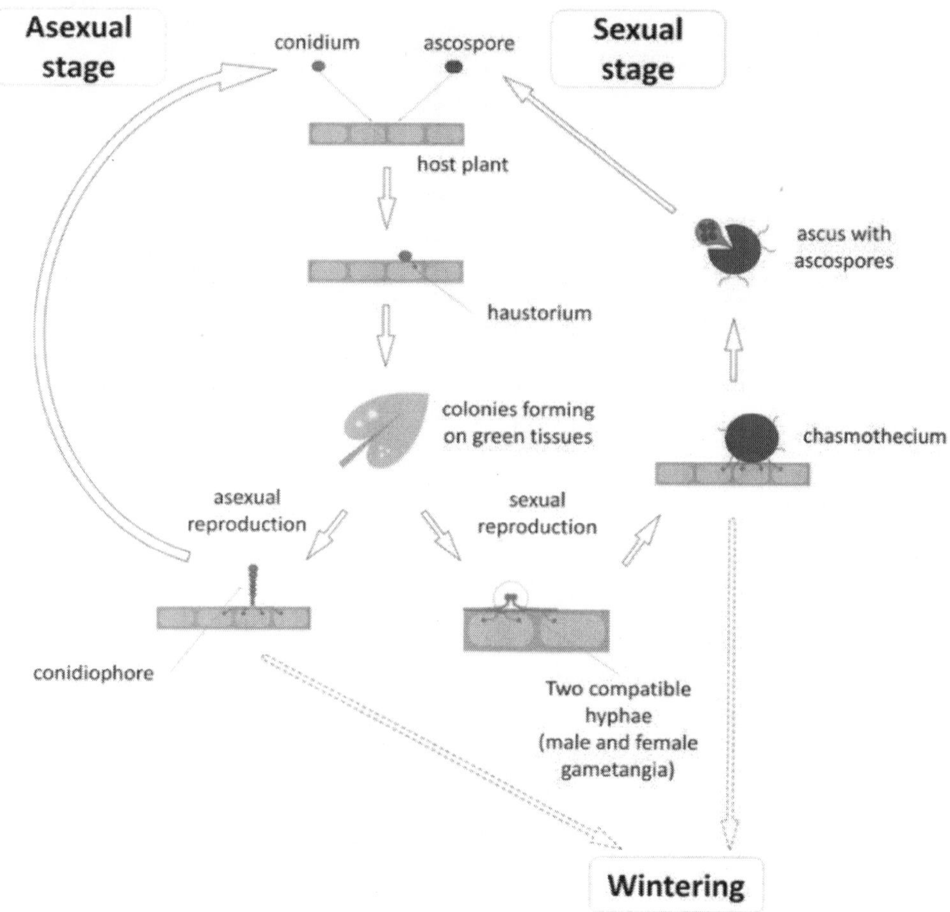

Figure 25.6: Disease Cycle of Powdery mildew of Sugarbeet (*Source*: Sulima and Zhukov, 2022).

The primary inoculum source is ovoid, hyaline, colourless conidia. The ideal conditions are 25 °C and 100 per cent relative humidity, but conidia can germinate on leaf surfaces in a variety of temperatures and relative humidities, including 0 per cent RH (Drandarevski, 1978). The conidia are wind-borne. When conidia germinate, they form a single tube that later develops into a septate, lobed appressorium that either pierces the host surface directly or sporadically via stomata that are open (Mukhopadhyay and Russell, 1979). Multiple short germ tubes or just one wandering aseptate tube are the products of abnormal germination (Mukhopadhyay and Russell, 1979). A lobed haustorium forms upon successful penetration (Hamacher, 1987). Subsequently, conidiation starts, more haustoria emerge, and elongating secondary hyphae ramify across the leaf surface. The ideal relative humidity for the growth of conidia, which are borne singly atop upright conidiophores, is between 30 and 40 percent (Drandarevski, 1978; Mukhopadhyay and Russell, 1979; Weltzien, 1963). This explains why locations with hot, dry temperatures experience higher levels of damage. Ascospores inside cleistothecia can arise by sexual reproduction between mildew strains that are compatible on the same leaf. Cleistothecium production is widespread in certain nations (Mamluk and Weltzien, 1973). Although the exact function of the sexual cycle in E. betae is yet unknown, it raises the possibility of overwinter survival and the emergence of new, perhaps more virulent mildew races (Byford and Bentley, 1976).The first step of infection–that is, the formation of working appressoria and their entry through the leaf surface–is arguably the most important in the life cycle of E. betae.

Management

1. Plant resistant varieties: A number of vegetable crop varieties have a resistance label that indicates varying degrees of resistance to powdery mildew.
2. After harvest, get rid of or plough over any sick plant material. Powdery mildew persists on plant debris, so getting rid of it can help prevent illness from emerging the next season. All afflicted plant parts, including leaves, branches, vines, and contaminated fruit, should be removed and destroyed. Plant debris that is diseased should not be added to compost piles.
3. Expand plant spacing. Well-placed plants will have improved airflow, which will lower the canopy humidity that can lead to infections with powdery mildew.
4. Morning irrigation will allow excess moisture to dry from the ground and leaves, reducing evening humidity. Where possible, switch to drip irrigation.
5. Spray sulphur 55.16 SC 300 ml/100 litre on one hectarecrop.

4. Rust

Economic Importance and Occurrence

Only the pycnial and aecial phases of *Puccinia* subnitens' life cycle infect sugar beetroot; the other phases (uredial and telial) infect other plants, including needlegrass (*Aristida* spp.) and saltgrass (*Distichlis* spp.). Inland saltgrass (*Distichlisspicata* L. Greene), a warm-season grass native to desert areas of the western United States, is the principal host known for *P. subnitens*. Additionally typical habitats for this plant include brackish, marshy places and heavily salinized soils. Because of its exceptional resistance to drought, inland saltgrass grows well on the highly alkaline soils found in the West and High Plains of North America. Significant *P. subniten* infections that happen in late August could result in yield decreases of at least 10 per cent.

Causal Organism

The heteroecious rust disease known as sugar beetroot seedling rust is brought on by PucciniasubnitensDietel, a macrocyclic pathogen (also known as *P. aristidae* Tracy).

Symptoms

Rust manifests as tiny, 1-2 mm diameter, elevated pustules that are brown or reddish-orange in colour on each side of the leaf.

Favourable Condition

The disease spreads when the weather is moist and temperatures range from 15 to 20°C. It reaches its peak severity when dew lingers for extended periods. Warm, dry weather stops the disease's progression, leaving behind sunken yellow spots on the leaves. In cases where the foliage is affected by rust and subjected to frost, the potential for greater damage to the plants is increased.

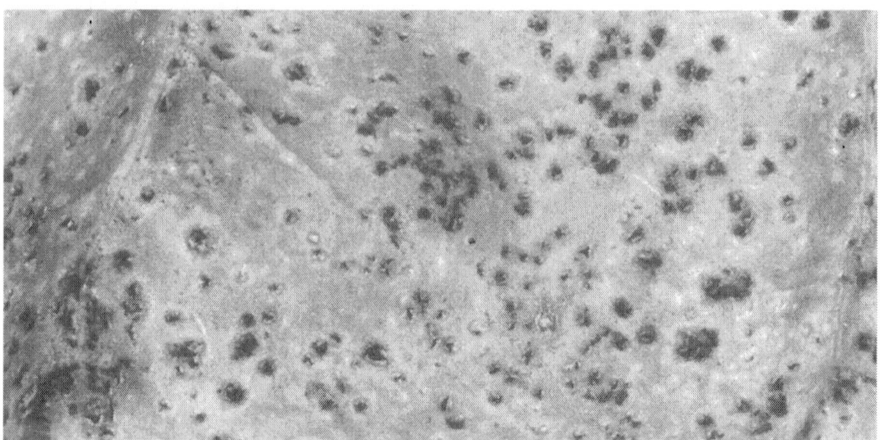

Figure 25.7: Rust of Sugarbeet (Source: fwi.co.uk).

Disease Cycle

Teliospores that overwintered on inland saltgrass begin to germinate from telial pustules in the early spring. They produce basidiospores, which the wind subsequently disperses. Sugar beets and a number of other alternative hosts are susceptible to infection by these basidiospores. The subsequent phases of pycnial (sometimes called spermagonial) and aecial spores are caused by basidiospore infections. Pycnial lesions appear as 2 to 5 mm in diameter, round, light yellow patches. Flask-shaped pycnia with spermatia and responsive hyphae embedded within these lesions are mainly found on the top leaf surface. After fertilisation, aecia grow straight from the pycnia and are usually seen on the underside of the leaf surface, directly beneath the pycnia. Aecia are clusters of spherical, yellowish-orange structures that, when they rupture and release spores, take on the shape of a cup. Circular arrangements are common for these structures. After that, freshly created aeciospores re-infect the inland saltgrass, starting fresh telia and uredia and finishing the pathogen's life cycle.

Management

1. Remove and destroy the plant debris that is contaminated.
2. Diseased plant should uproot immediately and destroy.
3. Destruction of weed.
4. Crop rotation should be adopted.
5. Applying Vitavax powder at a rate of 2.5 grammes per kilogramme of seed.
6. Foliar spray twice of PRIORI Gol delivered 33 per cent better control of the disease or mefentrifluconazole + fluxapyroxad after appearance of rust pastuals.

References

Byford, W.J. and Bentley, K. (1976). Cleistothecia of Erysiphe betae in England. *Trans. Br. Mycol. Soc.*, **67:** 544–545.

Drandarevski, C.A. (1969). Untersuchungenüber den echten Rübenmehltau Erysiphe betae (Vañha) Weltzien. III. Geophytopathologische Untersuchungen. *Phytopathol.*, **65**: 201–218.

Franc, G.D. (2010). Ecology and epidemiology of *Cercospora beticola*. In Cercospora leaf spot of sugar beet and related species, ed. R.T. Lartey, J.J. Weiland, L. Panella, P.W. Crous, and C.E. Windels, St. Paul, MN: *The American Phytopathological Society*, 7–19.

Hamacher, J. and Weltzien, H.C. (1983). Untersuchungenzu Resistenmechanismen von Zuckerrübengegenüberdem Echten Mehltau Erysiphe betae (Vañ.) Weltzien. *Med. Fac. Land. R. Gent*, **48**: 723–736.

Jacobsen, B.J. and Franc, G.D. (2009). Cercospora leaf spot. In: Compedium of Beet Diseases and Insects, 2nd ed, ed. R.M. Harveson, L.E. Hanson, and G.L. Hein, St. Paul, MN: *American Phytopathological Society*, 7–10.

Khan, M.F.R., and Khan, J.(2010). Survival, spore trapping, dispersal, and primary infection site for Cercospora beticola in sugar beet. In Cercospora leaf spot of sugar beet and related species,

Lawrence, J.S. and Meredith, D.S. (1970). Wind dispersal of conidia of *Cercospora beticola*. *Phytopathology*, **60**: 1076–1078.

Mamluk, O.F. and Weltzien, H.C. (1973). Untersuchungenüber die Hauptfruchtform des echtenRübenmehltaus, Erysiphe betae (Vañha) Weltzien. I. Die Fruchtkörperbildung in Abhängigkeit von Umweltfaktoren. *Phytopathol.*, **76:** 221–252.

Mcfarlane, J.S., Bardin, R. Snyder, W.C. (1954). An Alternaria leaf spot of the sugar beet. *Proc. Am. Soc. Sugarbeet Technol*, **8:** 241-245.

McKay, M.B. and Pool, V.W. (1918). Field studies of *Cercospora beticola*. *Phytopathology*, **8**: 119–136.

Mike's, V., Milat, M.L., Pugin, A. and Blein, J.P. (1994). *Cercospora beticola* toxins. Part VII: fluorometric study of their interactions with biological membranes. *Biochimica Biophysica Acta*, **1195**: 124–130.

Moser, H.S., Smith, G.A. and Martin, S.S. (1990). Saprophytic and gametophytic responses of sugar beet to two pathotoxins. *Crop Science*, **30**: 1–6.

Mukhopadhyay, A.N. and Russell, G.E. (1979). Light and scanning electron microscopy of sugar beet powdery mildew. *Trans. Br. Mycol. Soc.*, **72:** 316–319.

Munerati, O. (1932). Sull'incrociodellabarbabietolacoltivata con la beta selvaggiadella costa adriatica. *L'Industria SaccariferaItaliana*, **25:** 303–304.

Mutto, S., and D'Abra, V.(1981). Effects of CBT (*Cercospora beticola* toxin) on leaf cells of *Beta vulgaris* var. *saccharifera*. *Rivista di PatologiaVegetale*, **17:** 83–89.

Pal, V. and Mukhopadhyay, A.N. (1984). Study of the cellulolytic and pectolytic enzymes in nine biological forms of Cercospora beticolaSacc. *Acta Phytopathology Academy Science Hung*, **19**: 263–269.

Pool, V.W. and McKay, M.B. (1916). Climatic conditions as related to *Cercospora beticola*. *Journal Agricultural Research*, **6**: 21–60

Rathaiah, Y. (1977). Stomatal tropism of *Cercospora beticola* in sugarbeet. *Phytopathology*, **67**: 358–362.

Rees, R.W., Flood, J. Hasan, Y. and Cooper, R.M. (2007). Effect of inoculum potential, shading and soil temperature on root infection of oil palm seedlings by the basal stem rot pathogen *Ganoderma boninense*. *Plant Pathology*, **56**: 870–962.

Rossi, V., Meriggi, P. Biancardi, E. and Rosso, F.(2000). Effect of Cercospora leaf spot on sugar beet growth, yield and quality. In Advances in sugar beet research Vol. 2: *Cercospora beticola* Sacc. biology, agronomic influence and control measures in sugar beet, ed. M.J.C. Asher, B. Holtschulte, M. Richard Molard, F. Rosso, G. Steinruecken, and R. Beckers, 49–76. Brussels: International Institute for Beet Research.

Sharma, P., Kikkert, K. and Pethybridge, S. (2021). www.vegetables.cornell.edu 08/10/2023, www.vegetables.cornell.edu/pest-canagement/diseasefactsheets/alternaria-leaf-spot-of-table-beet.

Solel, Z. and Minz, G.(1972). Infection process of Cercospora beticola in sugar beet in relation to susceptibility. *Phytopathology*, **61**: 463–466.

Srivastava, S. N. (2004). Management of sugarbeet diseases. In Fruit and vegetable diseases. Dordrecht: *Springer*, pp. 307-355.

Srobarova, A. and Brillova, D. (1979). Study of the effect of Cercospora disease of beet on the polysaccharides of the host cell wall. *Pol'nohopodarstvo*, **25:** 878–886.

Steinkamp, M.P., Martin, S.S, Hoefert, E. G. Rupel. (1981). Ultra structure of lesions produced in leaves of Beta vulgaris by cercosporin, a toxin from *Cercospora beticola. Phytopathology*, **71**: 1272–1281.

Steinkamp, M.P., Martin, S.S. Hoefert, L.L. and Ruppel, E.G. (1979). Ultrastructure of lesions produced by *Cercospora beticola* in leaves of Beta vulgaris. *Physiolology of Plant Pathology*, **15**: 13–26.

Sulima, A. and Zhukov, V. (2022). Molecular Bases of Resistance to Powdery Mildew in Pea (*Pisum sativum* L.) and Other Legumes. *Plants*, **11**: 339.

Vereijssen, J., Schneider, J.H.M. and Termorshuizen, A.A.J. (2005). Root infection of sugar beet by *Cercospora beticola* in a climate chamber and in the field. *European Journal of Plant Pathology*, **112**: 201–210.

Weiland, J. and Koch, G. (2004). Sugar beet leaf spot disease (*Cercospora beticola* Sacc.). *Molecular Plant Pathology*, **5**(3): 157–166.

Weltzien, H.C. (1978). Geographical distribution. In: *The Powdery Mildews* (D. M. Spencer, ed.). London: Academic Press, pp. 39–49.

Index

A

B

F

G

H

I

J

K

L

R

S

T

U

V

W

X

Z